Springer Japan KK

H. Maeda · K. Edo · N. Ishida (Eds.)

Neocarzinostatin

The Past, Present, and Future of an Anticancer Drug

With 233 Figures, Including 7 in Color

Springer

Hiroshi Maeda, Ph.D., M.D.
Professor and Chairman
Department of Microbiology, Kumamoto University School of Medicine
2-2-1 Honjo, Kumamoto, 860 Japan

Kiyoto Edo, Ph.D.
Associate Professor and Director
Pharmacy, Fukushima Medical College Hospital
1 Hikariga-oka, Fukushima, 960-12 Japan

Nakao Ishida, M.D., Ph.D.
Professor Emeritus and Former President, Tohoku University
2-1 Seiryo-cho, Aoba-ku, Sendai, Miyagi, 980-77 Japan
President, Sendai Institute of Microbiology
2-5-22 Ichiban, Aoba-ku, Sendai, Miyagi, 980 Japan

ISBN 978-4-431-66916-6

Library of Congress Cataloging-in-Publication Data

Neocarzinostatin: the past, present, and future of an anticancer drug / H. Maeda, K. Edo, N. Ishida (eds.).
p. cm.
Includes bibliographical references and index.
ISBN 978-4-431-66916-6 ISBN 978-4-431-66914-2 (eBook)
DOI 10.1007/978-4-431-66914-2
1. Zinostatin. I. Maeda, H. (Hiroshi) II. Edo, K. (Kiyoto), 1950– . III. Ishida, N. (Nakao), 1923– .
[DNLM: 1. Zinostatin. 2. Antibiotics, Antineoplastic—chemistry. 3. Antibiotics, Antineoplastic—pharmacokinetics. 4. Neoplasms—drug therapy. QV 269 N438 1997]
RC271.Z55N46 1997
616.99'4061—dc21
DNLM/DLC
for Library of Congress 96-47101

On the frontcover: The cover illustrates the suppressive effects on metastatic liver cancer of portally administered SMANCS/Lipiodol. *Top*—control, *middle*—aqueous SMANCS, *bottom*—SMANCS/Lipiodol. *See* page 254.

This publication was supported by the Ministry of Education, Science, Sports and Culture of Japan, Grant-in-Aid for Scientific Research, Grant-in-Aid for Publication of Scientific Research Result; The Naito Foundation; and the Biodynamics Research Foundation.

Printed on acid-free paper

Originally published by Springer-Verlag Tokyo Berlin Heidelberg New York in 1997

Typesetting: Best-set Typesetter Ltd., Hong Kong

Preface

The 20th century has witnessed the great benefits of the development of antibiotics, which became a reality after World War II. More than 50 years ago I witnessed the miraculous therapeutic power of penicillin, when I was a student at the Tohoku University Medical School's Department of Bacteriology in Sendai, Japan. The late Dr. Kondo was a graduate student in the department at that time and developed the first crude penicillin preparation in Japan which was applied with dramatic results in two patients.

Although there was patient–family consent at that time, ethics committees, randomization, mutagenesis tests, distribution studies, purity-criteria, and pharmacokinetics were not yet in existence. Today, regulatory procedures have complicated the whole drug-approval process. For example, any new antibiotics that have been proven effective in laboratory studies against gram-negative bacteria, as might exist in deadly plague bacteria, must still undergo a long and enormously costly regulatory process before they can be introduced to benefit society, and before government insurance can be applied.

In spite of these circumstances, diseases like diabetes, hypertension, hypercholesterolemia, and osteoporosis, which have cases numbering from 500000 to 10000000 in Japan alone, provide great economic incentives for drug development. Before 1980, there was much enthusiasm in the pharmaceutical industry for the development of an anticancer drug. However, the soaring cost of drug development in recent years with no parallel profit caused the interest of pharmaceutical companies to wane. Any anticancer or antibacterial drugs must be approved for a specific type of cancer or bacteria before clinical use and before government insurance pays for its use (at least in Japan). Obtaining such government approval is far too costly for minor diseases, and it should be understood that many types of cancer fall into this category. There are about 400000 cancer patients annually in Japan, with about 30000 to 40000 (or fewer) cases per tumor type in tumors of the brain, esophagus, pancreas, kidney, bladder, skin, ovary, and skin and in leukemia and neuroblastoma. The effectiveness of neocarzinostatin—for example, in cancer of the bladder and brain—is established, but neocarzinostatin is not used as frequently as it should be nor is it widely advertised. The major reason is that these cases of minor tumors or cancer types do not provide an economic incentive for the pharmaceutical industry to pursue drug development. The problem is endemic in the field of cancer-drug development these days. The result is that few pharmaceutical companies are now as enthusiastic about cancer-drug development as they used to be. The cost–benefit issue is a reality, and it has created a growing disparity between scientific understanding and medical treatment. Economics is, more than ever before, the major pacesetter for new drug development. This is a problem we must solve.

Despite these practical industrial problems, scientific research on neocarzinostatin has given us the opportunity to witness the development and integration of multidisciplinary aspects of 20th-century science, including microbiology, biochemistry, NMR spectroscopy, X-ray crystallography, structural chemistry, pharmacology, immunology, and clinical practice in cancer therapy. On the 30th anniversary of the discovery of neocarzinostatin, Professor Maeda, Professor Edo, and I, as editors, believed that a comprehensive text dealing with this unique antitumor drug and including these diversified disciplines, should be in one volume as a landmark scientific development of this century.

The most critical requisite in antibacterial or anticancer chemotherapy is selective toxicity. This principle is straightforward for antibiotics, namely, selective toxicity to prokaryotic bacterial but not eukaryotic mammalian cells can be differentiated by unique molecular machinery or by structure in bacteria, and selective target molecules such as peptidoglycan do exist in bacteria. However, there are almost no such discrete molecules that differentiate normal cells from cancer cells at the cellular or subcellular level.

I realized in the late 1950s that the difference can be elucidated only by the in vivo screening model using animals that are tumor-bearing. When we screened culture filtrates of *Streptomyces* in vitro, we found that all cytotoxic compounds were detected, although most of these compounds did not exhibit antitumor activity in vivo. This line of screening effort led us to the discovery of neocarzinostatin. By the end of the 1960s, the strong antitumor effect of neocarzinostatin was confirmed at many leading institutions in the United States: the Southern Research Laboratory of the National Cancer Institute (NCI), Birmingham, Alabama; NCI, Bethesda, Maryland; Memorial Sloan-Kettering Cancer Institute, New York; and the Roswell Park Cancer Institute of Buffalo, New York. These highly reproducible antitumor effects of neocarzinostatin convinced my colleagues and me to pursue this very plausible challenge for a new class of antitumor drug development. Antibiotics with a protein nature were very unusual at that time, and how such protein could exert a potent action was a curiosity to me. The puzzle was solved by the discovery of the enediyne chromophore.

The importance of antitumor drug development for solid tumors using in vivo animal models increased as the enhanced permeability and retention (EPR) effect of macromolecules and lipids was elucidated by Professor Maeda about 10 years ago, as described in this monograph. The EPR effect can be explained by the unique anatomical and pathophysiological differences of tumor blood vessels and tissues, but it was not observed in the normal counterpart. It could be demonstrated only in the in vivo animal model, not by in vitro systems, as was shown later by a macromolecular conjugate of neocarzinostatin called SMANCS. Thus, tumor-selective toxicity by a tumor-targeting mechanism of drugs based on the EPR effect needs more attention.

As we have observed in many other areas of science, when we solve one problem, another one becomes visible. That has been my experience with neocarzinostatin. I am fascinated by the complexity of nature. Our quest for truth will continue in the next century.

The author acknowledges Drs. H. Maeda, C.B. Glaser, and H. Koprowski for their reading of and critical comments for this Preface.

NAKAO ISHIDA
Professor Emeritus

Contents

List of Contributors

AKAIKE, TAKAAKI (p 155)
Department of Microbiology, Kumamoto University School of Medicine, 2-2-1 Honjo, Kumamoto, 860 Japan

EDO, KIYOTO (p 1, 23)
Pharmacy, Fukushima Medical College Hospital, 1 Hikarigaoka, Fukushima, 960-12 Japan

HIRAMA, MASAHIRO (p 47)
Department of Chemistry, Graduate School of Science, Tohoku University, Aoba, Aramaki, Aoba-ku, Sendai, Miyagi, 980-77 Japan

ISHIDA, NAKAO (p 1, 269)
Sendai Institute of Microbiology, 2-5-22 Ichiban, Aoba-ku, Sendai, Miyagi, 980 Japan

KIM, KYOUNG-HEE (p 109)
Division of Chemistry and Chemical Engineering (147-75CH), California Institute of Technology, Pasadena, CA 91125, USA

KOBAYASHI, MAKIKO (p 167)
Department of Internal Medicine, University of Texas Medical Branch, 301 University Boulevard, Galveston, TX 77555-0835, USA

KOBAYASHI, YUJI (p 83)
Faculty of Pharmaceutical Sciences Osaka University, 3-3 Yamadaoka, Suita, Osaka, 565 Japan

KOIDE, YOSHIO (p 23)
Tohoku College of Medical Technology and Welfare, Kunimi, Sendai, Miyagi, 981 Japan

KONNO, TOSHIMITSU (p 227)
First Department Surgery, Kumamoto University School of Medicine, 2-2-1 Honjo, Kumamoto, 860 Japan

KUROMIZU, KENJI (p 3)
Laboratory of Chemistry, Fukuoka Dental College, 2-15-1 Tamura, Sawara-ku, Fukuoka, 814-01 Japan

MAEDA, HIROSHI (p 1, 3, 155, 187, 205, 227, 269)
Department of Microbiology, Kumamoto University School of Medicine, 2-2-1 Honjo, Kumamoto, 860 Japan

MASUDA, ETSUKO (p 187)
Molecular Medicine Research Labs, Institute for Drug Discovery Research, Yamanouchi Pharmaceutical Co., Ltd., 21 Miyukigaoka, Tsukuba, Ibaraki, 305 Japan

MATSUMOTO, SHIGEHIRO (p 269)
Kayaku Antibiotic Inc., 2-8-16 Funado, Itabashi-ku, Tokyo, 174 Japan

MEYERS, ANDREW G. (p 109)
Division of Chemistry and Chemical Engineering (164-30Cr), California Institute of Technology, Pasadena, CA 91125, USA

OHTSUKI, KENZO (p 129)
Laboratory of Genetical Biochemistry, Kitasato University, School of Allied Health Sciences, 1-15-1 Kitasato, Sagamihara, 228 Japan

ONO, YASUSHI (p 129)
Department of Microbiology, Nihon University School of Medicine, 30-1 Kamicho, Ohyaguchi, Itabashi-ku, Tokyo, 173 Japan

REES, DOUGLAS C. (p 109)
Division of Chemistry and Chemical Engineering (147-75CH), California Institute of Technology, Pasadena, CA 91125, USA

SATO, KEIZO (p 155)
Departments of Microbiology and Medicine I, Kumamoto University School of Medicine, 2-2-1 Honjo, Kumamoto, 860 Japan

SUZUKI, FUJIO (p 167)
Department of Internal Medicine, University of Texas Medical Branch, 301 University Boulevard, Galveston, TX 77555-0835, USA

TAKASHIMA, HIROYUKI (p 83)
International Research Laboratories Ciba-Geigy Japan, Ltd., 10-66 Miyuki-cho, Takarazuka, Hyogo, 655 Japan

Color Plates

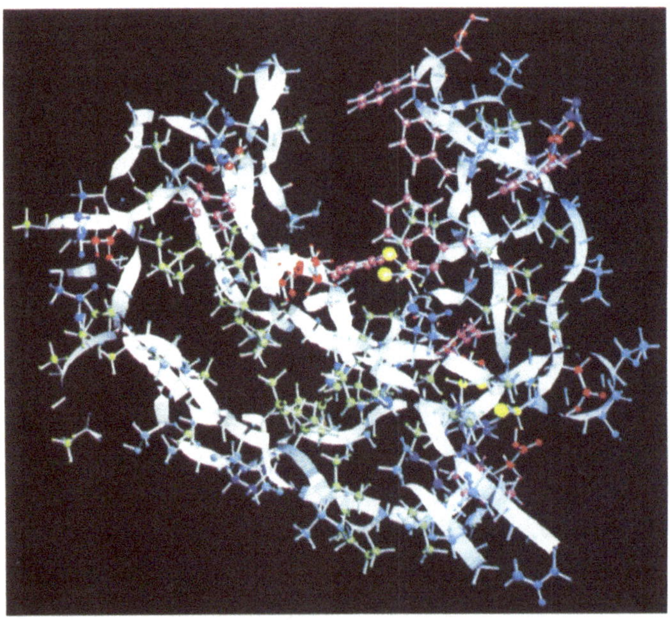

FIG. 13. Three-dimensional structure of protein moiety of holo-NCS determined by distance geometry calculations, with a ribbon drawing of the main chain and a ball-and-stick model drawing of the side chains (*p. 100*)

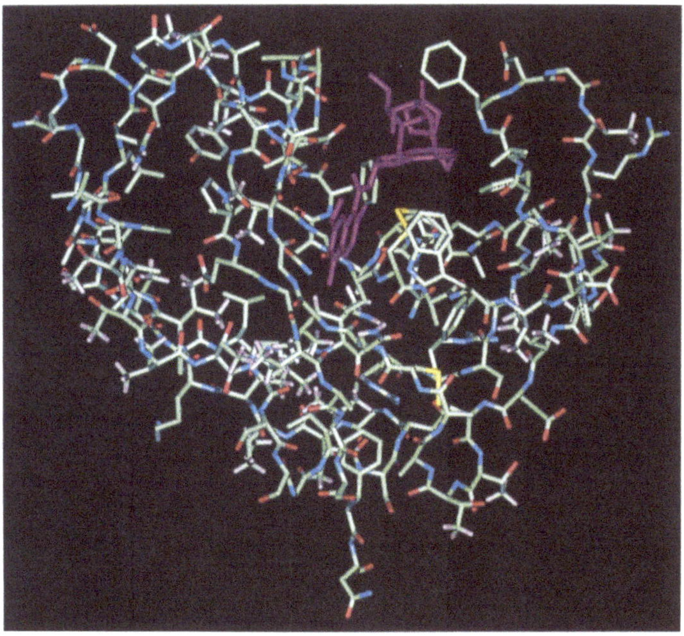

FIG. 14. Three-dimensional structure of holo-NCS determined by NMR data and restrained molecular dynamics calculations. The NCS-chrom molecule is displayed in *purple* (*p. 101*)

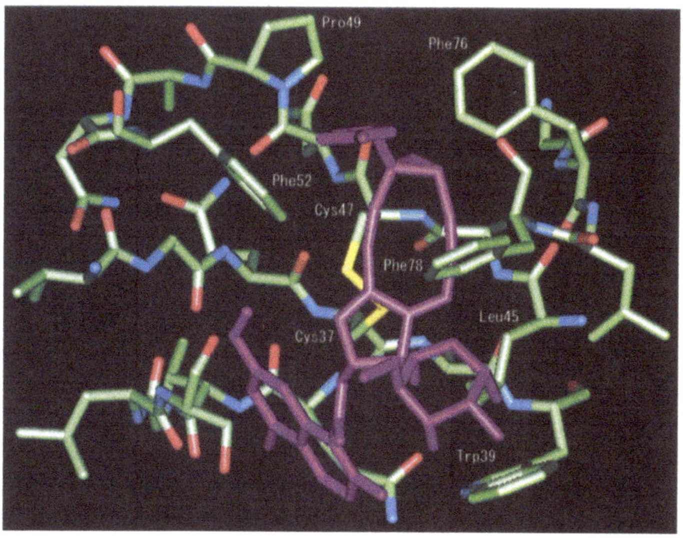

FIG. 15. NCS-chrom binding site of holo-NCS; the point of view is from top of Fig. 13. NCS-chrom (*purple*) and the residues from Val 34 to Trp 39, Leu 45 to Ser 53, Phe 76 to Phe 78, and Gln 94 to Leu 97 are indicated. The disulfide bond bridge between Cys 37 and Cys 47 is displayed in *yellow* (*p. 103*)

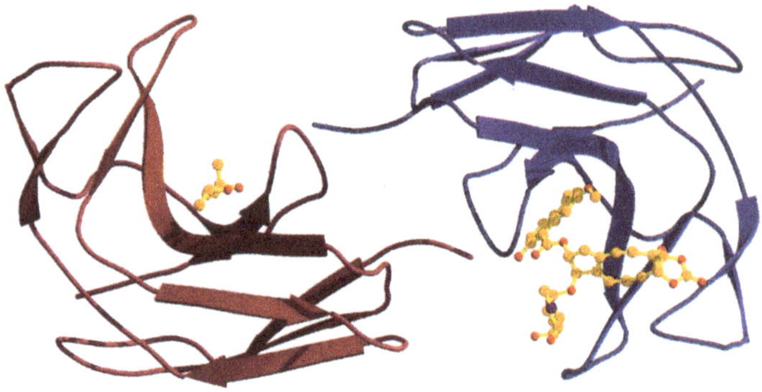

FIG. 2. Structures of two protein molecules in the asymmetric unit prepared with the program MOLSCRIPT [70]. The polypeptide chain is represented by a ribbons diagram: NCS-chrom and the 2-methyl-2,4-pentanediol (MPD) molecule are represented by ball-and-stick models. Holo-NCS (the *right* molecule) contains NCS-chrom, while apo-NCS (the *left* molecule) contains an MPD in the binding pocket (*p. 113*)

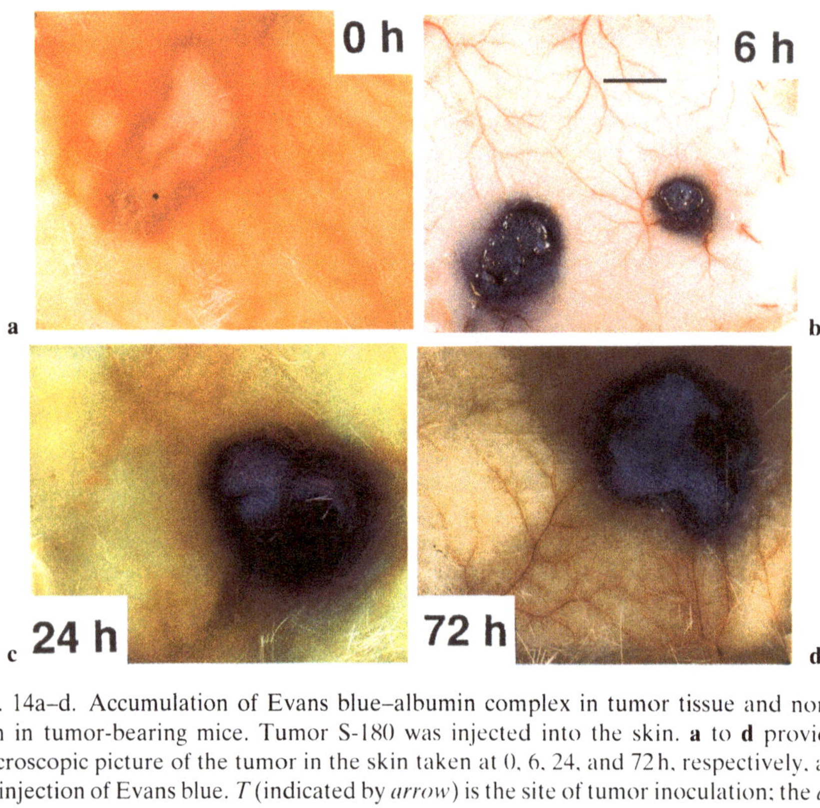

FIG. 14a–d. Accumulation of Evans blue–albumin complex in tumor tissue and normal skin in tumor-bearing mice. Tumor S-180 was injected into the skin. **a** to **d** provide a macroscopic picture of the tumor in the skin taken at 0, 6, 24, and 72 h, respectively, after i.v. injection of Evans blue. *T* (indicated by *arrow*) is the site of tumor inoculation; the *dark area* indicates the tumor, which became progressively bluer after injection. (From [18], with permission) (*p. 244*)

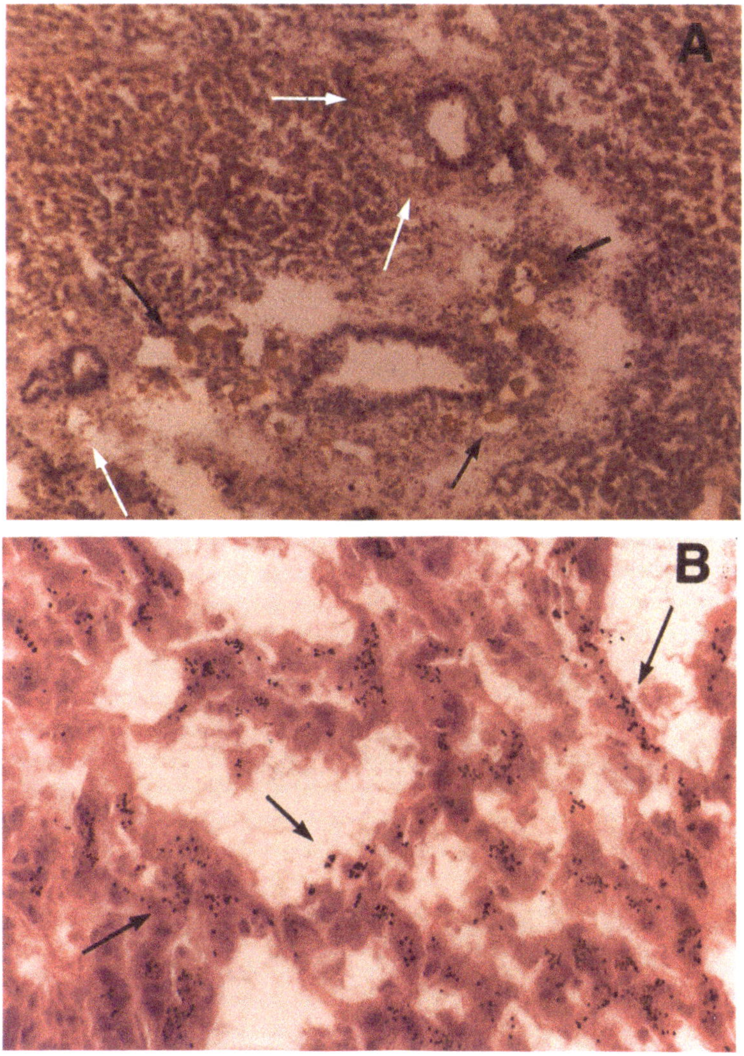

FIG. 21a,b. Microscopic examination of distribution of [^{14}C]-Lipiodol moiety in VX-2 carcinoma in rabbit liver. **a** Sudan III lipid staining in which cluster of lipid is seen in and out of the blood vessels (*black arrows*), although less intense radioactive microparticles are also seen in the interstitial space (*white arrows*). **b** Radioautograph shows *black dots* of Lipiodol throughout tumor tissue (*arrows* and elsewhere). (From [24], with permission) (*p. 253*)

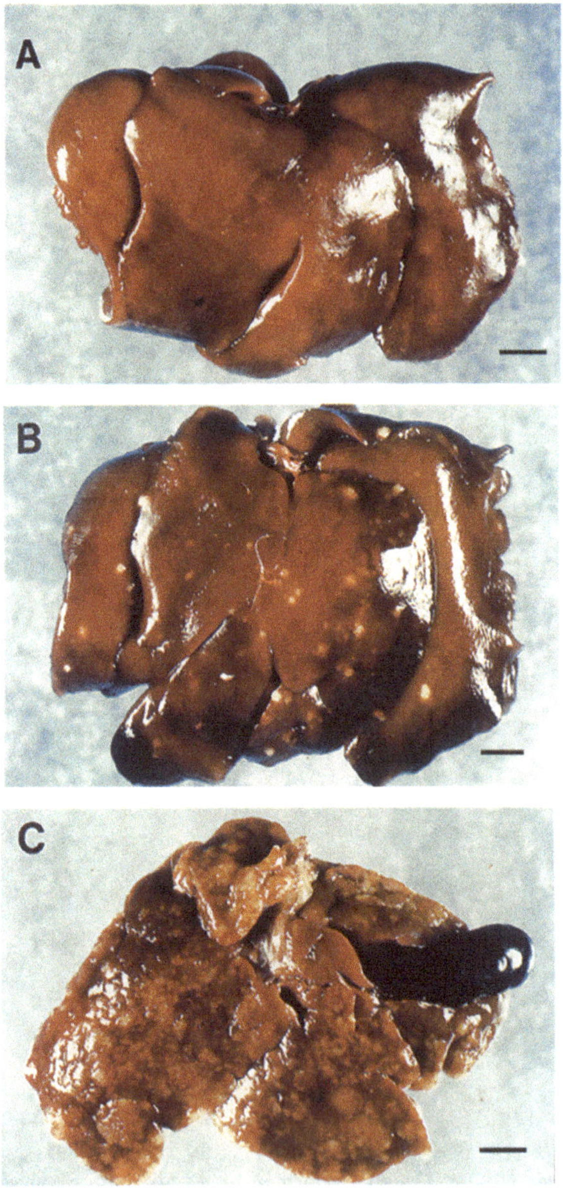

FIG. 22a–c. Suppression of metastatic liver tumor by SMANCS/Lipiodol against experimental metastatic liver tumor (VX-2) in rabbits. Tumor cells were inoculated via the portal vein, and the drugs were injected into the portal vein just before tumor inoculation. All liver specimens were obtained 12 days after inoculation. **a** SMANCS/Lipiodol, 0.4 mg/0.4 ml/kg. **b** SMANCS in 5% glucose (0.4 mg/0.4 ml/kg) followed by free Lipiodol as in **a**. **c** No drug control; tumor nodules number more than 500. (See Table 7.) *Bars*, 1 cm. (From [27], with permission) *(p. 254)*

Introduction

Hiroshi Maeda[1], Kiyoto Edo[2], and Nakao Ishida[3]

The year 1994 marked the 30h anniversary of the discovery of neocarzinostatin, with the first paper on the subject, by N. Ishida et al., appearing in the *Journal of Antibiotics* in 1965. In commemoration of that 30th anniversary, we organized a symposium held no November 22, 1994, in Kumamoto, Japan, entitled "Neocarzinostatin: Past, Present, and Future of Anticancer Drug Development."

Speakers at the symposium presented the molecular details, unique mode of action, pharmacological uniqueness, and clinical applications of neocarzinostatin, with reviews of the past progress, present status, and future outlook. In 1993 and 1994, most major journals, including *Science, Nature, American Scientist*, and the *Journal of the National Cancer Institute* (because of its anticancer action), featured commentaries on the unique properties of enediyne molecules, of which neocarzinostatin was a landmark. These years were also special for neocarzinostatin because its daughter molecule, or its derivative SMANCS, i.e., polymer (poly[styrene-co-maleic acid (SMA)]) conjugated neocarzinostatin (NCS), was approved as a recognized drug by the Japanese Ministry of Health and Welfare in September 1993. It became available for the treatment of hepatoma, one of the most difficult tumors to cure. Of further significance, this is the first polymer anticancer drug ever approved, and we are witnessing its great therapeutic efficacy, unprecedented at this time.

When the three-dimensional structure of neocarzinostatin was revealed in detail in 1993, we realized the amazing progress of science in the past 30 years. In the mid-1960s when neocarzinostatin was born, protein purification and even characterization were in the very early stages of infancy. An antibiotic that was proteinaceous was so unique that every experiment was something of a discovery. However, at the end of the 1970s a UV absorption profile (UV $\lambda_{max} = 330\,nm$) and

[1]Department of Microbiology, Kumamoto University School of Medicine, 2-2-1 Honjo, Kumamoto, 860 Japan
[2]Pharmacy, Fukushima Medical College Hospital, 1 Hikarigaoka, Fukushima, 960-12 Japan
[3]Sendai Institute of Microbiology, 2-5-22 Ichiban, Aoba-ku, Sendai, Miyagi, 980 Japan

other physicochemical and biological evidence suggested the presence of a non-protein component (chromophore) in neocarzinostatin which later was found to play a crucial role in the biological activities of the antibiotic. This prompted Ishida and his colleagues to make an effort to isolate the chromophore, but they soon faced unexpected difficulties due to its surprising fragility when exposed to light. Every purification procedure applied, including silica-gel column chromatography and preparative thin-layer chromatography, failed to recover the biological activity that was tested by inhibition of the growth of *Sarcina lutea*. It took almost 3 years to establish a method of purification in darkness. A graduate student, S. Katamine, and colleagues under the direction of Ishida at the Tohoku University School of Medicine performed all purification procedures in complete darkness and finally succeeded in purification of the nonprotein portion and subsequently characterized a partial structure of the chromophore in 1980. This was just the beginning of a great many efforts. It took 5 more years before Edo and colleagues from the Department of Bacteriology, Tohoku University School of Medicine, reported the entire structure in 1985. When they revealed the unique structure of the enediyne chromophore it was almost unbelievable initially because a 12-membered ring with two triple bonds did not fit easily into the conventional carbon skeleton model.

When we found that its activity is augmented enormously, i.e., more than 10000 fold, in the presence of cytochrome P-450 reductase in DNA breakage and radical generation, we realized that the complexity of nature is far beyond human comprehension.

The clinical benefits of neocarzinostatin are not well recognized, and its unique characteristics are not fully utilized for the treatment of such cancers as those of the urinary bladder. We are now forced to face an economic reality: that scientific validity and clinical benefits do not provide the pharmaceutical industry with adequate incentive in the case of rather small numbers of cancer patients, e.g., bladder cancer patients in Japan number about 20000 annually, compared with 10000000 or more patients with hypertension or hypercholesterolemia.

Regardless of the present clinical benefits of neocarzinostatin, we have come a long way and we are hopeful that it will be recognized in the future and that it will become a pacesetter in a very unique class of drugs. We are also hopeful that this class of drug will prove to be very powerful when its extremely potent cytotoxicity is harnessed and delivered to the target tumor cells, and will eventually help cure patients with cancer as demonstrated by SMANCS. We hope that this monograph will be useful for those scientists, pharmacologists, organic chemists, biochemists, and clinicians interested in neocarzinostatin and related subjects.

Chemical Structure of the Apoprotein of Neocarzinostatin

KENJI KUROMIZU[1] and HIROSHI MAEDA[2]

Summary. Purification and determination of the amino acid sequence of neocarzinostatin are described together with final assignments of the disulfide bridges and Asn/Asp residues. Deamidation of an asparagine residue during purification is also discussed. In addition to the spectral measurements of neocarzinostatin, precise determination of its secondary structure, which was obtained from multidimensional proton nuclear magnetic resonance spectroscopy and X-ray crystallography, is presented. Chemical modification for conformational analysis of the protein is also discussed.

Key words: Neocarzinostatin—Amino acid sequence—Apoprotein—Chromophore—Secondary structure

Introduction

Neocarzinostatin (NCS), isolated from the culture filtrate of *Streptomyces carzinostaticus* var. F-41 [1] is a member of the macromolecular antibiotic family of 10–20 kDa. The family includes a variety of proteinaceous antitumor antibiotics isolated from various species and subspecies of *Streptomyces* and *Actinomyces*: actinoxanthin (AXN) from *S. globisporus* 1131 [2,3], iyomycin from *S. phaeoverticillatus* [4], carzinostatin from *S. carzinostaticus* [5,6], plurallin from *S. pluricolorescens* [7], prunacetin A from *S. griseus* var. *purpureus* [8], macromomycin (MCR) from *S. macromomyceticus* [9], mitomalcin (MMC) from *S. malayensis* [10], largomycin (LRG) from *S. pluricolorescens* [11], sporamycin from *Streptosporangium* st. PO-357 [12], auromomycin (AUR) from

[1] Laboratory of Chemistry, Fukuoka Dental College, 2-15-1 Tamura, Sawara-ku, Fukuoka, 814-01 Japan
[2] Department of Microbiology, Kumamoto University School of Medicine, 2-2-1 Honjo, Kumamoto, 860 Japan

S. macromomyceticus [13], C-1027 antibiotic from *S. globisporus* C-1027 [14], kedarcidin (KED) from *Actinomycetes* st. 585–6 [15], and maduropeptin from *Actinomadura madurae* H710–49 [16]. All the macromolecular antibiotics are considered to be in a holo-form comprising the apoprotein and the nonprotein chromophore, except MCR, which is an apoprotein of AUR [13,17]. NCS is one of the most widely and extensively studied antitumor antibiotics.

NCS, with both antimicrobial and antitumor activities, consists of two components, a protein component (apo-NCS) and a nonprotein chromophore (Chr) [18–20]. Much evidence has suggested that nonprotein Chr is primarily responsible for the biological activity of NCS, while apo-NCS has no cytotoxic activity and is essential for the stability and delivery of Chr [21–24]. The unique chemical structure of Chr was finally determined by Edo et al. [25] using ^{1}H- and ^{13}C-NMR (nuclear magnetic resonance) spectroscopy (see chapter by K. Edo and Y. Koide, this volume). An elegant mechanism for NCS action has also been proposed by Myers [26].

The apoproteins of other chromoprotein antibiotics including AXN and AUR are similar to apo-NCS in size, having 108–113 amino acid residues, and they have extensive sequence homologies. X-ray crystallography shows that all exhibit very close three-dimensional (3-D) structures in these three apoproteins [27–30]. However, NCS cleaves preferentially at thymidine and less at adenosine sites in the DNA sequence, while AUR does so at the guanosine site [20,31]. It is interesting, however, that NCS-Chr does not bind to MCR properly, nor does AUR-Chr bind to apo-NCS [21,32]. Despite the related core structure in these chromophores and the close overall folding in the apoproteins, there are some distinct differences in 3-D structures of the apoproteins and in their mode of action.

The amino acid sequence of apo-NCS was first proposed in 1972 by Meienhofer et al. [33–35] and revised by several groups mainly using fast atom bombardment-mass spectrometry (FAB-MS) and gas chromatography-mass spectrometry (GC-MS) protein sequencing [36], manual and gas-phase sequence methods [37], two-dimensional (2-D)-NMR spectroscopy [38–41], and finally base sequence of the apoprotein gene [42]. The locations of two disulfide bonds were chemically determined by a conventional sequence manner [43]. Circular dichromism (CD), optical rotational dispersion (ORD), and infrared (IR) spectral studies as well as the prediction method of Chou and Fasman revealed the secondary structure of apo-NCS, having mainly β-pleated sheet and β-turn structures and a little α-helix [44–49]. More detailed and precise secondary structures of apo-NCS in solution were estimated by 2-D- and 3-D-NMR spectral analyses [38–41]. X-ray crystallographic studies of NCS in holo- and apo-states gave the detailed secondary and tertiary structures of apo-NCS and the structure-specific information relative to the binding site of Chr [30,50]. Extensive studies of holo-NCS by NMR and molecular modeling have clarified the interaction of the protein and nonprotein Chr [51–54].

Proteinaceous Active Entity of Neocarzinostatin

Early studies carried out in Ishida's laboratory at Tohoku University established the proteinaceous entity of NCS in the middle of the 1960s [1,55]. The situation regarding the so-called chromophore remained unclear at that time. The report by Maeda et al. stated on p. 258 [55] that "There may be an unidentified group in this molecule if not contaminant, while Molisch reaction is negative," based on the UV spectra at about 250–255 nm. Later it became apparent that they contain a class of prosthetic group called enediynes (see chapter by K. Edo and Y. Koide, this volume), which mediate radical-catalyzed DNA damage (see [24] and Chapters by K. Ohtsuki and Y. Ono, T. Akaike et al., respectively).

Before the active principle of NCS was established as proteinaceous in nature, a preliminary work by Shoji [6] indicated that carzinostatin (not neocarzinostatin), produced by a *Streptomyces* E-793, consists of two components, A and B. Although a methanol-soluble B component (a recent chromophore?) exhibits very weak activity, addition of water-soluble peptidic component A, which has also very weak activity, to the B component fortified the activity to a great extent. *Streptomyces carzinostaticus* var F-41, which produces NCS, is a mutant isolated from the strain *Streptomyces* E-793 (or later called *S. carzinostaticus*), and produced a single active principle referred to as NCS [1,55]. However, no relation between carzinostatin A/B and NCS was ever investigated further chemically, although they appear drastically different from each other.

Before the reproducible purification method of NCS was developed, reports of its specific biological activity, toxicity, or therapeutic activity varied widely. After 1965, a reproducible purification procedure was established as follow [55]: The culture filtrate of *Streptomyces carzinostacus* var. F-41 was filtered at 4°C, and the filtrate was concentrated in vacuo at or below 35°C; the concentrate was then placed in a dialysis tube, which was placed in the solid powder of ammonium sulfate at 4°C, allowing NCS to precipitate overnight or longer.

The content of the tube was centrifuged to collect the precipitates, which were dissolved in chilled deionized water, followed by dialysis against deionized water and lyophilization. The brown powder was then subjected to column chromatography on carboxymethyl (CM-) cellulose (pH gradient of 0.1 *M* acetate, pH 3.0–5.0) and Sephadex G-50. The fluffy pale brown material thus purified met the purity criteria satisfactory for this compound in those days, based on behavior on polyacrylamide gel electrophoresis, ultracentrifugation, and column chromatography (Sephadex, CM-cellulose) with a biologically active peak and on detection of a single amino-terminal amino acid (only Ala) [55]. Physicochemical constants are shown in Table 1.

It was found that NCS consists of all L-amino acids, contrary to many other peptide antibiotics [55], which frequently contain D-amino acids and are not synthesized via mRNA synthesis; in contrast, the synthesis of NCS in the cell is dependent on a normal protein biosynthesis.

TABLE 1. Physicochemical properties of neocarzinostatin (NCS)

1. Molecular weight, based on amino acid sequence: Protein portion, 11 076; chromophore, 659; total, 11 735
2. Terminal amino acids
   Amino-terminal: Ala
   Carboxyl-terminal: Asn
3. Cystine disulfide bridge: 2
4. Amino acid composition: All L-type, no cysteine (SH)
   Lys (1), His (0), Arg (3), Asp (8), Asn (4), Glu (2), Gln (3),
   Pro (4), Ser (11), Thr (13), Gly (15), 1/2Cys (4), Ala (18),
   Met (0), Val (12), Leu (6), Ile (1), Phe (5), Tyr (1), Trp (2): Total, 113
5. Diffusion constant D_{20w}: $1.4 \times 10^{-6} cm^2/s$
6. Sedimentation velocity: $S_{20w} = 1.44 S$
7. Partial specific volume: 0.713 ml/g
8. Isoelectric point (pI) by isoelectric focusing: 3.42 (NCS)
   3.40 (pre-NCS)[a]

[a] Preneocarzinostatin; see [34] and [55].

The apo-NCS consists of 113 amino acid residues with its unusually high content of hydrophilic amino acids. It contains two disulfide bonds, one tyrosine, and two tryptophan residues, but no histidine or methionine (see Table 1). It has been reported to be an acidic protein with an isoelectric point of 3.3–3.5 and a single-chain polypeptide with Ala and Asn as the amino- and carboxyl-terminal residues, respectively [55,56]. The protein in native form is highly resistant to the action of proteinases (e.g., trypsin and chymotrypsin) [55].

Primary Structure of Apo-Neocarzinostatin

To carry out sequence analysis, the peptide portion of apo-NCS has to be aligned to a single chain by cleaving the disulfide bonds. However, the reduction of disulfide bonds by various reducing agents was difficult to complete even in the presence of a high concentration of denaturing agents (8 *M* urea or 5 *M* guanidine hydrochloride) in an aqueous system [57]. Complete disulfide reduction was achieved by dithiothreitol in liquid ammonia [57,58].

The first studies to elucidate the chemical structure of apo-NCS were performed by Maeda, Glaser, and Kuromizu in the laboratory of Dr. Meienhofer at Children's Cancer Research Foundation and Harvard Medical School. (The early study of the amino acid sequence was initiated by Maeda et al. in Ishida's laboratory in 1967–1968, and the sequence of a fragment was reported at the 159th Meeting of the Japanese Antibiotic Research Association in Tokyo, January 26, 1968. The work was then continued in Boston after 1968.) In a preliminary experiment, partial sequences of the amino-terminal tetra residues and the carboxyl hexapeptide were determined using native NCS and heat-denatured and *S*-benzyl apo-NCS [59]. Tryptic digestion of tetra-*S*-

carboxymethylated (CM-) apo-NCS at pH 8.2 showed five fragments on peptide mapping. One spot fragment was free arginine. The other fragments were further digested by thermolysin, chymotrypsin, and pepsin if needed.

Purification of the peptides was carried out by combinations of preparative paper chromatography, paper electrophoresis, and gel filtration. Overlapping peptides of tryptic fragments were obtained from pepsin/chymotrypsin double digestion of native NCS. The isolated peptides were sequenced mainly by the subtractive and manual Edman dansyl procedures [60,61]. The amino-terminal 25 residues of the T2 fragment (corresponding to V21–L45; residue numbers are as seen in Fig. 4) were determined by use of the automated protein sequenator [62], while the other part had been determined by subtractive manual Edman degradation. Finally, the primary structure of apo-NCS consisting of 109 amino acids has been reported [33–35].

During the study on the primary structure of MCR, Gibson et al. [36] found that a comparison of the amino acid sequence between MCR and apo-NCS showed a considerable homology in the amino- and caboxyl-terminal regions but a discrepancy in the middle region corresponding to the T2 fragment of CM-apo-NCS. That is, some chymotryptic peptides in the T2 fragment were not consistent with those of the reported sequence in comparison with their molecular mass by FAB-MS (Fig. 1, Table 2). The sequence of the T2 fragment was further examined by FAB-MS of the thermolytic digest and by GC-MS of the partial acid hydrolysate. The other fragments were also sequenced by a combination of GC-MS and FAB-MS of partial acid and enzymatic hydrolysates.

TABLE 2. Peptides identified from enzymatic hydrolysis of T-2 (From [36], with permission)

MH^+	M_r	Uncorrected sequence [33]	Corrected sequence
Chymotrypsin			
375	374	Thr64–Arg66	Thr68–Arg70
603	602	None	Val40–Leu45
795	794	Ala55–Phe61	Ala46–Phe52
836	835	None	Asp33–Tep39
1078	1077	Val21–Tyr32	Val21–Leu32
1368	1367	None	Ser53–Leu67
Thermolysin			
374	373	Val21–Gly25	Val21–Gly25
838	837	Leu26–Asp33	Leu26–Asp33
1081	1080	None	Val55–Leu67
1181	1180	None	Val44–Ser54
1232	1231	None	Val40–Asn51
1283	1282	Leu26–Cys37	Leu26–Cys37
1540	1549	None	Leu26–Trp39
1553	1552	None	Val40–Ser54

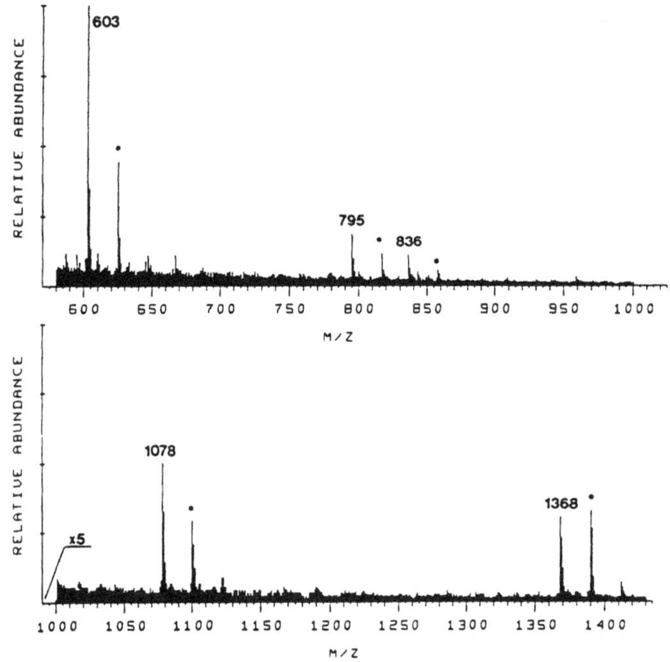

FIG. 1. Fast atom bombardment-mass spectrum of T-2 that had been hydrolyzed with α-chymotrypsin. The protonated molecular ions, MH^+, are labeled for each peptide. The ions with an asterisk correspond to the Na^+ adducts. (From [36], with permission)

The corrected primary sequence of apo-NCS with 113 amino acid residues has been reported [36]. The revised sequence involved reordering of two chymotryptic peptides (S53–L67 and A46–F52), the identification of di- and tripeptides (A59–D60 and A63–T65), the repositioning of Trp at position 39, and the assignment of the remaining Asx residues. Hirayama et al. [63] also reported the revised primary structure of apo-NCS using combinations of endopeptidases (trypsin, chymotrypsin, staphylococcus protease), carboxypeptidase Y digestions, and FAB-MS. The additional correction of replacement of Asp60 by Asn60 has been reported.

We also found that some peptic peptides from native NCS did not correspond with those expected from the previously reported sequence in the amino acid sequence and composition analyses. Therefore, we reexamined the primary structure of apo-NCS by conventional sequence determination using manual and gas-phase sequencing techniques [37]. Both CM- and tetra-*S*-aminoethylated (AE-) derivatives were obtained by reduction and alkylation of apo-NCS in liquid ammonia [58,64]. Tryptic peptides of CM- and AE-apo-NCS were fractionated by diethylaminoethyl (DEAE-) Sephacel chromatography (Figs. 2 and 3; peptides are designated by T1–T5 and TI–TIX, respectively) and further purified

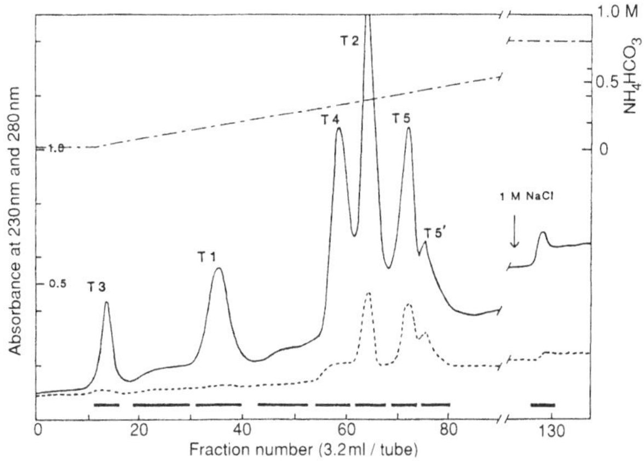

FIG. 2. Diethylaminoethyl (DEAE-) Sephacel column chromatography of tryptic peptides of carboxymethylated (CM-) apo-neocarzinostatin (apo-NCS). The tryptic hydrolysate of CM-apo-NCS (18mg) was dissolved in $0.2\,M\,NH_4HCO_3$, pH 8.5, adjusted with $6\,M\,NH_4OH$, and applied to a Sephacel column (1.2×25cm) equilibrated with the same solvent. Peptides were eluted with a linear gradient of 0.02–$1.0\,M\,NH_4HCO_3$, pH 8.5 (250ml each), at a flow rate of 28ml/h, and fractions of 3.2ml were collected. Absorbance at 230nm (*solid line*) and 280nm (*dashed line*) and concentration (*M*) of NH_4HCO_3 pH 8.5 (*dash/dot line*), are shown. Fractions indicated by *horizontal bars* were collected and rechromatographed (peaks T2, T4, and T5) on a DEAE-Sephacel column (0.9×25cm) with a linear gradient of appropriate concentration of NH_4HCO_3, pH 8.5. (From [37], with permission)

by rechromatography on the same column or reversed-phase high-performance liquid chromatography (HPLC). The T2 fragment consisting of 50 amino acids (V21–R70) was digested by α-chymotrypsin into seven subfragments (T2C1–T2C7), which were clearly separated by reversed-phase HPLC. Sequence analyses of tryptic peptides of both CM- and AE-apo-NCS as well as peptic peptides from native protein revealed a revised primary sequence of apo-NCS with 113 amino acid residues (Fig. 4). The revised sequence is essentially the same as that reported by Gibson et al. [36]. However, a minor but distinct difference is the assignment of amide to four Asp/Asn residues located at positions 41, 48, 51, and 103.

In the course of secondary structural elucidation of apo-NCS in solution, the primary structure was determined by several groups using 2-D-NMR spectroscopy [38–41]. Sequence-specific assignments for the peptide chain were made on the basis of the strong nuclear Overhauser enhancement (NOE) connectivities between the NH proton of a certain residue and the NH, $C_\alpha H$, or $C_\beta H$ proton of the preceding residue. All the results from the sequential NOE spectral analyses show that apo-NCS is a single-chain polypeptide consisting of 113 amino acids.

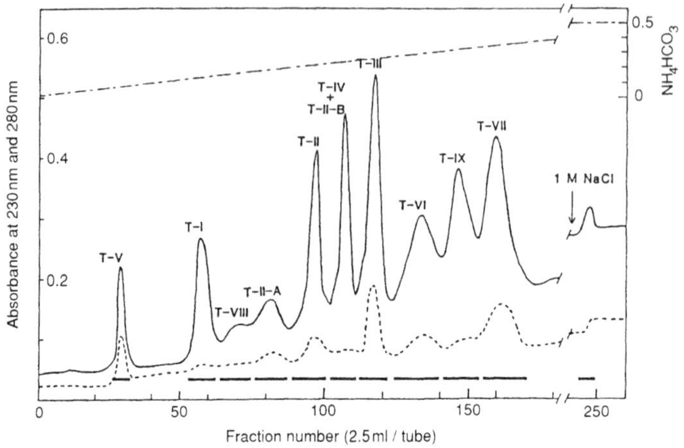

FIG. 3. Separation of tryptic peptides of aminoethylated (AE-) apo-NCS by DEAE Sephacel chromatography. The lyophilized tryptic digest of AE-apo-NCS (12 mg) was applied to a DEAE-Sephacel column (0.9 × 25 cm) equilibrated with 0.02 *M* NH_4HCO_3, pH 8.5. Elution was carried out with a linear gradient of 0.02–0.5 *M* NH_4HCO_3, pH 8.5 (125 ml each), at a flow rate of 18 ml/h, and fractions of 2.5 ml were collected. Absorbance at 230 nm (*solid line*) and 280 nm (*dashed line*), and concentration (*M*) of NH_4HCO_3 (*dash/dot line*) are shown. All the peaks underlined were separately lyophilized and subjected to high pressure liquid chromatography (HPLC) for further purification, except fraction T-V and T-VIII. (From [37], with permission)

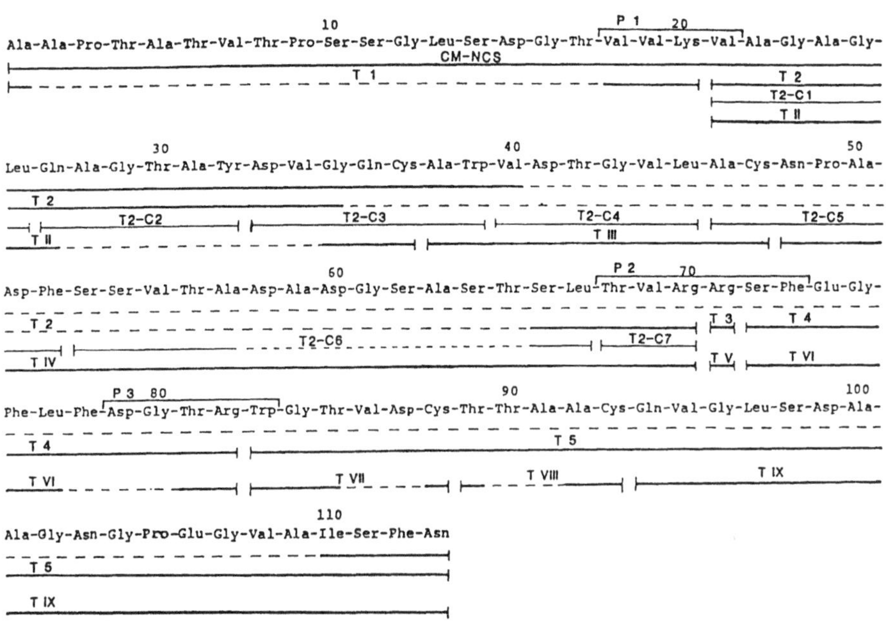

FIG. 4. Revised primary structure of neocarzinostatin (NCS). *Solid lines* indicate the residues identified by the sequence determination of the isolated peptides. (From [37], with permission)

Location of Disulfide Bridges and Assignment of Asparagine or Aspartic Acid

The location of disulfide bonds in apo-NCS was determined chemically [43]. The peptic and peptic/thermolytic peptides from native NCS were isolated by gel filtration and ion-exchange chromatography on a phospho-cellulose columm followed by further purification by use of reversed-phase HPLC. Cystine peptides were oxidized separately by performic acid at 0°C and further separated by HPLC into cysteic acid peptides. Sequence analyses of the isolated peptides revealed the location of disulfide bonds at Cys37–Cys47 and Cys88–Cys93 (Fig. 5). In this work we could reconfirm 84% of 113 amino acids in the protein including connective peptides of tryptic fragments. The definite location of the disulfide bonds was reconfirmed by NMR studies [38–41] and later by X-ray crystallography [30].

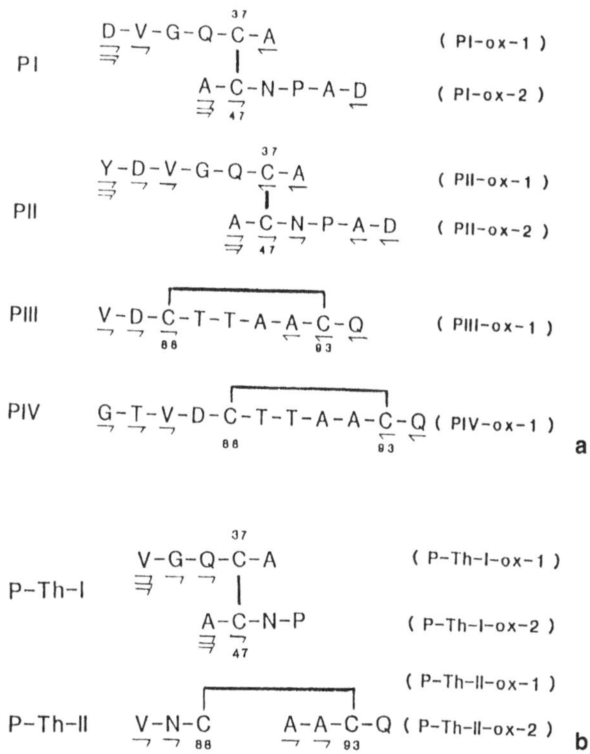

FIG. 5**a**,**b**. Amino acid sequence of cystine peptides obtained from the pepsin (**a**) and pepsin/thermolysin (**b**) digests of native NCS and their oxidized peptides. Amino acids are shown in the one-letter code. *Arrows* indicate dansylation (⇒), manual Edman degradation (⇁), and carboxypeptidase Y digestion (↽). (From [43], with permission)

During the course of purification and the sequence study, it became apparent that one tryptic peptide, T5, could separate into two components, T_{5a} and T_{5b}, which contain the same amino acid composition and sequence except one Asn/Asp issue [34]. In holo-NCS we have observed that a long-term dialysis or incubation at pH 3.0 in dilute acetic acid yielded more pre-NCS in quantity [65]. To clarify this issue we carried out extensive experiments and found one Asn is deamidated to become aspartic acid, as follows. The amide content in apo-NCS has been determined by analyzing Asn and Gln in the pronase-aminopeptidase M digests of CM-apo-NCS and carboxyl-modified apo-NCS in which the carboxyl groups were modified with water-soluble carbodiimide and [^{14}C]glycine methyl ester.

Preneocarzinostatin (PRE) was isolated from a crude NCS preparation by CM-cellulose column chromatography [34,55]. It was found that PRE contained 1 mole less Asn than NCS and that one Asn at position 103 in NCS was converted to Asp in PRE (Asn at position 83 [65] should be rectified to Asn at position 103). A time-dependent conversion of NCS to PRE at pH 3.2, 4°C or in 0.1 *M* acetic acid at 26°C was studied in two ways, that is, by quantitative determination of NCS and PRE by CM-cellulose chromatography and by following the release of free NH_3 during dialysis in an airtight container. Within experimental error, PRE was indistinguishable from NCS in amino acid content after acid hydrolysis, as well as in apparent molecular weight as determined by sodium dodecyl sulfate (SDS-) disk gel electrophoresis and also in amino- and carboxyl-terminal amino acid residues.

Both NCS and PRE shared a common antigenicity as determined by Ouchterlony's agar diffusion method using an antibody against NCS. Only a slight difference between the two in electrophoresis on a cellulose acetate membrane and on a peptide map of the tryptic digest was demonstrated. PRE, however, was completely devoid of biological activity. In addition to the chromatographic difference, a conformational difference was observed by CD spectroscopy; namely, an apparently looser structure of PRE was indicated by the shallowness of the trough in the 240- to 265-nm region. This interpretation was supported by the finding that digestions by pronase were more extensive with PRE than with NCS. These results indicate an important role of the single Asn of NCS in the biological activity, which is evidently governed by the conformation. This is an example of spontaneous deamidation of the protein after biosynthesis.

Sakata et al. [42] deduced the amino acid sequence of apo-NCS from the base sequence of the cloned gene. They cloned the apoprotein gene (ncs A) from the NCS-producing strain *S. carzinostaticus* subsp. *neocarzinostaticus* E-793. A DNA segment with 751 base pairs was obtained by the primer walking technique and sequenced by the dideoxynucleotide chain-termination method. The segment included an open reading frame for a precursor protein consisting of a leader peptide with 34 amino acids and the apoprotein. The amino acid sequence derived from the gene sequence is identical with our finding, except that a residue of Asp60 should be Asn, and provides the final picture of amide assignment in the NCS molecule. Ambiguous assignments of amide to Asx that have been reported

TABLE 3. Ambiguous assignments of Asp/Asn residues in Apo-NCS: the correct residues are shown in boldface type

References	Method[a]	Positions					
		41	48	51	60	87	103
Samy et al. 1983 [66]	—	N	D	N	D	N	D
Gibson et al. 1984 [36]	FAB-MS	N	D	N	D	**D**	D
Kuromizu et al. 1986 [37]	Gas phase	**D**	**N**	**D**	D	**D**	**N**
Hirayama et al. 1986 [63]	Ep/FAB-MS	**D**	D	N	**N**	N	D
Adjadj et al. 1990 [38]	2-D-NMR	**D**	D	**D**	**N**	**D**	**N**
Remerowski et al. 1990 [39]	2-D-NMR	**D**	D	N	**N**	N	D
Takashima et al. 1991 [41]	2-D-NMR	**D**	D	**D**	**N**	**D**	**N**
Gao and Burkhart 1991 [40]	3-D-NMR	**D**	**N**	**D**	**N**	**D**	**N**
Sakata et al. 1993 [42]	cDNA	**D**	**N**	**D**	**N**	**D**	**N**

[a] FAB-MS, Fast atom bombardment-mass spectroscopy; gas phase, gas-phase Edman degradation; Ep, endopeptidases; 2-D-/3-D-NMR, two- and three-dimensional NMR spectroscopy; cDNA, from DNA sequence of the apoprotein gene.

are summarized in Table 3. Comparison of the base sequence among the apoprotein genes of NCS, MCR, and C-1027 shows that homology is limited within the structural genes and that there is little homology in the flanking regions.

Sequence Homology in Apoproteins

MCR is an apoprotein of AUR, and its amino acid sequence with 112 amino acids was evaluated first by Samy et al. [66] using mainly GC-MS and FAB-MS. A revised primary sequence in which only one residue of Asp79 is replaced by Asn was deduced from the nucleotide sequence of the MCR apoprotein gene [67]. Amino acid seqences of AXN [3], C-1027 [68], KED [69], and MMC [40] have been elucidated in which the sequence of MMC apoprotein is the same as that of apo-NCS. The amino acid sequences of apoproteins reported are shown in Fig. 6 for their sequence comparison.

It is known that apo-NCS possesses a high degree of homology in amino acid sequence with the apoproteins of AXN and MCR. Indeed, apo-NCS has 47% sequence homology with MCR, 45% with AXN, 47% with C-1027, and 35% with KED. MCR shows 39% identity with AXN, 38% with C-1027, and 28% with KED. AXN is amazingly similar to the C-1027 protein and shows 97% sequence homology with C-1027. KED has less but considerable sequence homology and similarity in size with these antibiotic proteins. All four half-cystine residues are conserved in position. Two small loops bridged by disulfide bonds in apo-NCS are placed at similar positions in AXN, MCR, and C-1027.

In addition, Tyr at position 32 in apo-NCS is largely conserved among NCS, MCR, AXN, and C-1027 and is known to be buried in the interior of the molecule

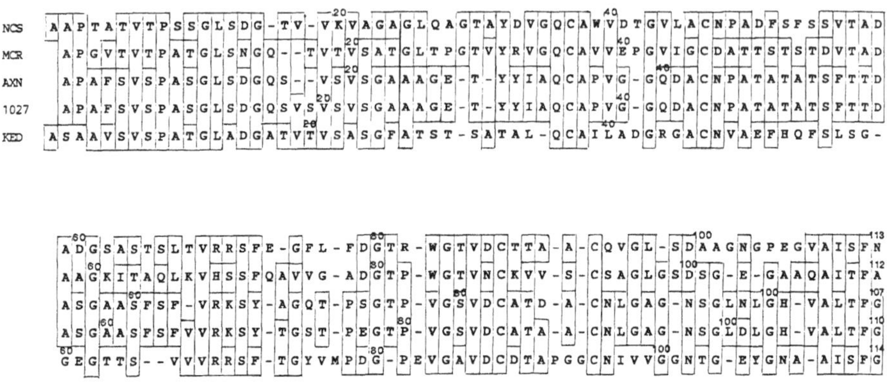

FIG. 6. Primary structure and sequence homology of the apoproteins of NCS [37], macromomycin (MCR) [66] actinoxanthin (AXN) [3], C-1027 antibiotic (1027) [68], and kedarcidin (KED) [69]. *Boxed regions* indicate the sequence homologies. The amino acid sequence of the apoprotein of mitomalcin is the same as that of apo-NCS [40]

in AXN, MCR, and NCS, as determined by crystallography [27–30]. However, no Tyr residue around position 32 is observed in the KED molecule. The partial sequence of amino-32 and carboxyl-4 residues of LRG has been estimated by Vandré et al. [70]. However, no sequence homology could be observed in any parts of other proteins. It is considered that LRG may belong to another category of the proteinaceous antitumor antibiotic because of its size (29300 daltons) and its amino acid composition (high content of Glx).

Secondary Structure of Apo-Neocarzinostatin

With regard to the secondary structure of apo-NCS Maeda et al. [44,45] have reported the peptide backbone conformation of apo-NCS based on plane-polarized IR and side-chain conformations on CD and ORD spectra. A high content of β-structure and very low content of α-helix were described [45]. A tight gross structure was suggested for native NCS, while an increased random conformation and a more symmetrical nature of aromatic side-chain chromophores were observed for the *S*-carboxymethylated derivative. The former explained the peculiar characteristics of native NCS, i.e., its resistance to disulfide bond reduction in an aqueous system and resistance to many proteolytic enzymes, as later pointed out by Samy et al. [47].

The presence of the unusual secondary structure of apo-NCS was expected by prediction analysis according to the method of Chou and Fasman [71,72] (K. Kuromizu, unpublished data). Saito et al. [48] have reported changes in CD bands of holo- and apo-states of NCS in the presence of SDS. They showed that the protein mainly consisted of β-structures by evidence of Gaussian analysis of Fourier-transform IR and curve-fitting analysis of CD spectra of apo-NCS and by a prediction of the secondary structure based on the amino acid sequence of the protein. The importance of the properties of the protein was suggested to be its hydrophobic nature for interaction between the chromophore and apo-NCS.

The very informative secondary structure of apo-NCS has been provided from proton 2-D- and 3-D-NMR analyses [38–41]. Adjadj et al. [38] reported that the sequence-specific resonance assignments of apo-NCS from the 2-D-NMR spectrum were obtained for the backbone protons of 111 of 113 residues of the protein, and that the secondary structure derived from the analysis of sequential connectivities corresponded to 10 β-strands separated by clearly identified loops and turns. They confirmed the presence of two disulfide bonds, Cys37–Cys47 and Cys88–Cys93, from the measurements of interstrand connectivities and slowly exchangeable amide protons, and indicated a global folding similar to those of AXN and MCR.

The similar secondary structural findings have been reported independently by Remerowski et al. [39] using 2-D-proton-NMR spectroscopy. A total of 98% of the main-chain and 77% of the side-chain resonances has been sequence-specifically assigned by use of information from coherence transfer experiments and from sequential and interstrand NOE connectivities. Gao and Burkhart [40] have developed semiautomated combined 2-D- and 3-D-proton-NMR spectroscopy and applied the method to use for more accurate assignments of overlapped 2-D cross-peaks in apo-NCS. They could assign 99% of the protons including those of the side chains, and identify about 1270 intra- and inter-residue proton–proton interactions in apo-NCS. Conformational analysis of NCS in the holo-state was carried out by Takashima et al. [41] using 2-D-proton-NMR spectroscopy. They discussed a β-barrel structure for the carrier protein in which the Chr was tightly enwrapped and thus shielded from exposure to the solvent. The detailed secondary structure of apo-NCS in solution, including the 3-D structure, is described in the chapter by H. Takashima and Y. Kobayashi (this volume).

Chemical Modification of Neocarzinostatin

Chemical and enzymatic modification of NCS was carried out to reveal the structure–activity relationships and to obtain the conformational information and at the same time the derivatives with improved chemotherapeutic properties. NCS possesses 2 free amino groups (1 at N-terminal Ala and the ε-amino group of Lys-20), 11 carboxyl groups (8 Asp, 2 Glu, and a C-terminus; see Fig. 4), and 1 Lys, 3 Arg, 1 Tyr, and 2 Trp residues accessible to site-specific chemical and enzymatic modification. The α-monodeaminated derivative produced by treatment of NCS with nitrous acid retained full antitumor activity and showed low toxicity in an experimental tumor model [73]. A promising therapeutic index of the derivative was shown by partial inhibition of tumor growth in a solid S-180 tumor, while native NCS did not affect the solid S-180 tumor even at the same dose.

Acylation of amino groups by treatment with fluorescein isothiocyanate [73], succinic anhydride [74], and 3-(4-hydroxyphenyl) propionic acid *N*-hydroxysuccinimide [75] has been reported, and the acylated derivatives retained biological activity; however, conversion of all carboxyl groups into carboxamides resulted in the loss of biological activity of the protein [75]. Unsuccessful attempts to derivatize the Tyr residue in NCS by *N*-bromosuccinimide resulted in the conclusion that the Tyr appeared to be buried in the interior of the native protein and inaccessible to the reagent [46].

Since modification of both amino groups of Ala 1 and Lys 20 did not affect biological activity of NCS [73,74], Maeda et al. prepared a poly(*s*tyrene-co-*m*aleic *a*cid) conjugated derivative of NCS via both amino groups (now referred to as SMANCS). Its in vivo antitumor effect was greatly improved as discussed in the chapter by Maeda and Konno.

The Arg-modified NCS and its 89-residue fragment (based on the sequence of Meienhofer et al. [33]) were obtained by reaction with 1,2-cyclohexanedione in borate buffer followed by trypsin digestion, and both were biologically inactive [47]. Removal of the protecting group regenerated the biologically active protein and 89-residue fragment having a conformation very close to that of the native protein in CD spectral properties. Selective Trp-oxidation in NCS was first observed by Maeda using *o*-nitrophenyl sulfenyl chloride as a modifier [76]. Subsequently, Samy et al. [46] reported selective Trp-oxidation with *N*-bromosuccinimide. The mono-Trp-oxidized NCS retained full biological activity, indicating that Trp in position 83 (for position numbers, see Fig. 4) does not play a role in the biological effects of NCS. The bis-Trp-oxidized product was totally inactive.

These results are consistent with the observation by CD spectral measurements, indicating that selective oxidation of the Trp-83 residue did not result in any conformational change, but oxidation of both Trp residues resulted in a change from β-pleated to random coil conformations. However, Edo et al. [49] have reported from titration analyses of NCS and apo-NCS by *N*-bromosuccinimide that selective oxidation of Trp residues was observed in NCS, retaining Trp-83 intact, and that both Trp residues were oxidized in apo-NCS. They have suggested that the chromophore directly interacts with the Trp-39 residue and that a β-sheeted loop containing the Trp-83 residue of apo-NCS changes the high-order structure on binding with the chromophore. They have also reported the isolation of a fragment consisting of the 43 carboxyl-terminal amino acids and its hydrophobic chromophore-binding ability [77]. The complex of the fragment with the chromophore was biologically active.

X-Ray Crystallography of Apo-Neocarzinostatin

An alternative technique for evaluation of the secondary and tertiary structures of macromolecular proteins is an X-ray crystallography that can elucidate the crystalline static nature of the molecule. Both secondary and tertiary structures of apo-NCS have been described by Teplyakov et al. [30] at 0.15-nm resolution. The

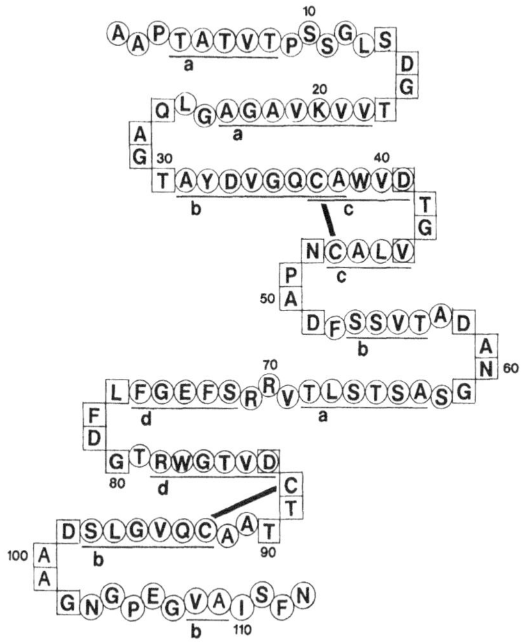

FIG. 7. Primary and secondary structures of apo-NCS. The amino acid sequence is identical to the one derived from cDNA of the apoprotein gene. *Heavy lines* indicate the position of disulfide bridges. The tetra-residues in *boxes* represent β-turns. The *underlined* residues shown by ***a*** form a three-stranded β-sheet, those by ***b*** a tetra-stranded β-sheet and those by ***c*** and ***d*** two-stranded β-sheets, respectively [30]. Note that there is little or no room for the helix structures

crystal structure of apo-NCS is similar to those of homologous proteins of AXN and MCR whose crystal structures have been determined at 0.25- and 0.16-nm resolution, respectively [27,28]. The stranded β-sheet structure and the hydrogen bonding network were in good agreement with those derived from 2-D- and 3-D-NMR spectroscopic analyses in solution. A more detailed description of the crystal structure of NCS, will be found in the chapter by K.-H. Kim et al. (this volume).

Conclusions

The amino acid sequence of apo-NCS as determined by chemical methods, NMR spectroscopy, and deduction from the apoprotein gene are now in good agreement (see, Figs. 4 and 7). There are structural homologies among other proteinaceous enediyne anticancer antibiotics. The protein portion of NCS stabilizes the chromophore in the holo-state by providing the hydrophobic binding site, and it is a unique carrier protein for the labile prosthetic group. The antibiotics possessing such extremely powerful cytotoxic activity as NCS still leave us many problems to be solved in drug targeting, formulation, and organic synthesis as well as biogenesis. The advantages of the unique properties of NCS in the clinical setting are discussed in separate chapters of this volume.

Acknowledgments. We are indebted to Dr. O. Abe, Fukuoka Dental College, for helpful discussions, and to Mrs. A. Izumi for her assistance with preparation of the manuscript.

References

1. Ishida N, Miyazaki M, Kumagai K, Rikimaru M (1965) Neocarzinostatin, an antitumor antibiotic of high molecular weight. Isolation, physicochemical properties and biological activities. J Antibiot (Tokyo) 18:68–76
2. Vichrova NM, Kruchkova TI, Preobrazhenskaya EV, Khokhlov AS (1957) Chemical investigation on the antibiotic actinoxanthin (in Russian). Antibiotiki 2:21–25
3. Khokhlov AS, Reshetov PD, Chupova LA, Cherches BZ, Zhigis LS, Stoyachemko IA (1976) Chemical studies on actinoxanthin. J Antibiot (Tokyo) 29:1026–1034
4. Nomura S, Yamamoto H, Matsumae A, Hata T (1964) Iyomycin, a new antitumor antibiotic from *Streptomyces*. I. Isolation and properties of iyomycin complex. J Antibiot (Tokyo) 17:104–111
5. Ishida N, Kumagai K, Miyazaki M, Ito Y (1960) Gann (Suppl) 51:56
6. Shoji J (1961) Preliminary studies on the isolation of carzinostatin complex and its characteristics. J Antibiot (Tokyo) 14:27–33
7. Ogawara H, Maeda K, Nitta K, Okami Y, Takeuchi T, Umezawa H (1966) An antibiotic, plurallin, consisting of a pluramycin-like prosthetic group and a glycoprotein. J Antibiot (Tokyo) 19:1–9
8. Arai T, Kushikata S, Takamiya K, Yanagisawa F, Koyama T (1967) Prunacetin A, an antitumor antibiotic from *Str. griseus* var. *purpureus* (syn. *Str. californicus*). J Antibiot (Tokyo) 20:334–343
9. Chimura H, Ishizuka M, Hamada M, Hori S, Kimura K, Iwanaga J, Takeuchi T, Umezawa H (1968) A new antibiotic, macromomycin, exhibiting antitumor and antimicrobial activity. J Antibiot (Tokyo) 21:44–49
10. McBride TJ, Axelrod M, Cullen WP, Marsh WS, Sadano CS, Rao KV (1969) Mitomalcin: a new proteinaceous antileukemic antibiotic produced by *Streptomyces malayensis* (nov. sp.). Proc Soc Exp Biol Med 130:1188–1190
11. Yamaguchi T, Furumai T, Sati M, Okuda Y, Ishida N (1970) Studies on a new antitumor antibiotic, largomycin. I. Taxonomy of the largomycin-producing strain and production of the antibiotic. J Antibiot (Tokyo) 23:369–372
12. Komiyama K, Sugimoto K, Takeshima H, Umezawa I (1977) A new antitumor antibiotic, sporamycin. J Antibiot (Tokyo) 30:202–208
13. Yamashita T, Naoi N, Hidaka T, Watanabe K, Kumada Y, Takeuchi T, Umezawa H (1979) Studies on auromomycin. J Antibiot (Tokyo) 32:330–339
14. Hu JL, Xue YC, Xie MY, Zhang R, Otani T, Minami Y, Yamada Y, Marunaka T (1988) A new macromolecular antitumor antibiotic, C-1027. I. Discovery, taxonomy of producing organism, fermentation and biological activity. J Antibiot (Tokyo) 41:1575–1579
15. Lam KS, Hesler GA, Gustavson DR, Crosswell AR, Veitch JM, Forenza S, Tomita K (1991) Kedarcin, a new chromoprotein antitumor antibiotic activity. J Antibiot (Tokyo) 44:472–478

16. Hanada N, Ohkuma H, Yonemoto T, Tomita K, Ohbayashi M, Kamei H, Miyai T, Konishi M, Kawaguchi H, Forenza S (1991) Maduropeptin, a complex of new macromolecular antitumor antibiotics. J Antibiot (Tokyo) 44:403–414
17. Kappen LS, Goldberg IH, Samy TSA (1979) Contrasts in the actions of protein antibiotics on deoxyribonucleic acid, structure and function. Biochemistry 18:5123–5127
18. Napier MA, Holmquist B, Strydom DJ, Goldberg IH (1979) Neocarzinostatin: spectral characterization and separation of a non-protein chromophore. Biochem Biophys Res Commun 89:635–642
19. Koide Y, Ishii F, Hasuda K, Koyama Y, Edo K, Katamine S, Kitame F, Ishida N (1980) Isolation of a non-protein component from neocarzinostatin (NCS) and their biological activities. J Antibiot (Tokyo) 33:342–346
20. Suzuki H, Miura K, Kumada Y, Takeuchi T, Tanaka N (1980) Biological activities of non-protein chromophores of antitumor protein antibiotics: auromomycin and neocarzinostatin. Biochem Biophys Res Commun 94:255–261
21. Kappen LS, Goldberg IH (1980) Stabilization of neocarzinostatin nonprotein chromophore activity by interaction with apoprotein and with HeLa cells. Biochemistry 19:4786–4790
22. Jung G, Kohnlein W (1981) Neocarzinostatin: controlled release of chromophore and its interaction with DNA. Biochem Biophys Res Commun 89:176–183
23. Kappen LS, Napier MA, Goldberg IH (1980) Roles of chromophore and apo-protein in neocarzinostatin action. Proc Natl Acad Sci USA 77:1970–1974
24. Goldberg IH, Kappen LS (1994) Neocarzinostatin; chemical and biological basis of oxidative DNA damage. In: Borders DB, Doyle TW (eds) Enediyne antibiotics as antitumor agents. Dekker, New York, pp 327–362
25. Edo K, Mizugaki M, Koide Y, Seto H, Furihata K, Otake N, Ishida N (1985) The structure of neocarzinostatin chromophore possessing a novel bicyclo [7.3.0]dodecadiyne system. Tetrahedron Lett 26:331–334
26. Myers AG (1987) Proposed structure of the neocarzinostatin chromophore-methyl thioglycolate adduct: a mechanism for the nucleophilic activation of neocarzinostatin. Tetrahedron Lett 28:4493–4496
27. Pletnev VZ, Kuzin AP, Trakhanov SD, Kostetsky PV (1982) Three-dimensional structure of actinoxanthin. IV. A 2.5-Å resolution. Biopolymers 21:287–300
28. Van Roey P, Beerman TA (1989) Crystal structure analysis of auromomycin apoprotein (macromomycin) shows importance of protein side chains to chromophore binding selectivity. Proc Natl Acad Sci USA 86:6587–6591
29. Sieker LC, Jensen LH, Samy TSA (1976) Neocarzinostatin: an antitumor protein, a preliminary X-ray diffraction study. Biochem Biophys Res Commun 68:358–362
30. Teplyakov A, Obmolova G, Wilson K, Kuromizu K (1993) Crystal structure of apo-neocarzinostatin at 0.15-nm resolution. Eur J Biochem 213:737–741
31. Takeshita M, Kappen LS, Grollman AP, Eisenberg M, Goldberg IH. (1981) Strand scission of deoxyribonucleic acid by neocarzinostatin, auromomycin, and bleomycin: studies on base release and nucleotide sequence specificity. Biochemistry 20:7599–7606
32. Kappen LS, Napier MA, Goldberg IH, Samy TSA (1980) Requirement for reducing agents in deoxyribonucleic acid strand scission by the purified chromophore of neocarzinostatin. Biochemistry 19:4780–4785

33. Meienhofer J, Maeda H, Glaser CB, Czombos J, Kuromizu K (1972) Primary structure of neocarzinostatin, an antitumor protein. Science 178:875–876
34. Maeda H, Glaser CB, Czombos J, Meienhoffer J (1974) Structure of the antitumor protein neocarzinostain. Purification, amino acid composition, disulfide reduction, and isolation and composition of tryptic peptides. Arch Biochem Biophys 164:369–378
35. Maeda H, Glaser CB, Kuromizu K, Meienhofer J (1974) Structure of the antitumor protein neocarzinostatin. Amino acid sequence. Arch Biochem Biophys 164:379–385
36. Gibson BW, Herlihy WC, Samy TSA, Hahm KS, Maeda H, Meienhofer J, Biemann K (1984) A revised primary structure for neocarzinostatin based on fast atom bombardment and gas chromatographic-mass spectrometry. J Biol Chem 259:10801–10806
37. Kuromizu K, Tsunasawa S, Maeda H, Abe O, Sakiyama F (1986) Reexamination of the primary structure of an antitumor protein, neocarzinostatin. Arch Biochem Biophys 246:199–205
38. Adjadj E, Mispelter J, Quiniou E, Dimicoli JL, Favaudon V, Lhoste JM (1990) Proton NMR studies of apo-neocarzinostatin from *Streptomyces carzinostaticus*. Sequence-specific assignment and secondary structure. Eur J Biochem 190:263–271
39. Remerowski ML, Glaser SJ, Sieker LC, Samy TSA, Drobny GP (1990) Sequential ^{1}H-NMR assignments and secondary structure of apo-neocarzinostatin in solution. Biochemistry 29:8401–8409
40. Gao XL, Burkhart W (1991) Two- and three-dimensional proton NMR studies of apo-neocarzinostatin. Biochemistry 30:7730–7739
41. Takashima H, Amiya S, Kobayashi Y (1991) Neocarzinostatin: interaction between the antitumor-active chromophore and the carrier protein. J Biochem (Tokyo) 109:807–810
42. Sakata N, Minamitani S, Kanbe T, Hori M, Hamada M, Edo K (1993) The amino acid sequence of neocarzinostatin apoprotein deduced from the base sequence of the gene. Biol Pharm Bull 16:26–28
43. Kuromizu K, Abe O, Maeda H (1991) Location of the disulfide bonds in the antitumor protein neocarzinostatin. Arch Biochem Biophys 286:569–573
44. Maeda H, Ishida N (1967) Conformational study of antitumor protein neocarzinostatin and a deaminated derivative. Biochim Biophys Acta 147:597–599
45. Maeda H, Shiraishi H, Onodera S, Ishida N (1973) Conformation of antibiotic protein, neocarzinostatin, studied by plane polarized infrared spectroscopy, circular dichroism and optical rotatory dispersion. Int J Pept Protein Res 5:19–26
46. Samy TSA, Atreyi M, Maeda H, Meienhofer J (1974) Selective tryptophan oxidation in the antitumor protein neocarzinostatin and effects on conformation and biological activity. Biochemistry 13:1007–1014
47. Samy TSA, Kappen LS, Goldberg IH (1980) Reversible modification of arginine residues in neocarzinostatin. Isolation of a biologically active 89-residue fragment from the tryptic hydrolysate. J Biol Chem 255:3420–3426
48. Saito K, Sato Y, Edo K, Akiyama Murai Y, Koide Y, Ishida N, Mizugaki M (1989) Characterization of secondary structure of neocarzinostatin apoprotein. Chem Pharm Bull (Tokyo) 37:3078–3082
49. Edo K, Saito K, Matsuda Y, Akiyama Murai Y, Mizugaki M, Koide Y, Ishida N (1991) Neocarzinostatin: selective tryptophan oxidation and neocarzinostatin-chromophore binding to apo-neocarzinostatin. Chem Pharm Bull (Tokyo) 39:170–176
50. Kim KH, Kwon BM, Myers AG, Rees DC (1993) Crystal structure of neocarzinostatin, an antitumor protein-chromophore complex. Science 262:1042–1046

51. Adjadj E, Quiniou E, Mispelter J, Favaudon V, Lhoste JM (1992) Three-dimensional solution structure of apo-neocarzinostatin from *Streptomyces carzinostaticus* determined by NMR spectroscopy. Eur J Biochem 203:505–511
52. Gao X (1992) Three-dimensional solution structure of apo-neocarzinostatin. J Mol Biol 225:125–135
53. Mohanty S, Sieker LC, Drobny GP (1994) Sequential ^{1}H-NMR assignment of the complex of apo-neocarzinostatin with ethidium bromide and investigation of protein-drug interactions in the chromophore binding site. Biochemistry 33:10579–10590
54. Ishiguro M, Imajo S, Hirama M (1991) Modeling study of the structure of the macromolecular antitumor antibiotic neocarzinostatin. Origin of the stabilization of the chromophore. J Med Chem 34:2366–2373
55. Maeda H, Kumagai K, Ishida N (1966) Characterization of neocarzinostatin. J Antibiot (Tokyo) 19:253–259
56. Maeda H, Koyanagi T, Ishida N (1968) C-Terminal amino acid of neocarzinostatin. Biochim Biophys Acta 160:249–251
57. Meienhofer J, Maeda H, Glaser CB, Czombos J (1972) In: Lande S (ed) Progress in peptide research. Gordon and Breach, New York, pp 295–306
58. Meienhofer J, Czombos J, Maeda H (1971) Reduction of disulfide bonds in peptides and proteins by dithiothreitol in liquid ammonia. J Am Chem Soc 93:3080–3081
59. Maeda H, Meienhofer J (1970) Amino- and carboxy-terminal sequence of neocarzinostatin, an antitumor polypeptide antibiotic. FEBS Lett 9:301–303
60. Konigsberg W (1967) Subtractive Edman degradation. In: Hire CHW (ed) Methods in enzymology, vol 11. Academic, New York, pp 461–469
61. Gray WR (1967) Sequential degradation plus dansylation. In: Hire CHW (ed) Methods in enzymology, vol 11. Academic, New York, pp 469–475
62. Glaser CB, Maeda H, Meienhofer J, Niall HD (1972) Partial sequence of the tyrosine region of neocarzinostatin. Experientia (Basel) 28:772–773
63. Hirayama K, Ando T, Takahashi R, Murai A (1986) A revised primary structure of neocarzinostatin using the combination of endopeptidase, carbxypeptidase Y and fast atom bombardment mass spectrometry. Bull Chem Soc Jpn 59:1371–1378
64. Kuromizu K, Meienhofer J (1972) Reactions in liquid ammonia. III. Complete aminoethylation of thiol groups in peptides and proteins by ethylenimine in liquid ammonia. J Biol Chem 247:5646–5647
65. Maeda H, Kuromizu K (1977) Spontaneous deamidation of a protein antibiotic, neocarzinostatin, at weakly acidic pH. J Biochem (Tokyo) 81:25–35
66. Samy TSA, Hahm KS, Modest EJ, Lampman GW, Keutmann HT, Umezawa H, Herlihy WC, Gibson BW, Carr SA, Biemann K (1983) Primary structure of macromomycin, an antitumor antibiotic protein. J Biol Chem 258:183–191
67. Sakata N, Kanbe T, Tanabe M, Hayashi H, Hori M (1989) Nucleotide sequence of the macromomycin apoprotein gene and its expression in *Streptomyces macromomyceticus*. J Antibiot (Tokyo) 42:1704–1712
68. Sakata N, Ikeno S, Hori M, Hamada M, Otani T (1992) Cloning and nucleotide sequencing of the antitumor antibiotic C-1027 apoprotein gene. Biosci Biotechnol Biochem 56:1592–1595
69. Hofstead SJ, Matson JA, Malacko AR, Marquardt H (1992) Kedarcin, a new chromoprotein antibiotic. II. Isolation, purification and physicochemical properties. J Antibiot (Tokyo) 45:1250–1254
70. Vandré DD, Zaheer A, Squier S, Montgomery R (1982) Largomycin: preparation, properties, and structure. Biochemistry 21:5089–5096

71. Chou PY, Fasman GD (1974) Prediction of protein conformation. Biochemistry 13:222–245
72. Chou PY, Fasman GD (1978) Empirical prediction of protein conformation. Annu Rev Biochem 47:251–276
73. Kumagai K, Maeda H, Ishida N (1966) Biological activity of neocarzinostatin and its derivatives. Antimicrob Agents Chemother 6:546–550
74. Maeda H (1974) Chemical and biological characterization of succinyl neocarzinostatin. J Antibiot (Tokyo) 27:303–311
75. Samy TSA (1977) Neocarzinostatin: effect of modification of side chain amino and carboxyl groups on chemical and biological properties. Biochemistry 16:5573–5578
76. Maeda H (1973) Chemical reactivity and state of tryptophan residues in neocarzinostatin under physiological condition. J Antibiot (Tokyo) 26:776–777
77. Edo K, Saito K, Akiyama Y, Mizugaki M, Koide Y, Ishida N (1986) Neocarzinostatin: carboxyl-terminal-43-peptide residue of apo-neocarzinostatin as the binding site of neocarzinostatin-chromophore. Chem Pharm Bull (Tokyo) 34:5180–5183

Neocarzinostatin Chromophore: Structure and Mechanism of DNA Cleavage

KIYOTO EDO[1] and YOSHIO KOIDE[2]

Summary. Neocarzinostatin (NCS) is a member of a family of chromoprotein antitumor antibiotics. It consists of two noncovalently bound components, a labile chromophore component (NCS-chrom) with biological activities such as DNA cleavage and a protein component that stabilizes NCS-chrom. NCS-chrom was the first of the enediyne antibiotics to be isolated and characterized. The chemical structure has been elucidated: a novel epoxybicyclo [7,3,0]dodecadienediyne system, an interconnecting highly unsaturated 12-member carbon ring subunit, bearing subunits 2-hydroxy-5-methyl-7-methoxyl-1-naphthoate, 2,6-dideoxy-2-methylamino-galactose, and ethylene carbonate. The mechanism for the activation of NCS-chrom highlights one of the crucial roles of the thiol in the mechanism of action of NCS. Nucleophilic attack by a thiol results in a Bergman-type electronic rearrangement to form a biradical species that cleaves DNA by abstracting hydrogen atoms from the deoxyribose sugar. Moreover, another pathway for the thiol-independent mechanism of action of NCS-chrom has been proposed. In this mechanism, the spirolactone cumulene is stereoselectively generated via an intramolecular Michael addition at C-12 by the enolate anion of naphthoate of NCS-chrom, resulting in the formation of the biradical.

Key words: Neocarzinostatin—Chromoprotein—Neocarzinostatin Chromophore—chemical structure—Enediyne core—DNA cleavage

Introduction

In general, the enediyne antibiotic family is classified into two groups. One group consists of a labile nine-membered enediyne-containing chromophore (chromoprotein's chromophore), noncovalently associated with its apoprotein

[1] Pharmacy, Fukushima Medical College Hospital, 1 Hikarigaoka, Fukushima, 960-12 Japan
[2] Tohoku College of Medical Technology and Welfare, Kunimi, Sendai, Miyagi, 981 Japan

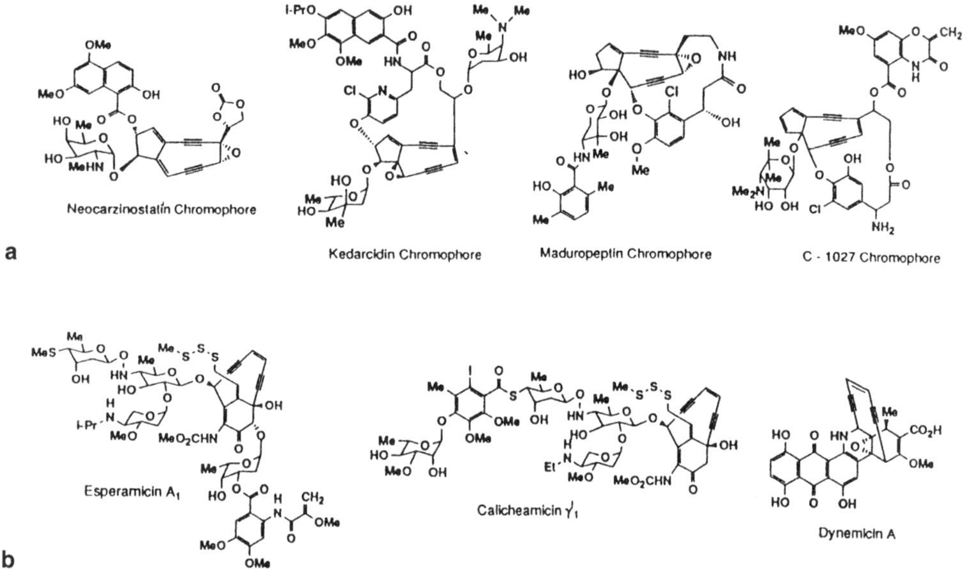

FIG. 1a,b. Families of the enediyne class of anticancer antibiotics. **a** Chromophores of chromoproteins (nine-membered enediyne-containing anticancer antibiotics). **b** Ten-membered enediyne-containing anticancer antibiotics

(Fig. 1a); the other group consists of ten-membered enediyne chromophoric compounds (Fig. 1b).

Enediyne chromoproteins are usually potent antitumor antibiotics with a unique molecular architecture. These chromoproteins include neocarzinostatin (NCS) [1–4], macromomycin [5], actinoxanthin [6,7], C-1027 [8–11], kedarcidin (KDC) [12–14], and maduropeptin (MDP) [15,16]. The ten-membered enediynes include esperamicin [17], calicheamicin [18], and dynemicin [19].

NCS is isolated from the culture filtrate of *Streptomyces carzinostaticus* var. F-41 [20,21]. It is composed of a protein component (apo-NCS) that consists of 113 amino acid residues [4,22–24] and a labile, nonprotein chromophore (NCS-chrom) [25–28]. Clinically, NCS has been used to treat patients with acute leukemia, gastric cancer, and pancreatic cancer in Japan since 1977 [29].

NCS was initially identified as a simple chain polypeptide (MW = 11 100) [20,30], capable of inhibiting DNA synthesis and inducing DNA degradation in cells [31,32]. The existence of a nonprotein chromophore possessing biological activity in the NCS molecule was derived as follows. First, it was found that the in vitro DNA-cleaving activity of NCS was enhanced, rather than inhibited, by adding agents that cause proteins to unfold or denature, such as organic solvents or urea, to the NCS solution [33]. Second, NCS was found to have ultraviolet-(UV-) visible absorption in the 300–360 nm range not assumed to occur with a simple protein [25]. Third, removal of this chromophoric materal by chromatog-

TABLE 1. Comparison of the antimicrobial spectra of neocarzinostatin chromophore (NCS-chrom) II, apo-NCS, and NCS-chrom II in the presence of apo-NCS and NCS

	Minimum inhibitory concentration (μg/ml)			
Test microorganisms	NCS-chrom II	Apo-NCS	NCS-chrom II[a] + apo-NCS	NCS
Staphylococcus aureus FDA209P	50	>100	1.56	6.2
Staphylococcus epidermidis IFO3726	25	>100	0.4	6.2
Sarcina lutea ATCC9341	12.5	>100	0.1	3.1
Streptococcus haemolyticus	25	>100	0.2	6.2
Streptococcus faecalis IFO12580	25	>100	0.4	6.2
Bacillus subtilis PCI219	25	>100	0.8	12.5
Bacillus cereus IFO3001	>100	>100	0.8	12.5
Bacillus licheniformis ATCC14580	>100	>100	3.1	12.5
Proteus vulgaris OX19	>100	>100	>100	>100
Proteus morganii FIO3848	>100	>100	>100	>100
Proteus rettgeri	>100	>100	>100	>100
Salmonella typhimurium	>100	>100	>100	>100
Salmonella paratyphi A1015	>100	>100	>100	>100
Salmonella typhi V1901	>100	>100	>100	>100
Shigella flexneri 2a	>100	>100	>100	>100
Shigella sonnei	>100	>100	>100	>100
Escherichia coli NIHJ	>100	>100	>100	>100
Pseudomonas aeruginosa GN83	>100	>100	>100	>100
Serratia marcescens IAM1022	>100	>100	>100	>100
Klebsiella pneumoniae GN350	>100	>100	>100	>100
Citrobactor freundii Ball07	>100	>100	>100	>100
Xanthomonas oryzae IAM1657	>100	>100	>100	>100
Candida albicans YU1200	>100	>100	>100	>100

[a] Minimum inhibitory concentration of NCS-chrom II on plate contained 10μg/ml of apo-NCS.

TABLE 2. Synergic action of NCS-chrom II and apo-NCS against Yoshida sarcoma

	MTD mg/kg[a]	MED mg/kg[b]
NCS-chrom II	>10.0	0.1
apo-NCS	>100.0	>10.0
NCS-chrom II + apo-NCS (1:10)	0.5	0.005
NCS	2.5	0.05

[a] MTD, maximum tolerated dose when given i.p.
[b] MED, minimum effective dose when given i.p.

raphy on Amberlite XAD-7 [34] and by gel filtration on Sephadex G-25 after UV-irradiation at 360nm resulted in total loss of biological activity [26,34].

In 1979, NCS-chrom was discovered and isolated from NCS by us and by others [25–28]. The accumulated evidence indicated that this NCS-chrom is responsible for biological activities such as DNA strand scission and the cytotoxic activities of

NCS against tumor and bacterial cells, while the apoprotein (apo-NCS) has no biological activities directly and stabilizes this chromophore, which is very labile, against UV light and heat [27,34,35] (Tables 1 and 2).

This chapter summarizes the elucidation of the structure of NCS-chrom and the mechanism of DNA cleavage by NCS-chrom at the molecular level.

Isolation and Physicochemical Properties of NCS-Chrom

Native NCS-chrom (epoxide form, NCS-chrom I) was extracted from lyophilized NCS powder with acetic acid in methanol [1]. Further, we extracted the NCS-chrom hydrogen chloride adduct (NCS-chrom II) [1,27] or the hydrogen bromide adduct (NCS-chrom III) [36] from purified NCS powder with hydrochloric or hydrobromic acid, respectively, in methanol. NCS-chrom III has been found to be a biologically active molecule and more stable than NCS-chrom I and NCS-chrom II to heat and UV light [36]. Three derivatives of NCS-chrom I, i.e., NCS-chrom II, NCS-chrom III, and the diol monomethyl ether, have been prepared by Goldberg et al. [37]. However, the monomethyl ether is biologically inactive.

Purified NCS-chrom II was obtained from NCS powder by extraction with methanolic hydrogen chloride followed by reverse-phase chromatography on a Wako-gel LC-ODS column (Fig. 2). The physicochemical properties of NCS-chrom II are summarized in Table 3.

The molecular formula of NCS-chrom I was established as $C_{35}H_{33}NO_{12}$ by fast atom bombardment-mass spectroscopy (FAB-MS), ^{1}H-NMR (nuclear magnetic

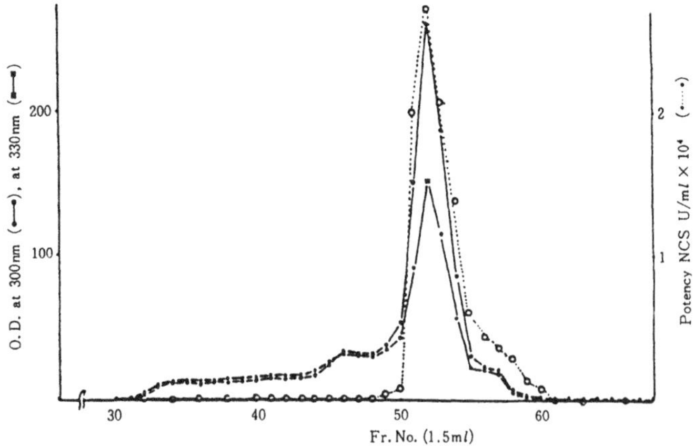

FIG. 2. Purification of NCS-chrom II by preparative high-pressure liquid chromatography (HPLC). Crude neocarzinostatin chromophore (NCS-chrom) II (1.3 g) in 70% MeOH (15 ml) was applied to a column (Wako-gel LC-ODS 15 × 300 mm) and eluted with 70% MeOH at flowrate of 1 ml/min; 1.5 ml of each fraction was collected. Each fraction was monitored by UV absorption at 300 nm (*dotted line, solid circles*), 330 nm (*dashed line, squares*), and potency against *Micrococcus luteus* (*dashed line, open circles*)

TABLE 3. Physicochemical properties of NCS-chrom II

MP	125°C (decomp)
$[\alpha]_D^{20}$	−171° (c 0.1, MeOH)
Solubility	MeOH soluble, water insoluble
Elemental analysis	
Found:	C, 53.41; H, 5.00; N, 1.67; Cl, 9.20
Calculated:	C, 53.44; H, 5.25; N, 1.78; Cl, 9.01 (for $C_{35}H_{34}NO_{12}Cl \cdot HCl \cdot 3H_2O$)
Color reaction	
Positive:	$KMnO_4$, diazo coupling, $AgNO_3$, Elson–Morgan
Negative:	Ninhydrin, biuret, Molish, Ehrlich

MP, Melting point.

resonance), and ^{13}C-NMR, indicating 20 degrees of unsaturation in the molecule. This formula is in agreement with that reported by Hensens et al. [38]. These data suggest that NCS-chrom I is the native form, while NCS-chrom II is a hydrogen chloride adduct.

Partial Structure of NCS-Chrom

We first proposed that NCS-chrom possesses 2-hydroxy-5-methoxy-7-methyl-1-naphthalenecarboxylic acid (NA′) [39]. Next, Goldberg et al. reported that NCS-chrom consisted of *N*-methyl-fucosamine (2,6-dideoxy-2-methylamino-galactose, MF) [40] and ethylene carbonate (EC) [41]. Further, partial structures of NCS-chrom were determined to be an aliphatic aromatic diacyl peroxide moiety (by us [42]) and a highly stranded epoxide (by Goldberg et al. [38]). Later, Shibuya et al. revised the structure of the naphthalene carboxylic acid moiety (NA′) to its stereoisomer, 2-hydroxy-7-methoxy-5-methyl-1-naphthalenecarboxylic acid (NA), on the basis of its total synthesis [43] (Fig. 3).

Spectroscopic Data of NCS-Chrom

No crystals were suitable for X-ray, so this powerful technique could not be used for structure elucidation of NCS-chrom containing its stable derivatives. It may be impossible to obtain adequate diffraction without decomposition even if it is possible to get suitable crystals of NCS-chrom. Most of the structural information for NCS-chrom was determined by NMR (two-dimensional homo- and heteronuclear correlation techniques), mass spectrometry (MS) [high-resolution measurement, fast atom bombardment (FAB), soft ionization techniques, and secondary ion (SI) analysis], infrared (IR), Fourier transform infrared (FT-IR), and ultraviolet (UV) spectrometry. Initially, we obtained much

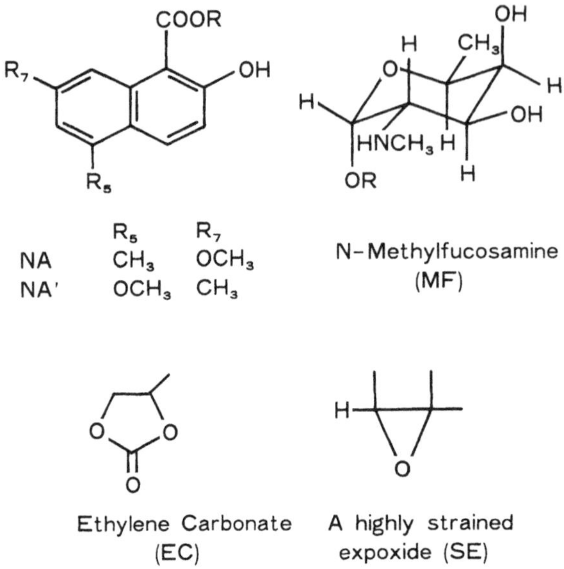

FIG. 3. Partial structure of NCS-chrom

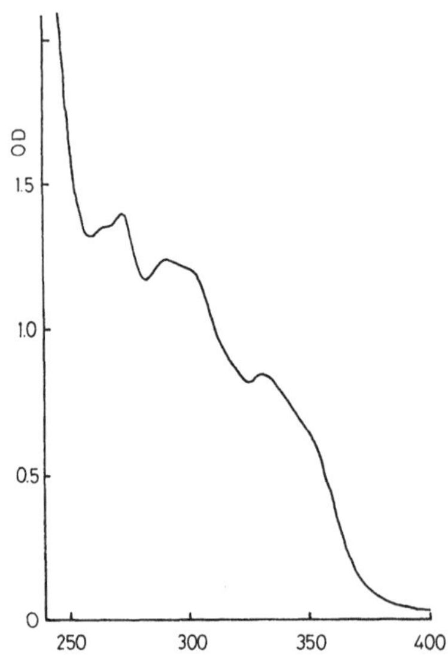

FIG. 4. UV absorption spectra of NCS-chrom II (70 μg/ml in methanol) measured by Hitachi spectrophotometer model 624 (Hitachi, Tokyo, Japan) in a 1-cm quartz cell

spectral data from NCS-chrom II because of its relative stability compared with NCS-chrom I.

Ultraviolet Absorption

NCS-chrom II has considerable absorption fine structure with maxima or shoulders at wavelengths of 228, 265, 272, 287, 300, 330, and 350 nm (log ε = 4.75, 4.23,

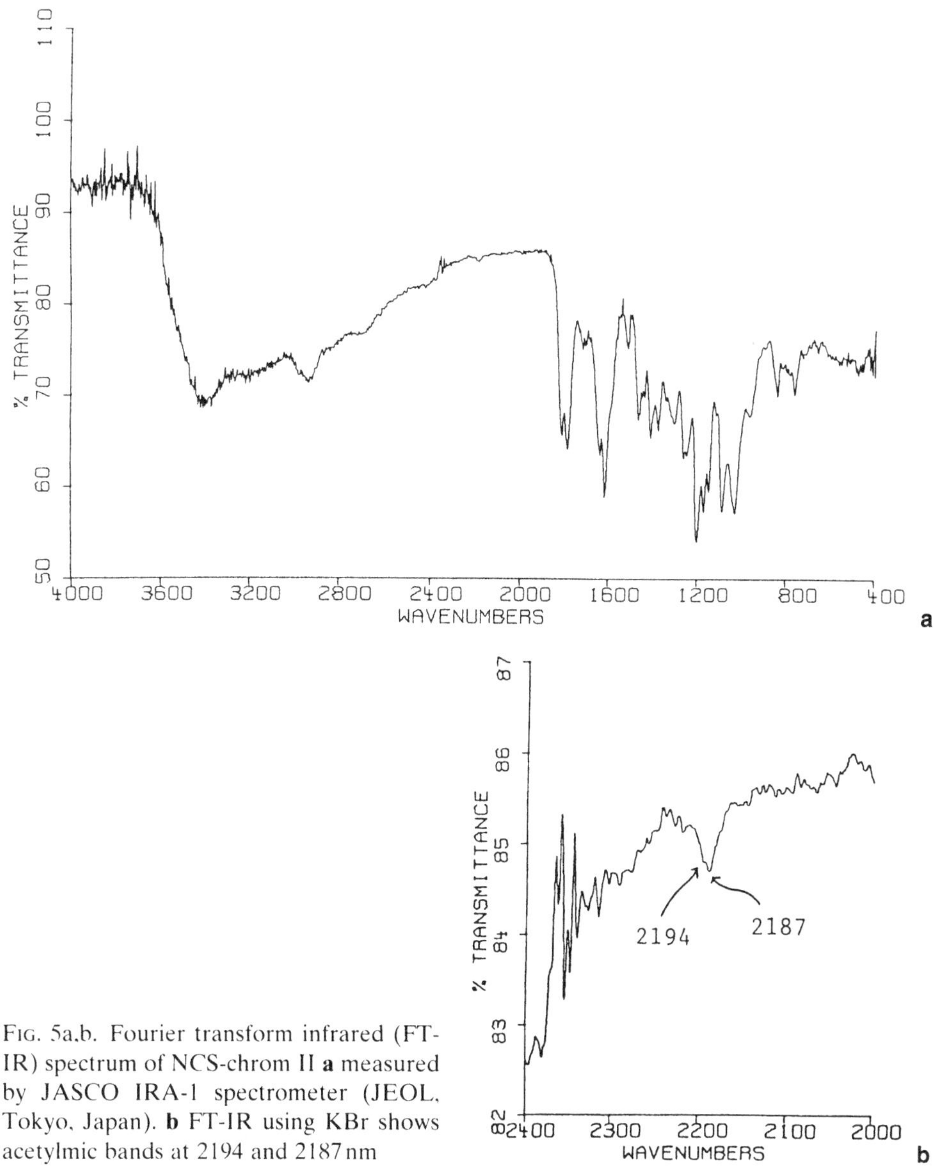

FIG. 5a,b. Fourier transform infrared (FT-IR) spectrum of NCS-chrom II **a** measured by JASCO IRA-1 spectrometer (JEOL, Tokyo, Japan). **b** FT-IR using KBr shows acetylmic bands at 2194 and 2187 nm

4.11, 4.11, 4.04, and 3.80, respectively) (Fig. 4). This UV spectrum was indicative of aromatic functionality with extended conjugation. The UV spectrum of NCS-chrom II has absorbance at 330 and 350 nm, which could be attributed to naphthalencarboxylate (NA), and at 272, 287 and 300 nm, which could result from the enediyne core moiety [44].

Infrared Spectrometry

The typical IR fingerprint of NCS-chrom II, composed of absorptions at 3400, 2960, 2922, 2850, 2080, 1811, 1783, 1715, 1638, 1613, 1520, 1463, 1409, 1380, 1300, 1260, 1200, 1172, 1150, 1087, 1030, 965, 843, and 765 nm, can be useful for identification purposes. Several bands characteristic of hydroxyl, aromatic ester, and α,β-unsaturated groups are found in the FT-IR spectrum (Fig. 5a). The FT-IR spectrum showed that the carbonate and naphthoate groups were preserved (1811, 1783, and 1638 cm^{-1}). Although no absorption of acetylenic groups in the range of 2300–2100 cm^{-1} is found in the IR spectra of NCS-chrom II, in the FT-IR spectra of NCS-chrom II (KBr tablet) twin acetylenic bands are observed at 2178 and 2194 cm^{-1} (Fig. 5b).

Mass Spectrometry

Using positive-ion [MH^+] FAB of NCS-chrom I and NCS-chrom II freshly extracted from NCS powder, strong peaks were obtained at m/z 660 for NCS-chrom I and at m/z 696 for NCS-chrom II, respectively. These values indicate that the molecular weights of NCS-chrom I and NCS-chrom II are 659 and 695, respectively. The structures of the molecular and ion fragment species at m/z 696 (MH^+), 482 (MH^+–$C_{13}H_{10}O_3$), 215 ($C_{13}H_{10}O$, NA), 160 ($C_7H_{14}O_3$, MF), and 88 ($C_3H_4O_3$, EC) were obtained by linked scan secondary ion mass spectrometry (SIMS) (Fig. 6). On the other hand, Hensens et al. reported high resolution mass spectrometry (HRMS) data of the trimethysilyl derivative of the $NaBH_4$ reduction product of NCS-chrom I (i.e., NCS-chrom I–methyl thioglycolate adducts) [41].

Nuclear Magnetic Resonance Spectrometry

To protect against polymerization, NCS-chroms were extracted directly with solvents for NMR such as $^{12}CD_3COOD$ or $^{12}CD_3OD$. The 100-MHz ^{13}C-NMR spectrum of NCS-chrom II taken in $^{12}CD_3OD$ indicated the presence of 35

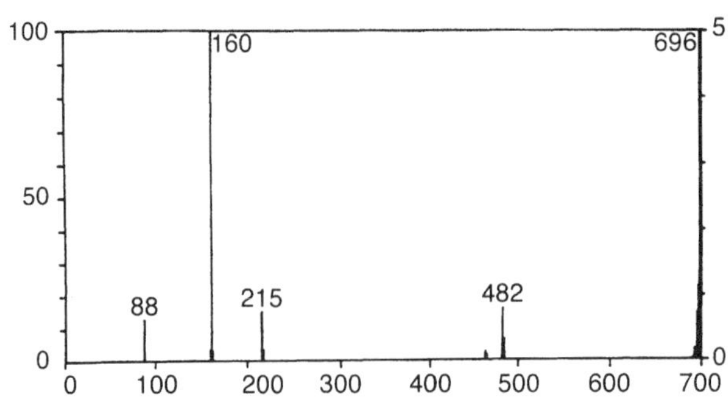

FIG. 6. Linked scan mass spectrometry (MS) of NCS-chrom II

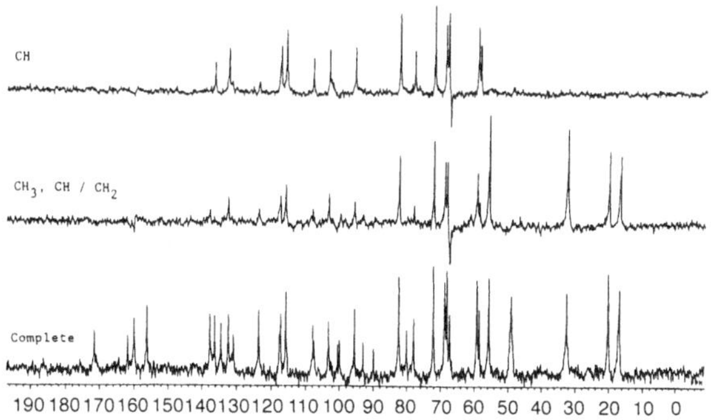

FIG. 7. 100-MHz ^{13}C-NMR (nuclear magnetic resonance) and INEPT (insensitive nuclei enhanced by polarization transfer) spectra of NCS-chrom II. NMR spectrum was measured by JNM-GX400 NMR spectrometer (JEOL)

carbons, classified as 4 methyls, 1 methylene, 15 methines, and 15 quaternary carbons on the basis of off-resonance decoupled and INEPT experiments (Fig. 7). The well-spaced 400-MHz ^{1}H-NMR spectrum taken in CD_3OD showed the presence of 29 nonexchangeable protons, which agrees with the number of protons counted by INEPT. All these proton signals were correlated to carbon resonances by selective proton decoupling [1] and the two-dimensional homodecoupling technique (Fig. 8 and Table 4).

Total Chemical Structure Elucidation of NCS-Chrom

Elucidation of the total chemical structure of NCS-chrom combined efforts of several laboratories [1,37–43,45–48]. Because of the lability of the 12-member carbon unit, it has not been isolated directly. Also, the high degree of unsaturation of the subunit has made completion of the structure assignment by NMR analysis difficult. In this section, we describe the total structure of NCS-chrom I as well as its hydrogen chloride adduct (NCS-chrom II).

Core Structure of NCS-Chrom

The remaining central core (12-membered carbon ring) must therefore consist of $C_{12}H_5$, which accounts for the remaining five sites of unsaturation in the parent molecule. After the proton and carbon signals attributable to the NA, MF, and EC moieties had been excluded, 5 protons and 12 carbons present in a very highly unsaturated moiety common to NCS-chrom I and NCS-chrom II remained to be assigned, and 5 methine and 7 quarternary carbons remained to be assigned to the C_{12} subunit. To reveal this moiety, NMR spectra were carefully analyzed to prove that both NCS-chrom I and NCS-chrom II have no aliphatic aromatic diacyl peroxide as proposed previously by us [37,42].

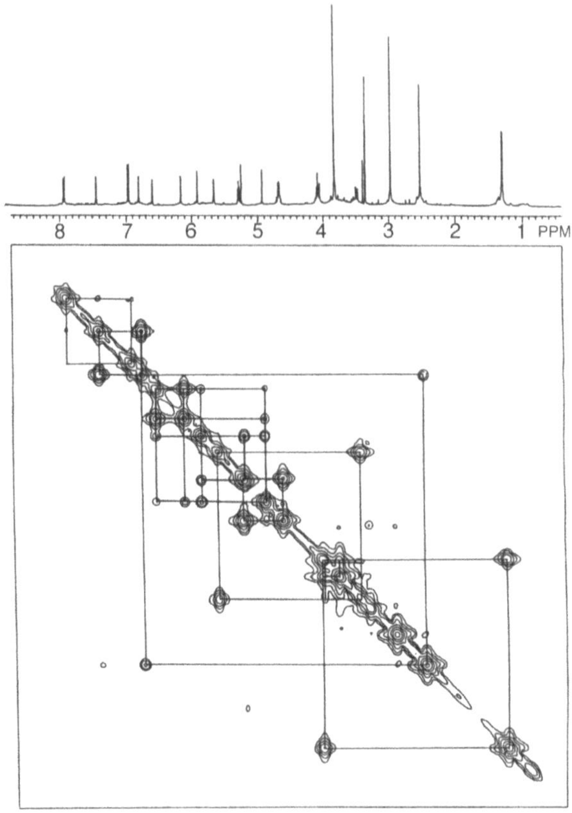

FIG. 8. Two-dimensional 400-MHz ^{1}H-NMR spectrum of NCS-chrom II

Because the coupling constants of H-10 in NCS-chrom II with H-8, -11, -12 and -1′ (anomeric proton of MF) were very small, distinction between vicinal and long-range couplings could not be made with certainty. However, nuclear Overhauser effect (NOE) experiments revealed the spatial proximity of H-10 with H-8, H-11, and H-1′. The magnitude of the coupling constant between H-11 and H-12 (J = 3.0 Hz) confirmed the vicinal relationship of these two protons. The ^{13}C-NMR data of NCS-chrom II indicated that C-8 (107.8 ppm) and C-12 (136.2 ppm) are *sp*^{2} carbons and that C-10 (82.5 ppm) and C-11 (82.6 ppm) are oxygenated methines. Thus, the relationship of these carbons is represented by either one of the two partial structure (a and b) shown in Fig. 9.

Because H-8 and H-12 are long-range coupled with each other (J = 1.2 Hz) without showing NOE enhancement, a and b must be extended to c and d, respectively (Fig. 9). However, accommodation of the remaining acetylenic units (vide infra) to the partial structure d would give a symmetrical structure, which could not account for the large ^{13}C-chemical shift differences between C-8 and C-12 and between C-1 and C-9. Thus, the partial structure d was excluded. Similar

TABLE 4. 100-MHz ^{13}C-NMR and 400-MHz ^{1}H-NMR spectral data of NCS-chrom I and NCS-chrom II

		NCS-chrom I (CD_3CO_2D, CD_3OD, 1:1)			NCS-chrom II ($^{12}CD_3OD$)		
		^{13}C(ppm)	$^{1}J_{C\text{-}H}$	H(ppm)	^{13}C(ppm)	$^{1}J_{C\text{-}H}$	H(ppm)
	1	129.8(s)			130.9(s)		
	2	87.5(s)			89.9(s)		
	3	97.6(s)			99.9(s)		
	4	63.8(s)			80.1(s)		
	5	55.2(d)	196.6	4.14(1H, brs)	58.5(d)	162.6	5.20(1H, brs)
	6	99.7(s)			100.5(s)		
	7	90.7(s)			93.0(s)		
	8	106.5(d)	168.8	5.82(1H, brs)	107.8(d)	169.5	5.89(1H, d, J = 1.2)
	9	160.2(s)			156.2(s)		
	10	82.2(d)	158.4	5.09(1H, brs)	82.5(d)	159.0	4.95(1H, brs)
	11	81.6(d)	156.7	6.25(1H, brs)	82.6(d)	150.0	6.14(1H, d, J = 3.0)
	12	139.4(d)	176.0	6.79(1H, brs)	136.2(d)	174.9	6.58(1H, dd, J = 3.0, 1.2)
	13	76.1(d)	160.0	4.93(1H, dd, J = 8.0, 5.0)	78.0(d)	162.6	5.25(1H, dd, J = 8.0, 5.5)
EC	14	68.0(t)	156.3	4.54(1H, dd, J = 5.0, 9.0)	67.3(t)	157.5	4.61(1H, dd, J = 5.5, 9.0)
				4.75(1H, dd, J = 9.0, 8.0)			4.64(1H, dd, J = 9.0, 8.0)
	15	155.6(s)			156.1(s)		
	1′	95.4(d)	168.7	5.75(1H, d, J = 3.0)	95.6(d)	174.0	5.63(1H, d, J = 3.8)
	2′	59.5(d)	144.6	3.65(1H, dd, J = 3.0, 10.5)	59.2(d)	142.5	3.43(1H, dd, J = 3.8, 11.0)
MF	3′	68.2(d)	143.8	4.21(1H, dd, J = 10.5, 2.5)	68.1(d)	142.5	3.99(1H, dd, J = 11.0, 3.0)
	4′	72.4(d)	146.2	3.90(1H, d, J = 2.5)	72.2(d)	142.5	3.76(1H, d, J = 3.0)
	5′	69.1(d)	136.4	4.07(1H, d, J = 6.5)	68.9(d)	144.6	4.07(1H, d, J = 6.5)
	6′	16.6(q)	ND	1.26(3H, d, J = 6.5)	16.6(q)	127.3	1.26(3H, d, J = 6.5)
	2′-NCH_3	32.7(q)	ND	3.02(3H, s)	32.3(q)	142.8	2.92(3H, s)
	1″	106.2(s)			107.4(s)		
	2″	136.2(s)			161.8(s)		
	3″	116.4(d)	164.4	7.03(1H, d, J = 9.0)	115.8(d)	163.8	6.98(1H, d, J = 9.5)
	4″	133.6(d)	158.4	8.10(1H, d, J = 9.0)	132.2(d)	159.0	8.00(1H, d, J = 9.5)
	5″	138.1(s)			137.6(s)		
NA	6″	117.6(d)	160.4	6.90(1H, brs)	117.3(d)	155.7	6.84(1H, d, J = 2.0)
	7″	160.4(s)			159.9(s)		
	8″	104.2(d)	160.4	7.76(1H, d, J = 2.0)	103.1(d)	160.5	7.51(1H, d, J = 2.0)
	4a″	124.0(s)			123.6(s)		
	8a″	134.9(s)			134.4(s)		
	1″-<u>C</u>OO	172.1(s)			171.4(s)		
	5″-<u>C</u>H_3	20.0(q)	ND	2.62(3H, s)	20.0(q)	127.3	2.56(3H, s)
	7″-O<u>C</u>H_3	55.9(q)	143.8	3.83(3H, s)	55.6(q)	143.4	3.83(3H, s)

NMR, Nuclear magnetic resonance; EC, ethylene carbonate; MF, *N*-methylfucosamine; NA, naphthalenecarboxylic acid; ND, not determined.

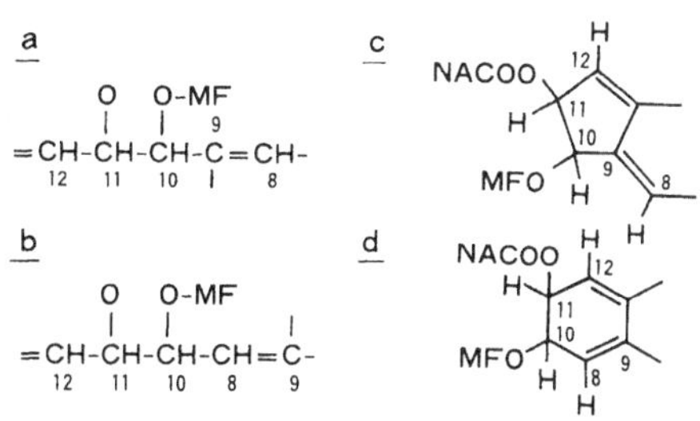

FIG. 9a–d. Common partial structures (***a–b***) for the unknown moiety in NCS-chrom I and NCS-chrom II

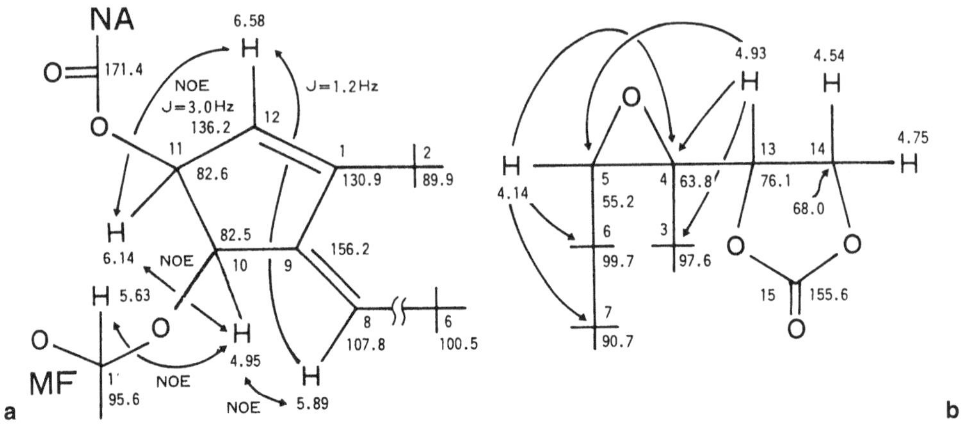

FIG. 10. Partial structure of NCS-chrom. **a** 2-Cyclopentene-1-ylidene (NMR data from NCS-chrom I). **b** Linkage of the epoxide and ethylene carbonate (NCS-chrom II). *Arrows* show long-range ^{13}C–H couplings

observations were made with NCS-chrom I (see Table 4). The relationship shown in Fig. 9c was confirmed by a long-range selective proton decoupling (LSPD) experiment irradiating H-12 of NCS-chrom II, which resulted in the collapse of the C-2, C-9, and C-10 resonances. When the resonances of chemical shifts of H-10 and H-11 of NCS-chrom II were irradiated, C-1′ and 1″-COO (171.4 ppm) were affected, respectively. On the basis of these results, we propose the 2-cyclopentene-1-ylidene unit indicated in Fig. 10a as the common partial structure for the unknown moiety in NCS-chrom I and NCS-chrom II. The relative stereochemical relationship between H-10 and H-11 is assumed to be trans because of the very small coupling between them.

The linkage of the epoxide and ethylene carbonate moieties was also established by NMR analysis. In the ^{13}C-NMR spectrum of NCS-chrom I, the very large coupling constant ($J_{C\text{-}H}$ = 196.6 Hz) of C-5 revealed the presence of the epoxide as reported by Hensens et al. [38]. Disappearance of the epoxide in NCS-chrom II was shown by comparison of ^{1}H-NMR and ^{13}C-NMR data of NCS-chrom I and NCS-chrom II (in NCS-chrom II, C-4-80.1 ppm, C-5-58.5 ppm, and $J_{C\text{-}H}$ = 162.6 Hz). The ^{13}C chemical shifts of the remaining carbons in NCS-chrom I and NCS-chrom II are very similar (see Table 4). These results suggest that NCS-chrom II is a hydrogen chloride adduct of NCS-chrom I at the epoxide moiety. When H-13 of the ethylene carbonate of NCS-chrom I was irradiated by an LSPD experiment, C-3 and both the epoxide carbons C-4 and C-5 were decoupled. On the other hand, when the epoxide proton H-5 was irradiated, the three quarternary carbons C-4, C-6, and C-7 were affected. Thus, the remaining partial structure has been elucidated as indicated in Fig. 10b.

The final problem is to connect partial structures a and b (Fig. 10). In the FT-IR spectrum (KBr) of NCS-chrom II, twin acetylenic bands at 2187 and 2194 cm^{-1} were observed (see Fig. 5b). In addition, there remained only four unassigned carbons in NCS-chrom II with the degree of unsaturation being five, which could be explained only by the presence of two acetylenic bonds and one ring structure. The relationship between structures a and b was examined by long-range spin decoupling and LSPD experiments. The epoxide methine proton H-5 of NCS-chrom I was long-range coupled with H-8. LSPD experimental results unequivocally suggested the connection pattern indicated in Fig. 11. The quaternary carbon resonances of NCS-chrom I were assigned to cyclic diacetylenic carbons in a highly strained system [49] by LSPD experimental data (C-2, C-3, C-6, and C-7, are 87.5, 97.6, 99.7, and 90.7, respectively). The structure in Fig. 11 was supported by the molecular and fragment ion species at m/z 696 (MH^+), 482 (MH^+-$C_{13}H_{10}O_3$), 215 ($C_{13}H_{10}O$, NA), 160 ($C_7H_{14}O_3$, MF), and 88 ($C_3H_4O_3$, EC) obtained by linked scan SIMS (see Fig. 6). From these results, the relative chemical structure of NCS-chrom I and its hydrogen chloride adduct (NCS-chrom II) were proposed as indicated in Fig. 11.

Stereochemistry of NCS-Chrom

The stereochemistry of the C-10 and C-11 substituents was shown to be trans but at C-4, C-5, C-10, C-11, and C-13 was not assigned [1]. The *N*-methylfucosamine appendage has been determined unambigously to be a D-sugar by the method of total synthesis [45]. The absolute and relative stereochemistry of NCS-chrom I was determined from spectroscopic data of the methyl thioglycolate adduct of NCS-chrom I by Myers et al. [46]. Thus, four diastereomers of model compounds, synthesized in nonracemic form, were tabular placed with the corresponding derivatives, and quantitative assessment of the data by computer analysis showed that the complete stereochemical assignment of **2** is 4R, 10R, 11S, 12S, and 13R. From these observations, it follows that the stereochemistry of NCS-chrom I is 4S, 5R, 10R, 11R, and 13R, unequivocally (Fig. 12).

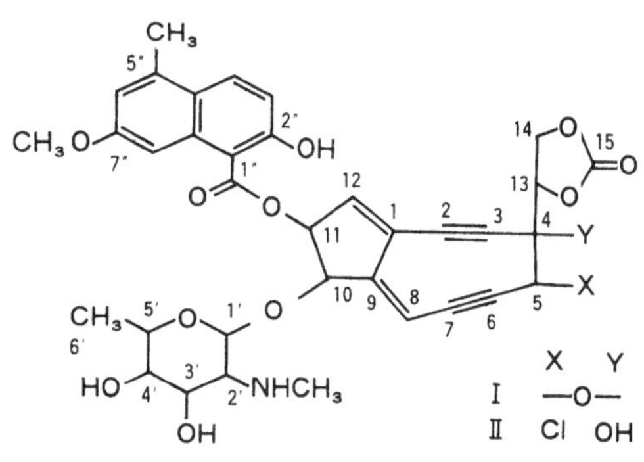

FIG. 11. The relative total structure of NCS-chrom I (native form) and NCS-chrom II (hydrochloride adduct form)

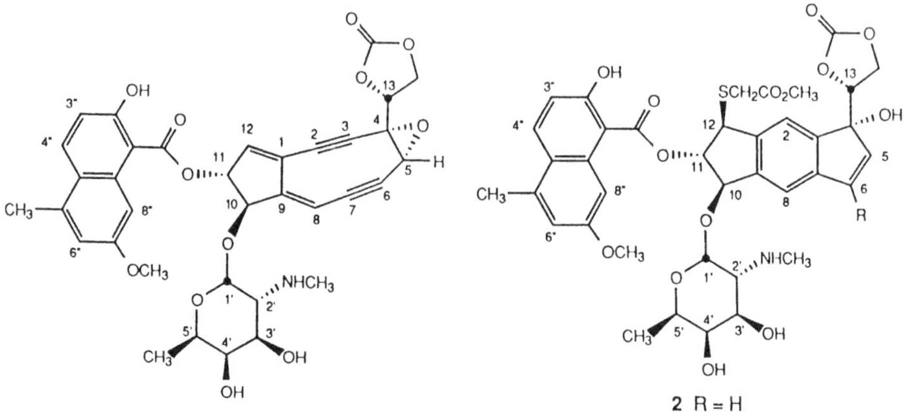

1

2 R = H

3 R = $SCH_2CO_2CH_3$

FIG. 12. Stereochemistry of NCS-chrom I (*1*) and structure of NCS-chrom I–methyl thiogycolate adducts (*2,3*)

Independent Findings in Support of the Proposed Structure

The synthesis of a bicyclo[7,3,0]dodeca-1,8-dien-2,6-diyne, the bicyclic core of NCS-chrom, was reported by Wender et al. [47]. ^{13}C-NMR data of the parent compound were in agreement with those reported for NCS-chrom I and its hydrogen chloride adduct (NCS-chrom II) (Table 5). The synthetic compound was found to be unstable at room temperature. The instability of NCS-chrom was proved by synthesis of the parent compound.

Futhermore, this unique structure has been supported by biosynthesis studies by Hensens et al. [48]. An acetate incorporation study on a culture of *Streptomyces carzinostaticus* suggested that the C_{12} naphthoic acid ring is derived from hexaketide (Fig. 13a), and the C_{14} cyclic carbonate/bicyclo[7,3,0] dodecadienediyne ring system appears to be derived from a minimum of eight head-to-tail coupled acetate units via the oleate-crepenynate biosynthetic pathway for polyacetylenes (Fig. 13b).

Mechanism of DNA Cleavage

The mechanism of DNA cleavage of NCS on tumor and bacteria cells was studied before the isolation of the NCS-chroms. However, DNA cleavage reactions induced by NCS-chrom have received only very preliminary investigation because information on the structure of NCS-chrom is limited. Therefore, the precise mechanism by which it effects DNA cleavage is not yet established at the molecular level. However, the proposed 12-member core subunit of NCS-chrom I suggests that the reactive intermediate could be a biradical produced from NCS-

TABLE 5. ^{13}C-NMR data for NCS-chrom I, NCS-chrom II, and bicyclic core

Carbon number / Solvent	NCS-chrom I (CD$_3$CO$_2$D, CD$_3$OD, 1:1)[b]	NCS-chrom II (CD$_3$OD)[b]	Bicyclic core[a] (THF-D$_8$)[c]
9	160.2	156.2	163.7
12	139.4	136.2	139.3
1	129.8	130.9	129.3
8	106.5	107.8	99.2
7	90.7	93.0	92.8
6	99.7	100.5	99.0
3	97.6	99.9	98.2
2	87.5	89.9	87.7
11	81.6	82.6	31.6
10	82.2	82.5	30.8
4	63.8	80.1	20.7
5	55.2	58.5	19.9

[a]
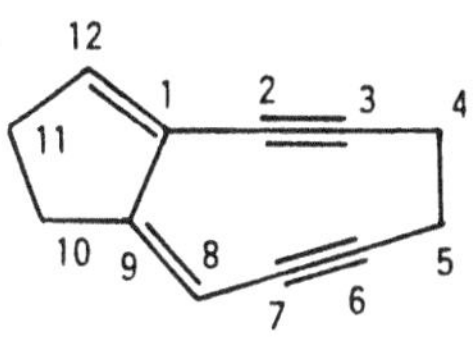

[b]Spectra recorded at room temperature.
[c]Spectrum recorded at −30°C to −35°C.

chrom I in a fashion analogous to that implicated in the mode of action of esperamicin/calicheamicin.

Generation of a Biradical of NCS-Chrom

The results of investigation of the effects of 2-mercaptoethanol (2-ME) and several organic solvents on the structural conformation and biological activities of NCS are important in understanding the inhibition of DNA synthesis in cultured mammalian cells as well as strand scission of DNA in vivo and in vitro [31,32].

Further, it is suggested that a free radical plays an important role in NCS-chrom-induced DNA strand scission [50]. Also, it was suggested that certain anticancer drugs such as bleomycin and adriamycin induce DNA strand scission through a mechanism involving the generation of free radicals such as a superoxide anion ($O_2\cdot^-$) and hydroxyl radical (OH·).

Generation of a free radical in DNA cleavage by NCS-chrom was confirmed by electron spin resonance (ESR) spectroscopy, using the spin-trapping technique. Figure 14 shows the ESR spectrum of free radicals formed by NCS-chrom II in the presence of 2-ME and *N-tert*-butyl-α-phenylnitrone (BPN). This spectrum has the parameters $g = 2.006$, $a^N = 15.6\,G$, and $a^H_\beta = 4.1\,G$. These values suggest that the radicals are differnt from those known for the OH· or $O_2\cdot^-$ spin adduct of

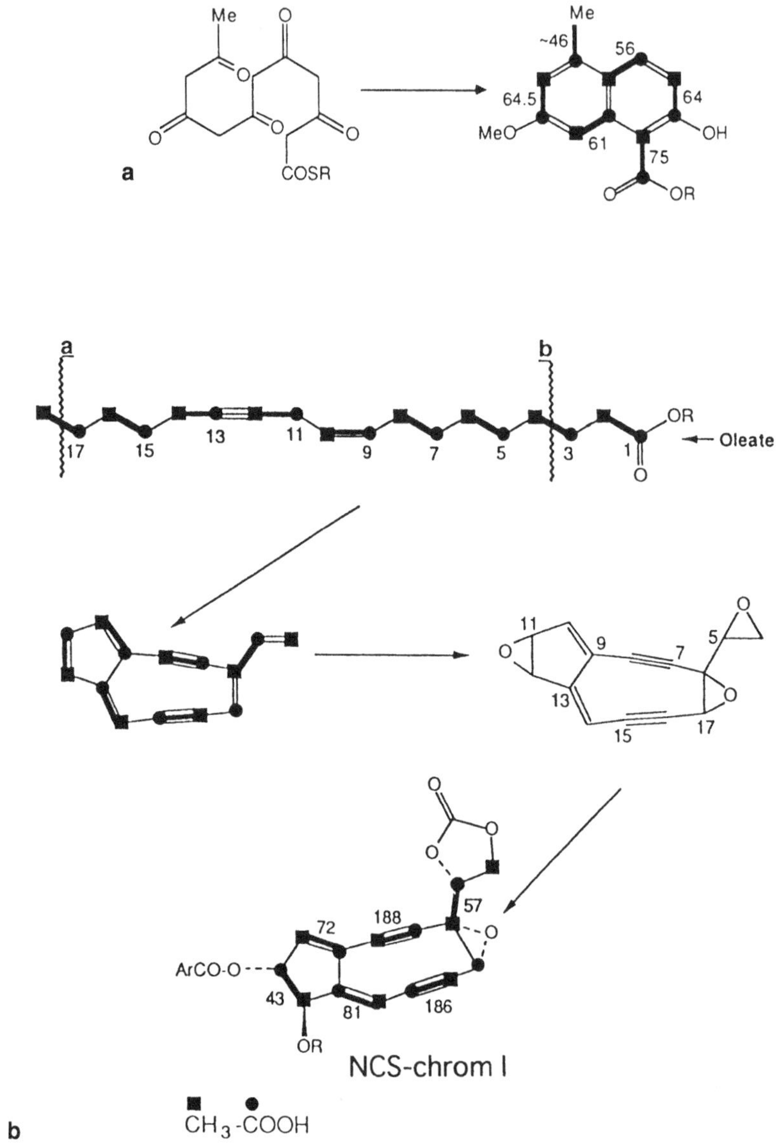

FIG. 13a,b. Proposed biosynthetic pathway of NCS-chrom I. **a** Naphthoic acid ring of NCS-chrom I from labeling pattern with ^{13}C-acetate. **b** Enediyne core structure of NCS-chrom I from labeling pattern with ^{13}C-acetate. *Squares*, CH_3; *Circles*, COOH. $^1J_{cc}$ (in hertz) indicated for each acetate pair

BPN. Of the free radical scavengers, superoxide dismutase (SOD) and D-mannose had little effect but α-D-tocopherol was effective in reducing NCS-chrom-induced radical formation. These effects of various agents on radical formation induced by NCS-chrom parallel those on DNA degradation induced

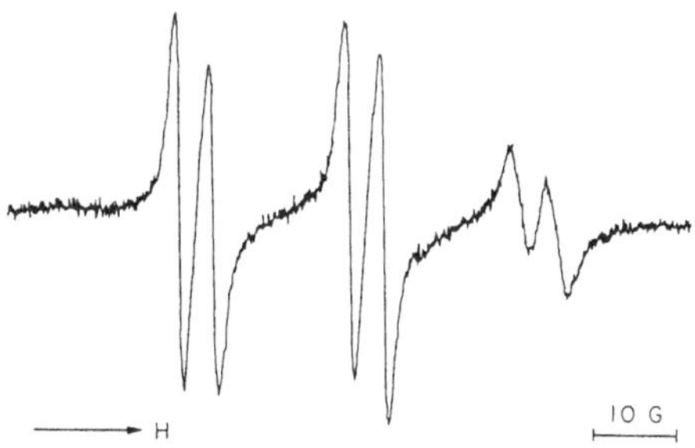

FIG. 14. Electron spin resonance (ESR) spectrum from NCS-chrom II (500 μg/ml) in the presence of 2-mercaptoethanol (2-ME, 5 m*M*) and saturated *N-tert*-butyl-phenylnitorone (BPN). The conditions of ESR spectroscopy were microwave power, 20 mW; modulation amplitude, 1G; time constant, 1 s; scan time, 8 min

by NCS-chrom. The foregoing data strongly suggested that the generation of a radical plays an important role in the DNA cleavage activity of NCS-chrom [50]. This same observation, that NCS-chrom induced radical formation, was reported by Sheridan and Gupta [51].

On the other hand, the side-chain model compounds of NCS-chrom, such as the synthetic methyl ester of naphthalenecarboxylate, synthetic D-methyfucosamine, and propyrene carbonate modeled as an ethylene carbonate moiety, have no biological activity. Further, the characteristic change in absorption between native NCS-chrom and its degradation product occurred at 274, 292, and 302 nm (see Fig. 4), caused by the dienediyne core. Thus, the biological activities of NCS-chrom can be traced to the bicyclic core composed of an oxygenated enediyne by the spectral data and biological activities for NCS-chrom and related compounds [44].

NCS-Chrom-Induced Cleavage Chemistry

NCS-chrom I undergoes rapid irreversible reaction with thiols to produce a species that is capable of cleaving DNA under aerobic incubation. The structure of NCS-chrom I-methyl thioglycolate adduct, a key compound of the DNA cleavage mechanism, was elucidated from the data of ^{1}H-NMR and MS of Hensens et al. [38] by Myers [52]. Myers proposed the precise mechanism for NCS-chrom I activation by thiols at the molecular level [53,54] (Fig. 15). Cycloaromatization of the enediyne core of NCS-chrom I, on nucleophilic attack by thiol at C-12, is proposed to proceed via formation of a cumulene intermediate **3** to a biradical species **4** of NCS-chrom I. In this mechanism, nucleophilic attack by thiol at C-12 and the epoxide ring opening generate a cumulene intermediate **2,** which cyclizes to form the indacene biradical with radical centers at C-2 and C-6 (**3**). The rearrangement is closely analogous to the enediyne-benzenediyl rearrangement (Bergman-type rearrangement). The biradical then abstracts hydrogens from DNA to form the reduced chromophore. The intermediate (**3**) was detected by ^{1}H-NMR at −70°C [54].

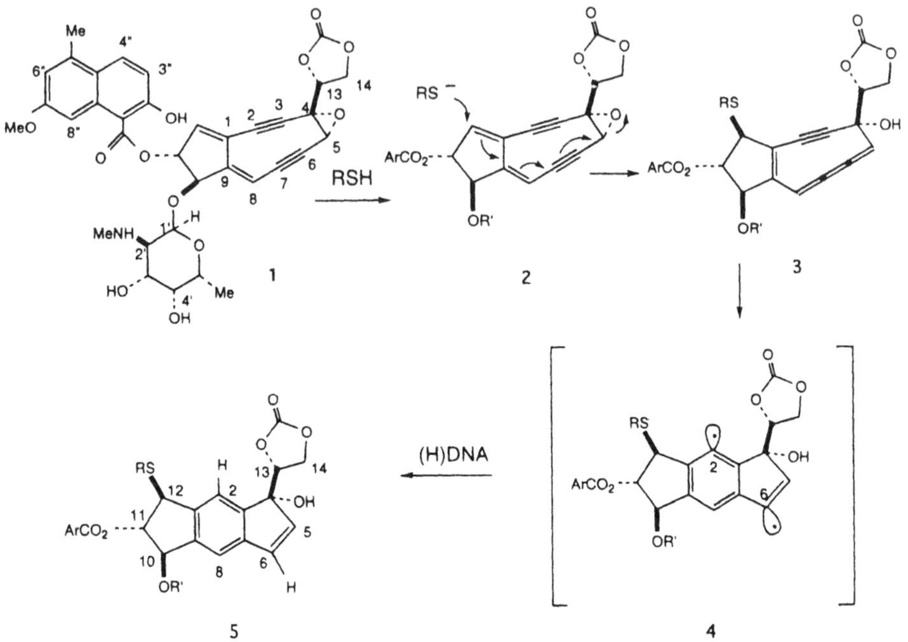

FIG. 15. Thiol activation of NCS-chrom I to the biradical species. *RSH*, Glutathione

Goldberg et al. [55] recently proposed that thiol-independent mechanisms of NCS-chrom I-induced cleavage of bulged DNA are catalyzed by an alkaline condition initially (Fig. 16). After the enolate at C-1″ of naphthoate of NCS-chrom I attacks at C-12, the spirolactone cumulene intermediate 1c (intramolecular Michael addition) is generated stereoselectively at the epoxide ring opening. Then, 1c intermediates rearrange to form the indacene biradical 1d (Bergman-type rearrangement) similar to the thiol-dependent NCS-chrom I-induced DNA cleavage mechanism. The radical at C-2 is then quenched by C-8″ of naphthoate to the final product 3 in the presence of bulged DNA. This base-catalyzed novel mechanism is undergone with DNA bearing a specific bulge conformation structure in the absence of thiol.

Role of Moieties of NCS-Chrom in DNA Cleavage

NCS-chrom binds in the minor groove of duplex DNA by intercalation of its naphthoate moiety and the specific electrostatic recognition between the negatively charged sugar–phosphate backbone of DNA and a positively charged *N*-methyl sugar moiety. These two attachment sites are viewed as serving as anchors for proper placement in the minor groove of the highly strained epoxybicyclo[7,3,0]dodecadiyne system close to the C-5 of deoxyribose [56].

Furthermore, it has been proposed by Myers et al. that the amino group plays a role in thiol activation because the carbohydrate residue of the NCS-chrom–

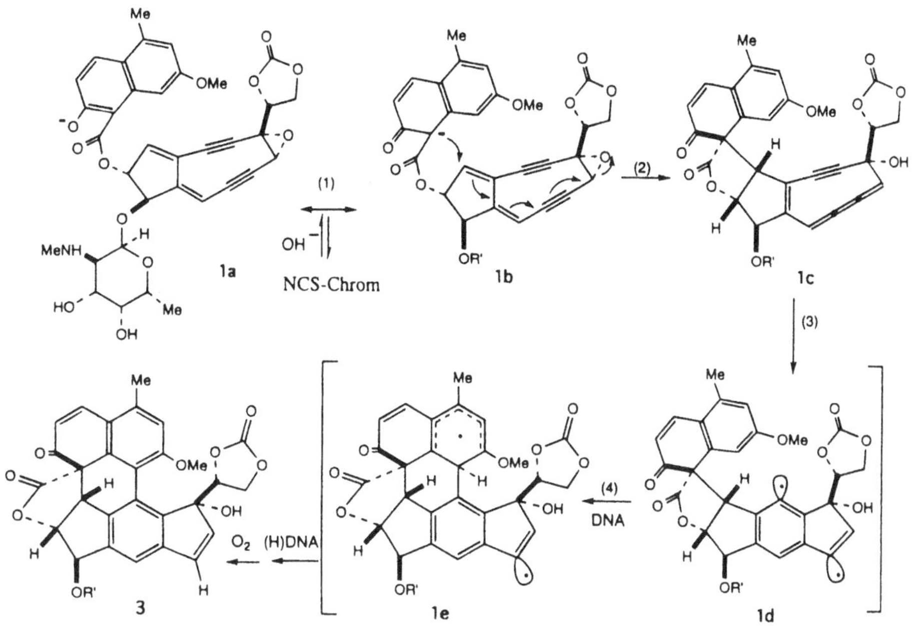

FIG. 16. Thiol-independent NCS-chrom I-induced cleavage of DNA

thioglycolate adduct adopts a chair conformation with the amino group oriented above C-12 approximating the van der Waals diameter of a sulfur atom [57]. On the other hand, the role of epoxide in NCS-chrom is activation of the thiol-dependent DNA cleavage (see Fig. 15). The cyclic carbonate moiety takes part in the passage of the NCS-chrom through cellular and nuclear membranes [41,58].

Concluding Remarks

The first of the bicyclic enediyne antibiotics for which the structure has been elucidated is NCS. NCS was initially identified as a simple acidic protein antitumor antibiotic with the ability to inhibit DNA synthesis and induce the degradation of DNA in cells. Until almost 15 years after the discovery of NCS, it was not appreciated that the biological activities of NCS was not derived from the protein (Apo-NCS) but rather from a nonprotein chromophore, NCS-chrom I.

Using a combination of NMR spectroscopic techiques, mass spectrometry, and preparation of relative stable derivatives, the unprecedented structure of NCS-chrom was elucidated to the epoxybicyclo[7,3,0]dodecadienediyne structure element bearing 2-hydroxy-5-methyl-7-methoxy-1-naphthoate, 2,6-deoxy-2-methylamino-galactose, and ethylene carbonate. This unprecedented structure is supported by synthesis of the core subunit and NCS-chrom I biosynthesis study.

NCS-chrom is endowed with unprecedented structural properties that enable it to aromatize to generate a biradical in the presence of thiols.

Early studies on the mechanism of action of NCS-chrom, before the elucidation of the enediyne character of its structure, suggested that the active form of the drug was a radical species. The precise mechanism of DNA cleavage by NCS-chrom has been proposed. After activation by thiols, the NCS-thiol adduct collapses to form a putative biradical that can abstract a hydrogen from the deoxyribose of DNA.

In conclusion, the unprecedented structure elucidation of NCS-chrom led to clarifying its instability and its elegant mechanism of DNA cleavage at the molecular level.

References

1. Edo K, Mizugaki M, Koide Y, Seto H, Furihata K, Ōtake N, Ishida N (1985) The structure of neocarzinostatin chromophore possessing a novel bicyclo[7,3,0]dodecadyne system. Tetrahedron Lett 26:331–334
2. Goldberg IH, Kappen LS (1995) Neocarzinostatin: chemical and biological basis of oxidative DNA damage. In: Borders BD, Doyle TW (eds) Enediyne antibiotics as antitumor agents. Dekker, New York, pp 327–362
3. Hensens OD, Goldberg IH (1989) Mechanism of activation of the antitumor antibiotic neocarzinostatin by mercaptan and sodium borohydride. J Antibiot (Tokyo) 42:761–768
4. Sakata N, Minamitani S, Kanbe T, Hori M, Hamada M, Edo K (1993) The amino acid sequence of neocarzinostatin apoprotein deduced from the base sequence of the gene. Biol Pharm Bull 16:26–28
5. Chimura H, Ishizuka M, Hamada M, Hori S, Kimura K, Iwanaga J, Takeuchi T, Umezawa H (1968) A new antibiotic, macromomycin, exhibiting antitumor and antimicrobial activity. J Antibiot (Tokyo) 21:44–49
6. Khokhlov AS, Cherches BZ, Reshetov PD, Smirnova GM, Sorokina IB, Prokoptzeva TA, Koloditskaya TA, Smirnov VV, Navashin SM, Fomina IP (1969) Physicochemical and biological studies on actinoxanthin, an antibiotic from *Actinomyces globisporus* 1131. J Antibiot (Tokyo) 22:541–544
7. Khokhlov AS, Reshetov PD, Chupova LA, Cherches BZ, Zhigis LS, Stoyachenko IA (1976) Chemical studies on actinoxanthin. J Antibiot (Tokyo) 29:1026–1034
8. Hu J, Xue YC, Xie MY, Zhang R, Otani T, Minami Y, Yamada Y, Marunaka T (1988) A new macromolecular antitumor antibiotic, C-1027. I. Discovery, taxonomy of producing organism, fermentation and biological activity. J Antibiot (Tokyo) 41:1575–1579
9. Otani T, Minami Y, Marunaka T, Zhang R, Xie MY (1988) A new macromolecular antitumor antibiotic, C-1027. II. Isolation and physico-chemical properties. J Antibiot (Tokyo) 41:1580–1585
10. Zhen YS, Ming XY, Yu B, Otani T, Saito H, Yamada Y (1989) A new macromolecular antitumor antibiotic, C-1027. III. Antitumor activity. J Antibiot (Tokyo) 42:1294–1298
11. Sugimoto Y, Otani T, Oie S, Wierzba K, Yamada Y (1990) Mechanism of action of a new macromolecular antitumor antibiotic, C-1027. J Antibiot (Tokyo) 43:417–421

12. Lam KS, Hesler GA, Gustavson DR, Crosswell AR, Veitch JM, Forenza S, Tomita K (1991) Kedarcidin, a new chromoprotein antitumor antibiotic. I. Taxonomy of producing organism, fermentation and biological activity. J Antibiot (Tokyo) 44:472–478
13. Hofstead SJ, Matson JA, Malacko AR, Marquardt H (1992) Kedarcidin, a new chromoprotein antitumor antibiotic. II. Isolation, purification and physico-chemical properties. J Antibiot (Tokyo) 45:1250–1254
14. Leet JE, Schroeder DR, Langley DR, Colson KL, Huang S, Klohr SE, Lee MS, Golik J, Hofstead SJ, Doyle TW, Matson JA (1993) Chemistry and structure elucidation of the kedarcidin chromophore. J Am Chem Soc 115:8432–8443
15. Hanada M, Ohkuma H, Yonemoto T, Tomita K, Ohbayashi M, Kamei H, Miyaki T, Konishi M, Kawaguchi H (1991) Maduropeptin, a complex of new macromolecular antitumor antibiotics. J Antibiot (Tokyo) 44:403–414
16. Schroeder DR, Colson KL, Klohr SE, Zein N, Langley DR, Lee MS, Matson JA, Doyle TW (1994) Isolation, structure determination, and proposed mechanism of action for artifacts of maduropeptin chromophore. J Am Chem Soc 116:9351–9352
17. Golik J, Dubay G, Groenewold G, Kawaguchi H, Konishi M, Krishnan B, Ohkuma H, Saitoh K, Doyle TW (1987) Esperamicins, a novel class of potent antitumor antibiotics. 3. Structures of esperamicin A_1, A_2, and A_{1b}. J Am Chem Soc 109:3462–3464
18. Lee MD, Dunne TS, Chang CC, Ellestad GA, Siegel MM, Morton GO, McGahren WJ, Borders DB (1987) Calichemicins, a novel family of antitumor antibiotics. 2. Chemistry and structure of calichemicin γ_1^I. J Am Chem Soc 109:3466–3468
19. Konishi M, Ohkuma H, Matsumoto K, Tsuno T, Kamei H, Miyaki T, Oki T, Kawaguchi H, vanDuyne GD, Clardy J (1989) Dynemicin A, a novel antibiotic with the anthraquinone and 1,5-diyn-3-ene subunit. J Antibiot (Tokyo) 42:1449–1452
20. Ishida N, Miyazaki K, Kumagai K, Rikimaru M (1965) Neocarzinostatin, an antitumor antibiotic of high molecular weight. J Antibiot Ser A (Tokyo) 18:68–76
21. Shoji J (1961) Preliminary studies on the isolation of carzinostatin complex and its characteristics. J Antibiot Ser A (Tokyo) 14:27–33
22. Maeda H, Glaser CB, Kuromizu K, Meienhofer J (1974) Structure of the antitumor protein neocarzinostatin. Arch Biochem Biophys 164:379–385
23. Gibson BW, Herlihy WC, Samy TSA, Hahm KS, Maeda H, Meienhofer J, Biemann K (1984) A revised primary structure for neocarzinostatin based on fast atom bombardment and gas chromatographic-mass spectrometry. J Biol Chem 259:10801–10806
24. Kuromizu K, Tsunasawa S, Maeda H, Abe O, Sakiyama F (1986) Reexamination of the primary structure of an antitumor protein, neocarzinostatin. Arch Biochem Biophys 246:199–205
25. Napier MA, Holmquist B, Strydom DJ, Goldberg IH (1979) Neocarzinostatin: spectral characterization and separation of a non-protein chromophore. Biochem Biophys Res Commun 89:635–642
26. Iseki S, Koide Y, Ebina T, Ishida N (1980) Biological activities and physicochemical properties of pre-neocarzinostatin and UV-irradiated neocarzinostatin. J Antibiot (Tokyo) 33:110–113
27. Koide Y, Ishii F, Hasuda K, Koyama Y, Edo K, Katamine S, Kitame F, Ishida N (1980) Isolation of a non-protein component and a protein component from neocarzinostatin (NCS) and their biological activities. J Antibiot (Tokyo) 33:342–346
28. Suzuki H, Miura K, Kumada Y, Takeuchi T, Tanaka N (1980) Biological activities of non-protein chromophores of antitumor protein antibiotics: auromomycin and neocarzinostatin. Biochem Biophys Res Commun 94:255–261

29. Maeda H (1981) Neocarzinostatin in cancer chemotherapy (review). Anticancer Res 1:175–186
30. Meienhofer J, Maeda H, Glaser CB, Czombos J, Kuromizu K (1972) Primary structure of neocarzinostatin, an antitumor protein. Science 178:875–876
31. Ono Y, Watanabe Y, Ishida N (1966) Mode of action of neocarzinostatin: inhibition of DNA synthesis and degradation of DNA in *Sarcina lutea*. Biochim Biophys Acta 119:46–58
32. Ohtsuki K, Ishida N (1975) Neocarzinostatin-induced breakdown of deoxyribonucleic acid in HeLa-S3 cells. J Antibiot (Tokyo) 28:143–148
33. Kappen LS, Goldberg IH (1979) Mechanism of the effect of organic solvents and other protein denaturants on neocarzinosatin activity. Biochemistry 18:5647–5653
34. Napier MA, Kappen LS, Goldberg IH (1980) Effect of nonprotein chromophore removal on neocarzinostatin action. Biochemistry 19:1767–1773
35. Ohtsuki K, Ishida N (1980) The biological effect of a nonprotein component removed from neocarzinostatin (NCS). J Antibiot (Tokyo) 33:744–750
36. Edo K, Akiyama-Murai Y, Saito K, Mizugaki M, Koide Y, Ishida N (1988) Hydrogen bromide adduct of neocarzinostatin chromophore: One of the stable derivatives of native neocarzinostatin chromophore. J Antibiot (Tokyo) 41:1272–1274
37. Lee SH, Goldberg IH (1988) Role of epoxide in neocarzinostatin chromophore stability and action. Mol Pharmacol 33:936–401
38. Hensens OD, Dewey RS, Liesch JM, Napier MA, Reamer RA, Smith JL, Schönberg GA, Goldberg IH (1983) Neocarzinostatin chromophore: presence of a highly strained ether ring and its reaction with mercaptan and sodium borohydride. Biochem Biophys Res Commun 113:538–547
39. Edo K, Katamine S, Kitame F, Ishida N, Koide Y, Kusano G, Nozoe S (1980) Naphthalenecarboxylic acid from neocarzinostatin (NCS). J Antibiot (Tokyo) 33:347–351
40. Schönberg GA, Dewey RS, Hensens OD, Liesch JM, Napier MA, Goldberg IH (1980) Neocarzinostatin: chemical characterization and partial structure of the non-protein chromophore. Biochem Biophys Res Commun 95:1351–1356
41. Napier MA, Goldberg IH, Hensens OD, Dewey RS, Liesch JM, Schönberger GA (1981) Neocarzinostatin chromophore: presence of a cyclic carbonate subunit and its modification in the structure of other biologically active forms. Biochem Biophys Res Commun 100:1703–1712
42. Edo K, Ito M, Ishida N, Koide Y, Ito A, Haga M, Takahashi T, Suzuki Y, Kusano G (1982) Peroxy acid as a partial structure of neocarzinostatin (NCS) chromophore. J Antibiot (Tokyo) 35:106–110
43. Shibuya M, Toyooka K, Kubota S (1984) Synthesis of the naphthalenecarboxylic acid derivative obtained from neocarzinostatin (NCS): a structure revision. Tetrahedron Lett 25:1171–1174
44. Koide Y, Ito A, Edo K, Ishida N (1986) The biologically active site of neocarzinostatin-chromophore. Chem Pharm Bull (Tokyo) 34:4425–4428
45. Edo K, Akiyama Y, Saito K, Mizugaki M, Koide Y, Ishida N (1986) Absolute configuration of the amino sugar moiety of the neocarzinosatin chromophore. J Antibiot (Tokyo) 39:1615–1619
46. Myers AG, Proteau PJ, Handel TM (1988) Stereochemical assignment of neocarzinostatin chromophore. Structures of neocarzinostatin chromophore–methyl thioglycolate adducts. J Am Chem Soc 110:7212–7214

47. Wender PA, Harmata M, Jeffery D, Mukai C, Suffert J (1988) Studies on DNA-active agents: the synthesis of the parent carbocyclic subunit of neocarzinostatin chromophore A. Tetrahedron Lett 29:909–912
48. Hensens OD, Giner JL, Goldberg IH (1989) Biosynthesis of NCS chrom A, the chromophore of the antitumor antibiotic neocarzinostatin. J Am Chem Soc 111:3295–3299
49. Kloster-Jensen E, Wirz J (1975) 1,5-Cyclooctadiyne. Preparation and reactivity. Helv Chim Acta 58:162–177
50. Edo K, Iseki S, Ishida N, Horie T, Kusano G, Nozoe S (1980) An electron spin resonance study of a spin adduct of the non-protein component (NPC) of neocarzinostatin. J Antibiot (Tokyo) 33:1586–1589
51. Sheridan RP, Gupta RK (1981) Electron spin resonance detection of free radicals in the mercaptan-activation and UV-inactivation of neocarzinostatin. Biochem Biophys Res Commun 99:213–220
52. Myers AG (1987) Proposed structure of the neocarzinostatin chromophore–methyl thiogycolate adduct; a mechanism from the nucleophilic activation of neocarzinostatin. Tetrahedron Lett 28:4493–4496
53. Myers AG, Cohen SB, Kwon BM (1994) DNA cleavage by neocarzinostatin chromophore. Establishing the intermediacy of chromophore-derived cumulene and biradical species and their role in sequence-specific cleavage. J Am Chem Soc 116:1670–1682
54. Myers AG, Proteau PJ (1989) Evidence for spontaneous, low-temperature biradical formation from a highly reactive neocarzinostatin chromophore-thiol conjugate. J Am Chem Soc 111:1146–1147
55. Yang CF, Stassinopoulos A, Goldberg IH (1995) Specific binding of the biradical analog of neocarzinostatin chromophore to bulged DNA: implications for thiol-independent cleavage. Biochemistry 34:2267–2275
56. Goldberg IH (1987) Free radical mechanisms in neocarzinostatin-induced DNA damage. Free Radical Biol Med 3:41–54
57. Kim KH, Kwon BM, Myers AG, Rees DC (1993) Crystal structure of neocarzinostatin, an antitumor protein-chromophore complex. Science 262:1042–1046
58. Napier MA, Holmquist B, Strydom DJ, Goldberg IH (1981) Neocarzinostatin chromophore: purification of the major active form and characterization of its spectral and biological properties. Biochemistry 20:5602–5608

Synthesis and Chemistry of Nine-Membered Cyclic Enediyne Chromophores of Chromoprotein Antitumor Antibiotics

Masahiro Hirama

Summary. Recent progress in the synthesis and chemistry of the nine-membered cyclic enediyne chromophores of the chromoprotein antitumor antibiotics, neocarzinostatin, C-1027, and kedarcidin, is reviewed.

Key words: Neocarzinostatin—Enediyne—Chromophore—Synthesis—C-1027—Kedarcidin

Introduction

Neocarzinostatin (NCS) is a potent antitumor antibiotic isolated from the culture of *Streptomyces carzinostaticus* by Ishida et al. [1]. NCS is a complex composed of an unstable chromophore (**1**) (Fig. 1) [2,3] and an apoprotein of 11000 daltons (11 kDa) secreted simultaneously [4–6]. The chromophore (**1**) is very labile to heat, light, and pH levels above 6, but it is stabilized substantially when bound to the apoprotein. Thus, the apoprotein serves as a stabilizer and a carrier of the unstable chromophore (**1**). Their binding structure and the mechanism of stabilization are of major significance in connection with molecular recognition and protein transport, and these issues were clarified quite recently [7–13]. The stabilization mechanism of other so-called chromoprotein antibiotics, C-1027 (chromophore **2**) [14,15] and kedarcidin (chromophore **3**) [16] (Fig. 1), has also been proposed [17].

Another intriguing issue is the mode of action in connection with those of enediyne antitumor antibiotics (for reviews, see [18–22]; also [23–25]. Goldberg [2] and Edo [3] elucidated that **1** is fully responsible for the biological activity of NCS, and that a thiol activates **1** to generate a carbon radical, which abstracts hydrogen from DNA and leads to DNA scission under aerobic conditions. Myers proposed a chemical mechanism for the action of **1**: thiol addition to C-12, and a

Department of Chemistry, Graduate School of Science, Tohoku University, Aoba, Aramaki, Aoba-ku, Sendai, Miyagi, 980–77 Japan

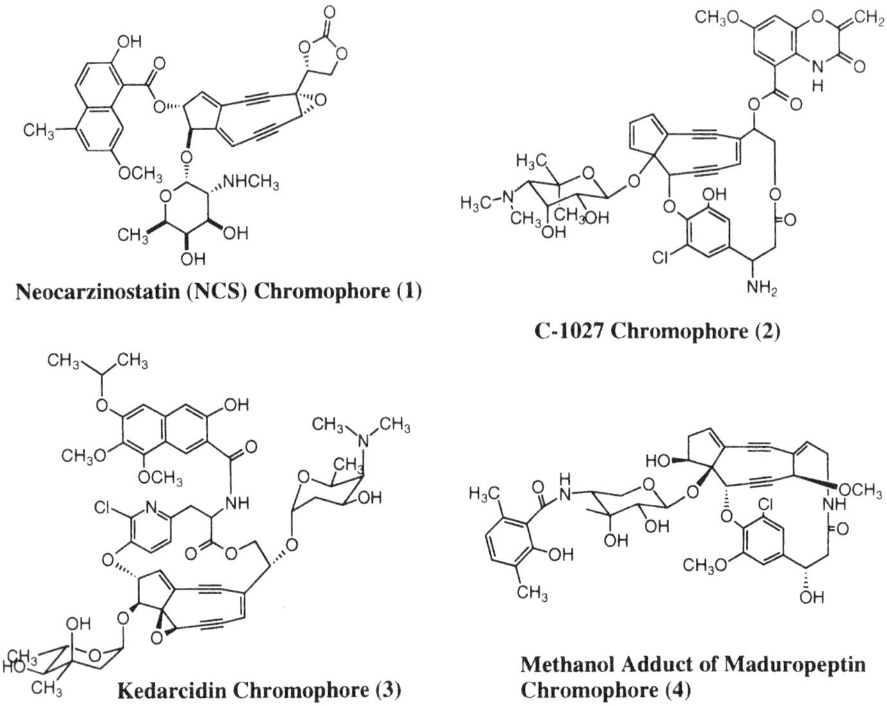

FIG. 1. Structures of chromophores of chromoprotein antitumor antibiotics

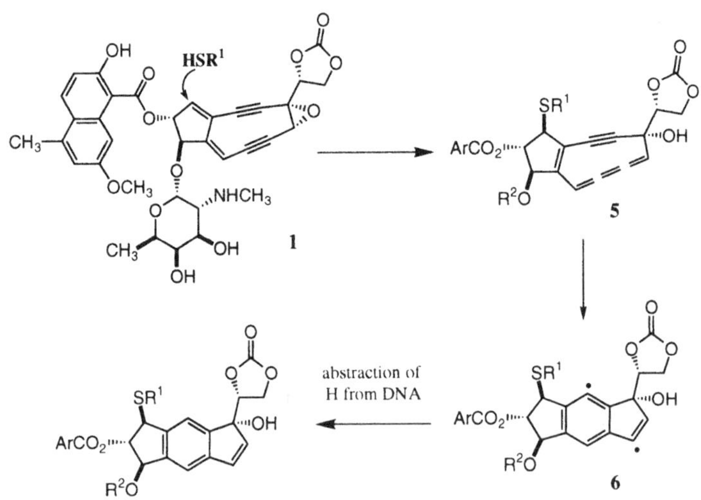

FIG. 2. A proposed mechanism for action of the NCS (neocarzinostatin) chromophore

concomitant opening of the epoxide initiates the cascade of reactions that lead to the generation of the dehydroindacene diradical (**6**) through Masamune–Bergman-type cyclization [18,26,27] of an enynecumulene intermediate (**5**) (Fig. 2) [28–31].

The first half of this chapter reviews our mechanistic and synthetic studies referring to the following questions and issues of the NCS chromophore [18–22]: (1) Is the highly strained nine-membered ring essential for the thiol-triggering aromatization? (2) Since there is no precedent for a conjugated enynecumulene system, synthesis of such systems is necessary to understand their chemical behaviors. (3) Is there any other mode of activation of **1**? (4) How does the apoprotein (**2**) recognize **1** and prevent it from decomposing? (5) Is the design and synthesis of new DNA-cleaving molecules related to **1** a suitable objective to pursue for attaining low molecular weight, sequence-specific DNA cleavers?

In the latter half of this chapter, the recent synthetic studies and the chemistry of nine-membered cyclic enediyne chromophores, including the C-1027 and kedarcidin chromophores, are reviewed [17,21,32]. Since the structure of NCS chromophore was determined as a highly strained bicyclo[7,3,0] dodecadienediyne system **1** by Edo et al. [3] in 1985, this novel class of the enediyne anticancer antibiotics has been rapidly emerging. In 1987 the structures of two new families, the calicheamicins (**7**) and the esperamicins (**8**), that have ten-membered cyclic enediyne core structures were reported by the Lederle and Bristol-Myers groups, respectively [33–36]. Konishi et al. (of Bristol-Myers,

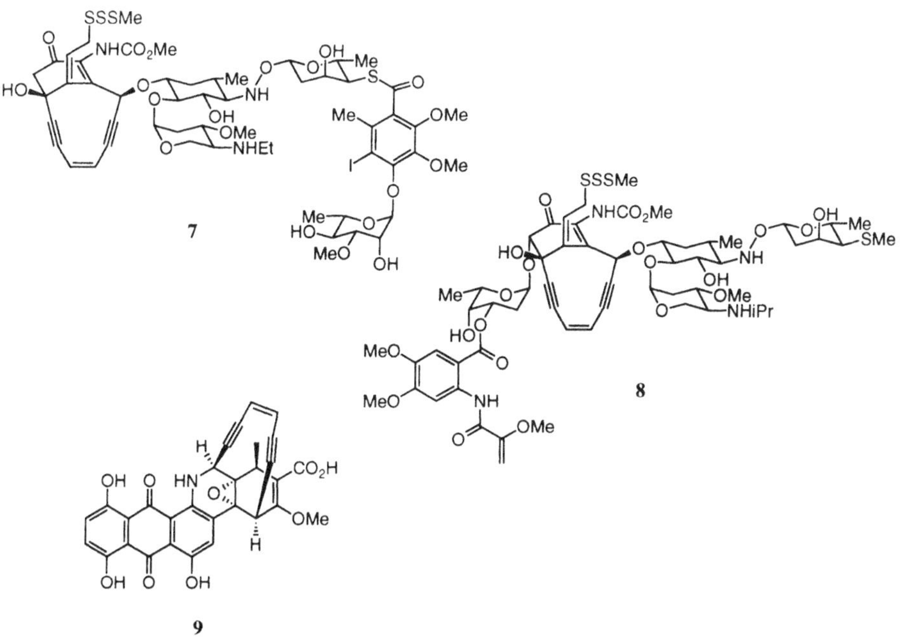

7

8

9

FIG. 3. Structures of ten-membered cyclic enediyne antitumor antibiotics

Japan) also reported the fourth member, the dynemicins (**9**), in 1989 [37] (Fig. 3). Recently, three novel structures of kedarcidin (**3**) [16], maduropeptin (**4**) [38], and the C-1027 chromophore (**2**) [14,15], which are very closely related to the structure of the NCS chromophore (**1**), have been determined by the Bristol-Myers Squibb and Taiho Pharmaceutical Co. groups, respectively (see Fig. 1). These complexes—NCS, kedarcidin, C-1027, and maduropeptin—constitute another novel family of the chromophore-apoprotein complexes, the so-called chromoprotein antibiotics, and moreover their chromophores possess the highly strained bicyclo[7.3.0]dodecadiyne, i.e., the nine-membered cyclic enediyne core structures. The total syntheses of calicheamicin (**7**) [39–42] and dynemicin (**9**) [43,44] were accomplished recently. As several comprehensive reviews on the synthetic studies of ten-membered enediynes and the related mimics have appeared [18–25], the latter half of this chapter focuses on and reviews the syntheses and the chemistry of the extremely strained and labile nine-membered enediynes of NCS, kedarcidin, and C-1027, which elude total synthesis [17,21,32].

Synthesis and Chemical Properties of Acyclic Enynecumulenes

To address the issues implicit in the first and second questions, we synthesized an acyclic cross-conjugated diene-diyne system (**10**) in a straightforward manner and treated it with thiol in acetonitrile [45]. Under carefully controlled conditions, we isolated a vinylogous S_N2' reaction product, enynecumulene (**11**), in moderate yield, although interestingly a formally direct substitution product (**12**) was formed exclusively when a large excess of thiol or dimethylsulfuride (DMSO) was used as solvent. The ^{13}C-NMR (nuclear magnetic resonance) spectrum strongly supported the cumulene structure for **11**, which was stable at room temperature in the absence of oxygen (Fig. 4) [45].

The next question we concerned ourselves with was whether **11** would undergo the Masamune–Bergman-type cyclization [26,27]. Heating at 80°C in 1,4-cyclohexadiene produced a styrene derivative (**13**). When the reaction was carried out in cyclohexadiene-d_8, two deuteriums were incorporated at the relevant positions, which clearly indicates that such an enynecumulene system has an

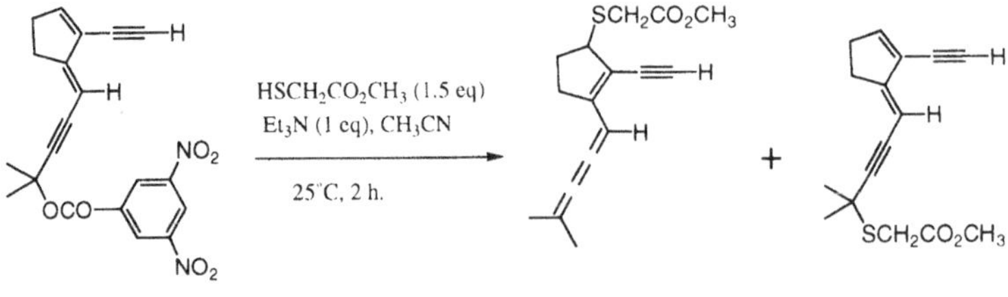

10 49% **11** 14% **12**

FIG. 4. Synthesis of acyclic enynecumulene

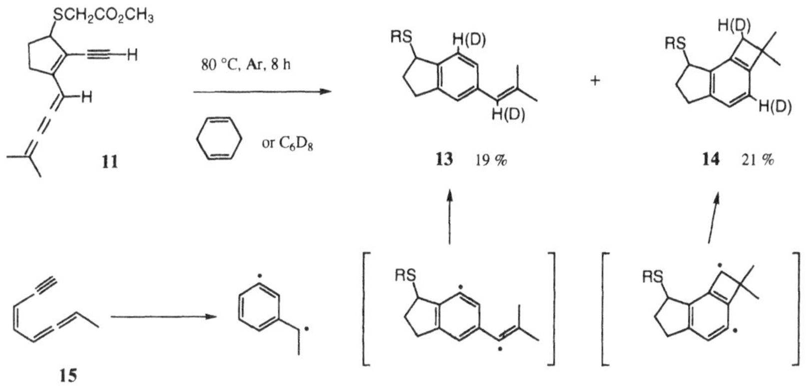

FIG. 5. Cycloaromatization of acyclic enynecumulene

ability to undergo the cycloaromatization leading to the generation of σ,σ-diradicals. The first-order disappearance rate of **11** is approximately 30 times slower than that of the related enynallene (**15**) (Fig. 5) [46,47]. In addition, benzocyclobutane (**14**) formation turned out to be another major reaction pathway in this conformationally flexible system [45]. Brückner has supplemented another mechanistic rationale for the formation of **14** [48].

Synthesis, Cycloaromatization, DNA Cleaving, and Biological Activities of Ten-Membered Ring Analogs

Synthesis

In connection with the first question, a cyclic, less strained ten-membered analog is an another intriguing system. Noteworthy in the syntheses of the ten-membered ring analogs (**17–19**) is the palladium-mediated effective internal cyclization of vinyl bromide-alkynyl stannane (**16**) [49]. Attempts at direct cyclization of **15** without the stannyl group failed and instead resulted in the formation of dimeric oxidative coupling products. The final oxidation and concomitant dehydration proceeded smoothly by Swern oxidation [50,51] (Fig. 6).

Radical-Induced Cycloaromatization

It was demonstrated that analogs (**17–19**) can undergo thiol-initiated aromatization at room temperature as well as thiyl or hydroperoxy radical-induced cycloaromatization [50,51]. The less strained dienediyne system seems to have a marked tendency to react with radical species at C-13, leading to a delocalized radical intermediate such as **20**. Smooth formation of an unstable epoxide **21** from **19** with molecular oxygen was rationalized on the basis of formation of the radical intermediate **20** [52] (Fig. 7).

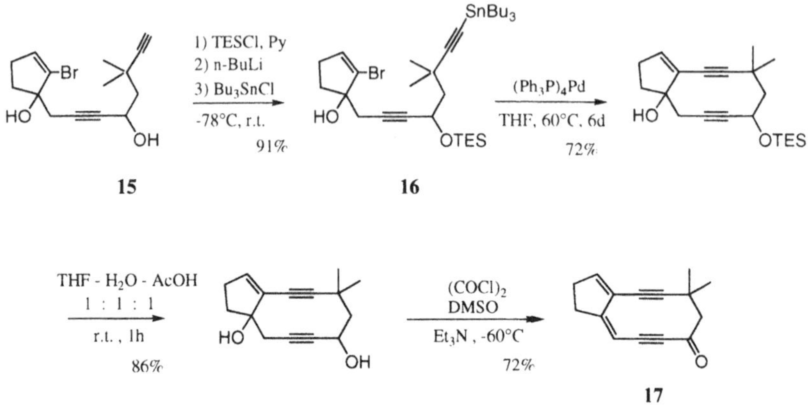

FIG. 6. Synthesis of ten-membered ring analogs. *TES*, triethylsilyl

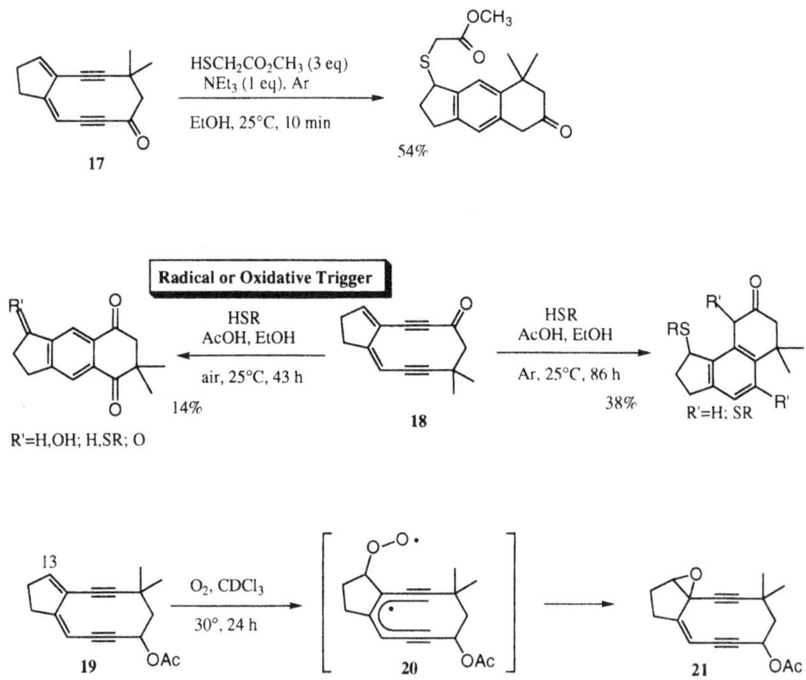

FIG. 7. Radical-induced cycloaromatization of ten-membered analogs

New Initiation Modes of Cycloaromatization of the NCS Chromophore

Radical-Triggering Pathway

The radical-induced aromatization demonstrated in the reaction of ten-membered analogs (**17–19**) has also been observed in the aromatization of NCS chromophore (**1**). Formation of ketone **22** (~10% yield) was confirmed by decomposition of **1** in methanolic acetic acid in the presence of air and methyl thioglycolate at –30°C in the dark [53] (Fig. 8). This experiment suggested that the aromatization of **1** triggered by a hydroperoxy radical formed from thiol and molecular oxygen would become important under physiological conditions when nucleophilic thiols such as glutathione are insufficient.

New Mode of Triggering

On the other hand, we suggested an alternative activation pathway of **1** (Fig. 8) based on model studies using **19** [51]. Saito et al. [54,55] have found that aromatization product **23** related to this new activation mode was produced as a major product when **1** was incubated with 2-mercaptoethanol in the presence of apoprotein in aqueous solution, although it is likely to be an ionic pathway and therefore will not partake in the DNA cleavage mechanism.

Second Generation of NCS Chromophore Analogs

Analogs **17** and **18** did not exhibit DNA cutting or cytotoxic activities, possibly because they lack a hydrophilic or DNA-binding group. Therefore, we designed the second-generation models (**24–27**) to improve these activities (Fig. 9) [56–58]. Our basic strategy for designing new DNA-cleaving molecules was to combine the DNA-cleaving unit with some DNA-binding group.

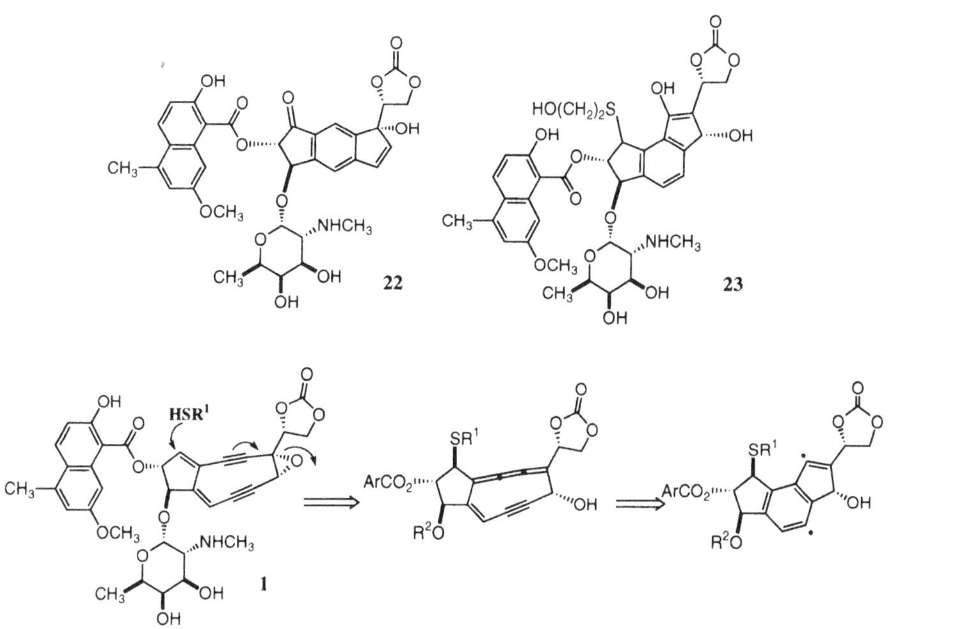

FIG. 8. An alternative activation pathway of the NCS chromophore

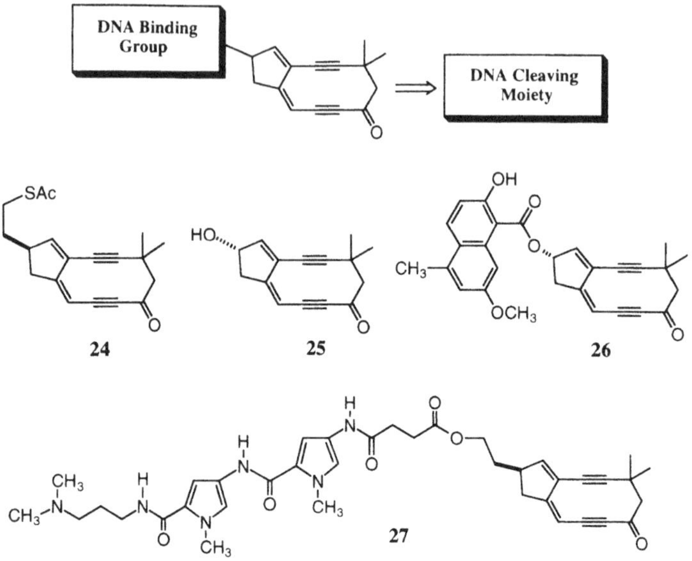

FIG. 9. Second generation of NCS chromophore analog

A New Model Equipped with an Intramolecular Trigger

First, we synthesized a new model (**24**) equipped with an intramolecular nucleophile, a design idea contrived from the related potent antitumor antibiotics calicheamicin [33,34] and esperamicin [35,36]. Methanolysis using potassium carbonate in the presence of 1,4-cyclohexadiene at room temperature was complete within 30 min in an argon atmosphere to give the reductive cycloaromatization product through the intramolecular addition of thiolate [56]. Deuterium incorporation was observed at C-2 in the presence of cyclohexadiene-d_8. This internal trigger model (**24**) showed an appreciable activity toward *Micrococcus luteus* without addition of external thiol.

New Functional Analogs

The naphthoate conjugate **26** is of interest, particularly in connection with two issues. First, it would exhibit a T-selective DNA cleavage similar to that of **1** [2]. If so, it should provide evidence that the naphthoate moiety is responsible for T-specific DNA scission by **1** through its specific recognition of the T residue by intercalation [59,60]. Second, the NCS apoprotein would serve as a carrier of this lipophilic molecule (**26**), as our recent NMR studies of the three-dimensional structure of the NCS complex (see following) have shown that the naphthoate moiety is responsible for the specific binding of the chromophore [7–10]. The enantiomerically pure alcohol (**25**) and naphthoate conjugate (**26**) were then synthesized. Their thiol triggered cycloaromatization, which proceeded smoothly at room temperature in ethanol in the presence of 1 equivalent of triethylamine [57].

The alcohol (**25**) clearly showed antibiotic and cytotoxic activities against *Bacillus licheniformis* and K562 leukemia cells, respectively. On the other hand, the activities of naphthoate (**26**) alone were quite low, but in the presence of apoprotein the mixture exhibited the distinctive activities. This is likely to result from the complex formation and transport function of the apoprotein. Similar results were again observed in the DNA cleavage experiments. The naphthoate **26** showed some sequence-selective DNA cleavage activity, namely cutting G of 5′-GGT and T of 5′-GCT, but only in the presence of the apoprotein [19,57]. Because this T selectivity is characteristic of the selectivity of **1**, it implies that the naphthoate moiety should play an important role for the site-selective DNA cleavage of **1**. On the other hand, remarkable G specificity was found for alcohol **25**. This was surprising because **25** cannot be an intercalator, nor does it possess an apparent DNA binding moiety. Recent detailed DNA cleavage experiments suggest that G specificity for **25** might arise from its chemical reactivity (alkylation mechanism), rather than the molecular recognition of a guanine base, followed by DNA cutting via a radical mechanism (Gombuchi et al., unpublished results).

A New Hybrid Molecule Containing a Minor Groove Binder

The antibiotic netropsin is known to be a strong minor groove binder to the AT-rich regions of DNA through hydrogen bonding and van der Waals interactions. Thus, we anticipated that the netropsin conjugate (**27**) would function as an effective AT-specific DNA cleaver. Synthesis of optically pure **27** has been achieved [58]. Because **27** possesses an internal amine, it undergoes the thiol-triggering cycloaromatization quickly without addition of an external amine as expected. We compared the DNA-cleaving abilities of synthetic molecules **24**, **25**, and **27** on supercoiled pBR322 DNA. Agarose gel electrophoresis clearly showed that netropsin conjugate (**26**) cleaved supercoiled DNA-form I to give a nicked circular form II and then further to form a linear form III in a highly efficient manner [58].

Apoprotein–Chromophore Binding Structure and Stabilization Interactions

Solution Structure and Binding Elements

Our recent study of the three-dimensional solution structure of the NCS complex utilizing two-dimensional (2D) NMR and distance geometry calculations revealed the elements of the specific binding and the stabilization interactions between unstable **1** and apoprotein [7–13]: (also see the chapter by H. Takashima and Y. Kobayashi, this volume). This is the first time that the chromophore-binding structure has been determined in a member of the chromoprotein antibiotic family. A hydrophobic pocket with approximate dimensions of $12 \times 9 \times 9$ Å is the binding site of **1**. Apoprotein accepts the naphthoate moiety of **1** deeply into this pocket. The C-5″–C-7″ site of the naphthoate is in proximity to the bottom of the pocket where the main chains of three strands (Val34-Gly35-Gln36, Leu97-Gly96-Val95, and Gly107-Val108) are exposed. The carbocyclic core is located on the Cys37–Cys47 disulfide bond, and the aminosugar and the carbonate moieties are facing outward. There seems to be no aromatic stacking interaction with the naphthalene ring, such as that predicted by computer modeling. Instead, several other principal interactions are indicated. The oxygen atoms of C-7″–O and C-2″–O of the naphthoate of **1** are close enough to the OH of Ser98 and NH of Trp39, respectively, to produce a hydrogen bond. The C-5″ methyl is approximately in van der Waals contact with the β-methylene of Gln94. The upfield shift of the C-7″–OCH_3 resonance by 1 ppm upon binding most likely results from the diamagnetic anisotropy of the nearby aromatic ring of Phe52, which supports a possible CH–π interaction [61].

On the other hand, the C2–C3 triple bond of the core is just above the sulfur atom of Cys47 and is covered perpendicularly by the aromatic ring of Phe52. The aromatic ring of Phe78 is surrounding the other side of the core, the C6–C7 triple bond. It can be concluded therefore that Phe78 and Phe52, as well as Cys37–Cys47 disulfide bond, must be of importance for holding the carbocyclic core moiety of **1** [7–13].

Recent X-ray crystallographic study [62] (also see the chapter by K.H. Kim et al., this volume) of the NCS holoprotein structure showed a different conformation of the Phe78 aromatic ring from that of our solution structure [10]. Thus, we have carried out further analyses combining the crystallographic data of the apoprotein (**2**) [63] and the NMR nuclear Overhauser effects (NOEs) observed between **2** and apoprotein [10]. These data suggest that the aromatic moiety of Phe78 changes significantly in its conformation upon binding of **1** and that this moiety in the holoprotein is still rather flexible in solution. We concluded that Phe78 in the crystal structure of the holoprotein exists as one of the stable conformers, while Phe78 is dynamically undergoing multiple conformations in the solution. Although the overall shape of the binding pocket of the apoprotein is similar to the related chromoproteins, auromomycin [64] and actinoxanthin [65], the side chains of the residues along the binding pocket are different in size and functionality with the exception of the Cys37–Cys47 disulfide. Because the NCS chromophore (**1**) is a poor ligand for auromomycin apoprotein [64], it is very likely that the residues described previously are crucial for chromophore-binding specificity. In particular, Phe78 and Phe52 are not conserved in other chromoprotein antibiotics. The interactions between the core acetylenic bonds and the Phe52 and Phe78, in a similar manner to an aromatic edge-to-face interaction, should be important for the specific binding of the core moiety of **1** in the holoprotein. Because Phe78 has multiple conformers covering the core, it should play an important role in release of the chromophore (**1**) and in the protection of **1** in the holoprotein [8,10].

Confirmation and Evaluation of Binding Elements

Synthesis of Ligands

The binding structure indicates that the naphthoate moiety of **1** is essential for its specific and strong binding to the apoprotein, and that the several interactions between the functional groups on the naphthoate ring and the side chains of

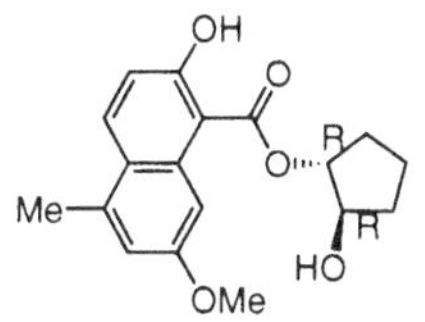

28 : Ka = 1.3 x 10^4 M^{-1}
ΔG = 5.6 kcal / mol

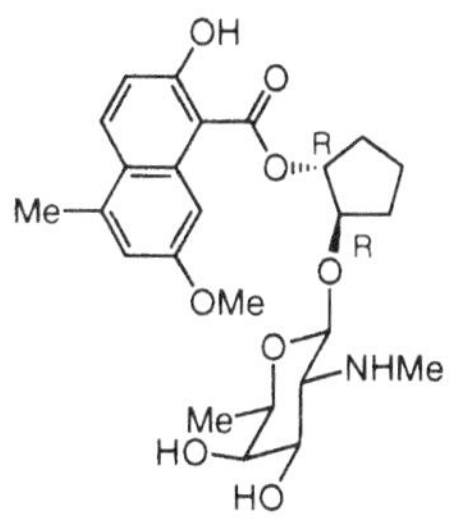

29 : Ka = 2.0 x 10^4 M^{-1}
ΔG = 5.9 kcal / mol

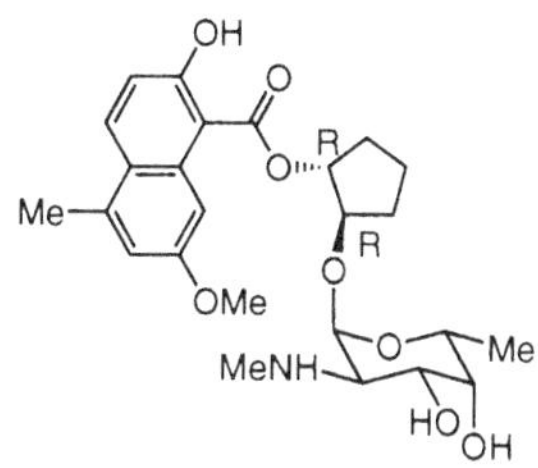

30 : Ka ≈ 1.1 x 10^5 M^{-1}
ΔG ≈ 6.9 kcal / mol

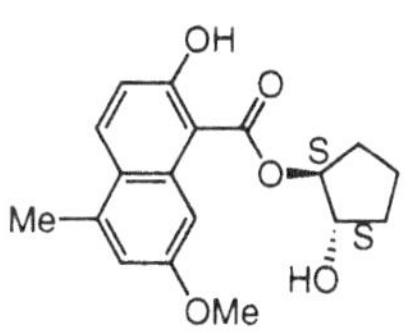

31: Ka = 1.5 x 10^4 M^{-1}
ΔG = 5.7 kcal / mol

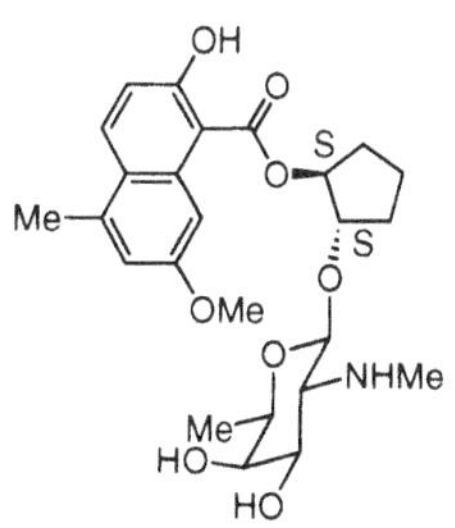

32 : Ka ≈ 1.6 x 10^5 M^{-1}
ΔG ≈ 7.1 kcal / mol

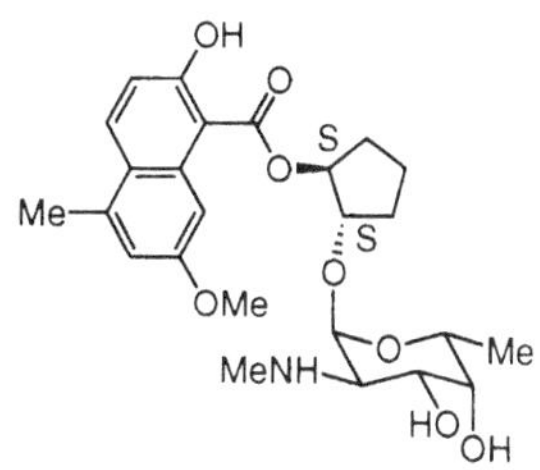

33 : Ka = 2.7 x 10^4 M^{-1}
ΔG = 6.1 kcal / mol

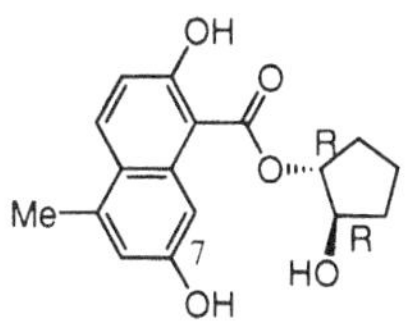

34 : Ka = 4.2 x 10^3 M^{-1}
ΔG = 5.0 kcal / mol

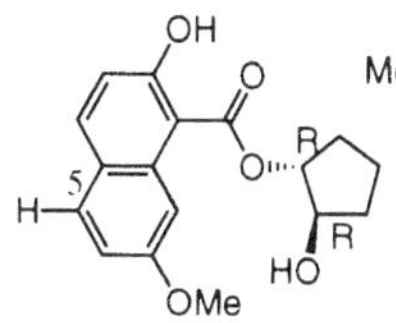

35 : Ka = 4.0 x 10^3 M^{-1}
ΔG = 4.9 kcal / mol

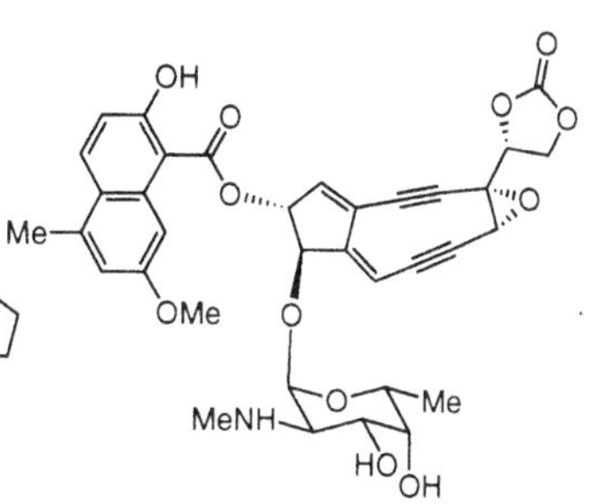

1 : Ka ≈ 3.0 x 10^6 M^{-1}
ΔG ≈ 8.9 kcal / mol

FIG. 10. Binding constants of simple model compounds to apoprotein, *Ka*, association constant residues appear to be responsible for such an important role [7–13]. To confirm the importance of the naphthoate moiety of **1** for the specific binding, we synthesized simple analogs (**28–35**) (Fig. 10) and carried out experiments on their binding to the apoprotein [66]. On this occasion, we have developed a simple and versatile route to the naphthoic acid moiety (**38**) of **1**, which is useful to synthesize the related 2-hydroxy-1-naphthoic acid derivatives [66]. Activation of thioglycoside with NIS/TfOH or NIS/TESOTf in CH_2Cl_2 accomplished the glycosidation in reasonable yields, although the reaction lacked stereocontrol. The β-glycosides **29** and **32** were obtained stereoselectively from **28** and **31**, respectively [66]. A stereoselective synthesis of the α-glycosides, **30** and **33**, has been recently developed, and their binding constants were determined (Kaneko et al., unpublished results).

Association Constant

The binding constants of the model compounds (**28–35**), obtained by fluorescence quenching titration, supported the foregoing idea that the naphthoate moiety is a necessary design feature for sustaining this binding (see Fig. 10) [7–13]. While the bindings of the naphthoates **28** and **31** are still 200 times weaker than the natural chromophore (**1**), their binding energies reach approximately 64% of the total binding energy of **1**. The binding affinity of the demethylated analog **35** is approximately one-third that of **28**, supporting the van der Waals contact between C-5–CH_3 and the β-methylene of Gln94. While the demethylated diol **34** also showed a lower affinity than **28**, **34** should be able to retain the hydrogen bonding with Ser98. Thus, the $\boldsymbol{K_a}$ difference between **34** and **28** appears to correspond to the loss of a CH-π interaction between C-7–OCH_3 and the nearby aromatic ring of Phe52. This CH-π interaction could be then estimated as approximately –2.8 KJ/mol [66]. The $\boldsymbol{K_a}$ of the β-glycoside **29** is close to that of **28**, suggesting that the wrong β-stereochemistry has only negligible effects on the binding. That the affinities of **30** and **32** are approximately 10 times higher than those of **29** and **33**, respectively, is interesting. These results suggest that the contribution of the glycoside moiety would not be so large (~20%) and that the space for the glycoside may be loose. The rest of the binding elements (20%) may arise mainly from the interactions between the carbocyclic core and the hydrophobic amino acid residues such as Phe52 and Phe78, as well as Cys37–Cys47 disulfide, as discussed under *Solution Structure are Binding Elements*.

Stabilization Interactions

Chemical Stability

The conformation of bound chromophore (**1**) in the complex showed some interesting features. The sugar portion hangs over the five-membered ring of the core with the hydrophobic side down, thus satisfying the exo-anomeric effect [7,8]. The methylamino group is forced to come closer to C-12 (4.3 Å), partly because of the salt bridge with Asp33, compared with the conformation of uncomplexed **1** (5.4 Å). The naphthoate stays away from the core to fit into the hydrophobic pocket of the apoprotein, thus allowing the core to lie on the disulfide bridge, while in the stable conformations of the free form it lies just below the core to become more compact. The side chains of Phe52, Phe78, and Asp33, and the Asp99–Asn103 loop, cover the reactive center C-12 as well as the acetylenic bonds. These steric hindrances in C-12 and the acetylenic carbons should be a major stabilizing factor for **1** [7–11].

Thermal Stability

The naphthoate moiety was found to be shifted away from the core so as to fit into the pocket of the apoprotein, so that the plane of the carbocyclic core was located on the Cys37–Cys47 disulfide bridge. The distance between the core plane and the disulfide bridge was close enough (about 3.4 Å) to have an orbital interaction. The sulfur atom of Cys37 was situated at 3.4 Å from the center of the C-12–C-1 double bond, and the sulfur atom of Cys47 was near the acetylenic carbon C-2. The disulfide bond was parallel to the core plane. Therefore, HOMO (the dienediyne π system)-LUMO (vacant sulfur *d* orbital) or LUMO (dienediyne)-HOMO (sulfur lone pair) interaction would stabilize the core π system. Because this disulfide bridge is conserved in all chromoprotein antibiotics, such an interaction with the strained enediyne chromophore may be common in this family [7–11].

Photochemical Stability

The stability of **1** is significantly reduced under ambient light. However, the apoprotein protects **1** against photodecomposition [2]. The central pathway of photochemical decomposition of the analog (**36**) proved to be a fragmentation reaction of a naphthoate ester, a reaction that is anologous to a Norrish type II split, leading to the formation of an unstable fulvene derivative (**37**) and the naphthoic acid (**38**) [12]. Very recently, photodecomposition of **1** proved to give a product **39** together with **38** [13]. The photo-induced naphthoate migration product (**39**) of **1** leading to the formation of cumulene (**41**) and biradical (**42**) would be an alternative central pathway (Fig. 11).

Thus, the naphthoate carbonyl of **1** should be excited by light and would thus undergo photoreactions such as the foregoing naphthoate migration and Norrish type II fragmentation. Such processes can compete with the reversible abstraction of the hydroxyl H to form the enol (**40**). Two stabilizing factors to prevent **1** from undergoing photodecomposition were indicated by the three-dimensional structure of the NCS complex [4]. First, several binding interactions between the naphthoate and binding-site residues could quench the excited state. Second, in the bound chromophore the naphthol proton is fixed very closely (2.0 Å) to the ester carbonyl. There must be a strong intramolecular hydrogen bonding in the bound conformation, which should promote the hydroxyl H 1,5 shift leading to a photo-enol equilibrium and subsequent quenching of the excited state.

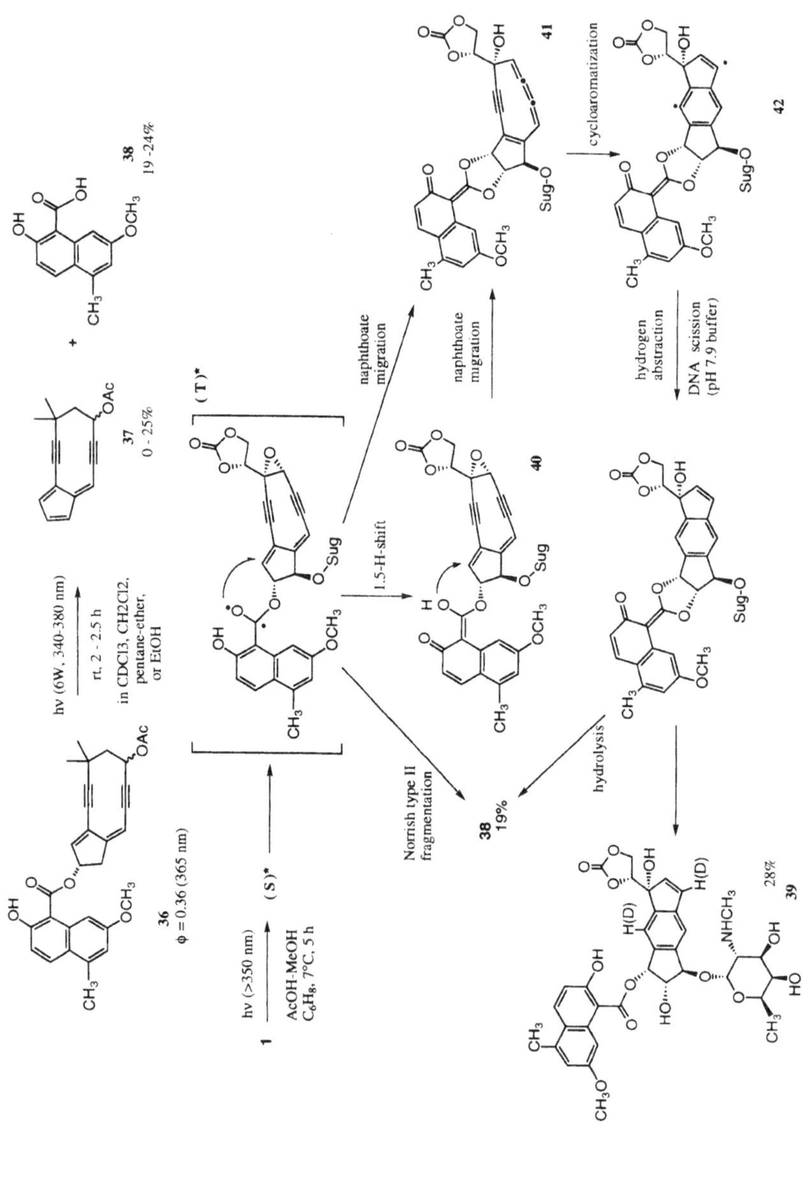

FIG. 11. Photodecomposition of analog and NCS chromophore

FIG. 12. Structure of a possible protein–chromophore conjugate (C-1027)

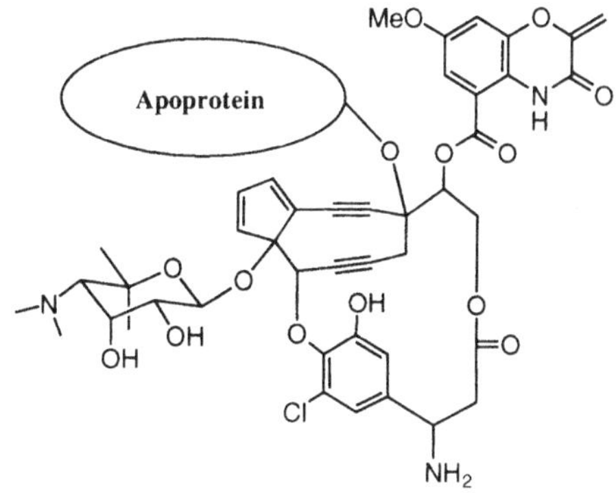

43

Synthetic Studies of NCS Chromophore and Nine-Membered Ring Analogs

Very recently, chromophores of the potent antitumor antibiotic chromoprotein C-1027 (**2**) [14,15] and kedarcidin (**3**) [16] have been shown to possess a highly strained bicyclo[7.3.0]dodecadiyne core structure. Because such nine-membered 3-ene-1,5-diyne systems are highly labile to undergo cycloaromatization at ambient temperature [14–16,67,68], a specific mechanism that prevents spontaneous aromatization of **2** and **3** should be identified in the holoprotein. Although some noncovalent stabilization interactions have been suggested between **2** and the apoprotein, they have not been verified [7–15]. The most simple alternative might be to bond them covalently as a protein conjugate **43** (Fig. 12) [32]. This strategy, which masks the 3-ene-1,5-diyne system **2** as a 1,5-diyne **43**, is also a fascinating approach from the perspective of design and synthesis of related DNA-cleaving molecules [68,69,70]. After surveying the preceding examples of syntheses of the highly strained bicyclo[7.3.0]dodecadiyne system, we describe here our general and efficient route, a fine tuning of the extremely facile Cope rearrangement of nine-membered cyclic 1,5-diynes [32], and further the successful synthesis of enediynes, **86** and **90**, as models of **2** and **3**, respectively, as well as their equilibration with a *p*-benzyne biradical and the kinetic stabilization [17].

First Synthesis of the Carbocyclic Core Structure of Chromophore

The first outstanding contribution in the synthesis of unprecedented bicyclo-[7.3.0]dodeca-1,8-diene-2,6-diyne core structure of NCS chromophore **1** was made by Wender et al. in 1988 [71]. They presented a general strategy (Fig. 13) applicable to the synthesis of NCS chromophore analogs: (i) the introduction of alkyne units through 1,2-addition of propargyl Grignard reagent to 2-

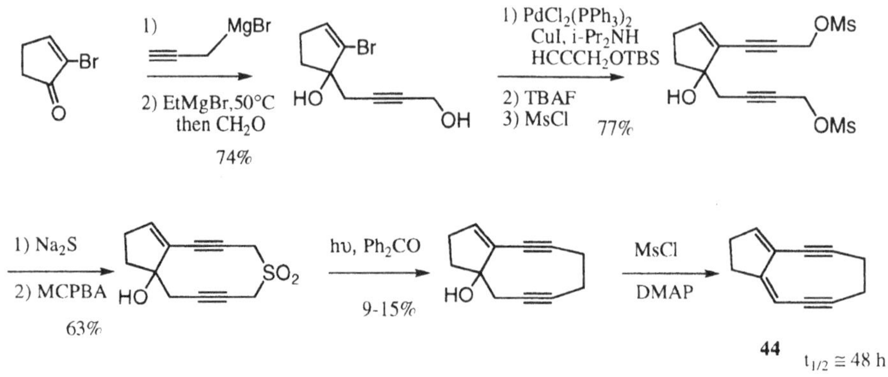

FIG. 13. The first synthesis of the parent carbocyclic core of NCS chromophore. *MCPBA*, meta-chloroperoxybenzoic acid; *DMAP*, p-dimethylaminopyridine

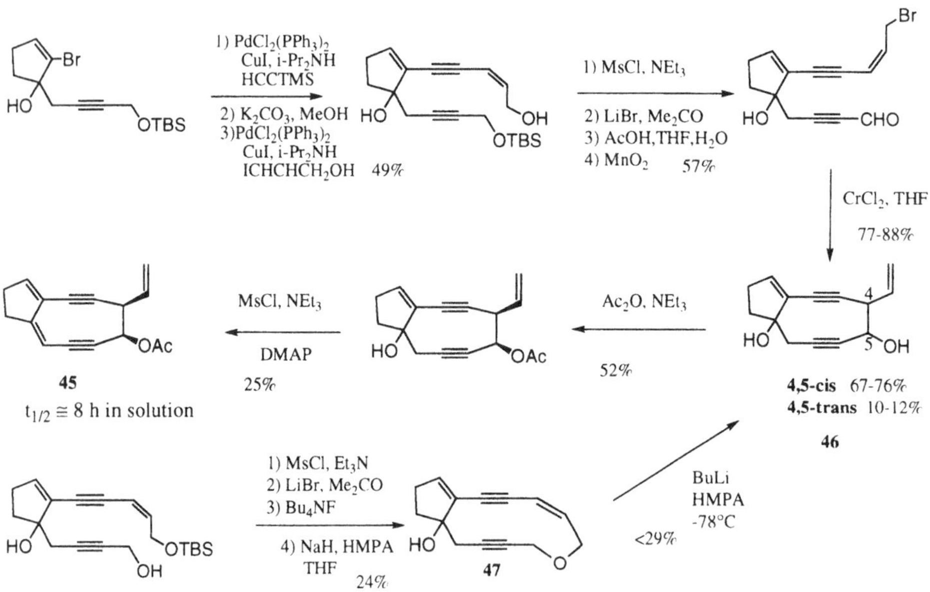

FIG. 14. Synthesis of elaborated bicyclic core analog of NCS chromophore.

bromocyclopentenone and through palladium-catalyzed coupling of vinyl bromide with a terminal alkyne; (ii) the cheletropic ring contraction entry to the highly strained nine-membered ring system; and (iii) a successful strategy and methodology for introducing the crucial 8,9-double bond at the latest stage of the synthesis. The parent hydrocarbon **44** was found to be very unstable at room temperature, suffering from polymerization upon removal of solvent. Its ^{13}C-NMR data provided corroborative support for the structure of **1**.

Synthesis of the more elaborated bicyclic core of **1** was again reported by Wender et al. in 1990 [72,73]. They achieved a novel strained-ring synthesis via the internal chromium-mediated closure, presumably through a 13-membered macrometallocyclic chelate (Fig. 14). The functionalized analog **45** was even more labile than the parent hydrocarbon **44**.

Macroring Contraction Approach Using [2,3]-Wittig Rearrangement

Another approach to **46** through [2,3]-Wittig rearrangement of **47** was also reported by Wender et al., but they found it unsatisfactory in both yield and purity (Fig. 14) [72,73]. At the same time, Krebs et al. and Takahashi et al. independently took the same macroring contraction approach. The former group synthesized 12-membered oxacycloalkenediyne **48** and reported that the attempted [2,3]-Wittig rearrangement of **48** was unsuccessful (Fig. 15) [74].

However, Takahashi et al. invented a reaction system adequate for this rearrangement (Fig. 16) [67]. They found the protection of the hydroxy group of **49** as a silyl ether was essential for the success of this rearrangement. Thus, 4,5-*cis*-**50** and **51** were obtained stereoselectively from **49** (R=H) and **49** (R=CH_3), respectively, in a reasonable yield. Compound **51** was converted to NCS chromophore analog **52**, which failed to undergo the thiol-triggering cycloaromatization because of the low leaving group ability of acetate. To improve the electrophilicity, the hydroxyl group of **51** was oxidized to form the ketone analog **53**. Without isolation of labile **53**, addition of methyl thioglycolate at –30°C then resulted in the production of a 1:1 mixture of diastereomeric benzenoid **54** in analogy with ten-membered analogs [8,50] (see following).

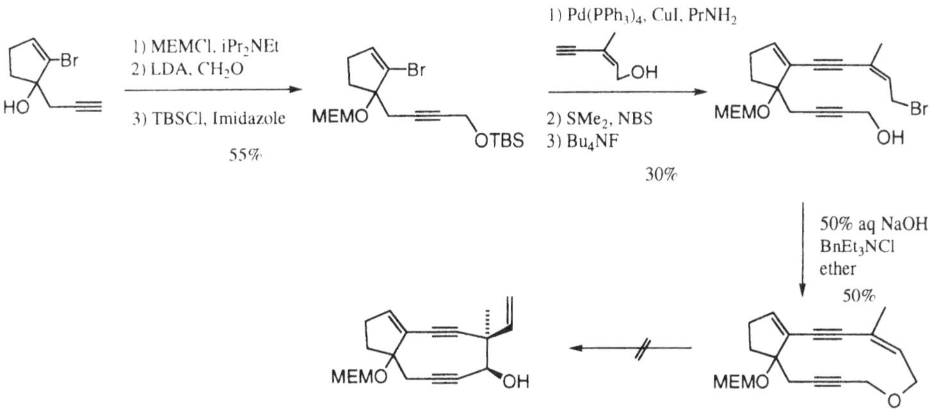

48

FIG. 15. Synthetic approach to bicyclic core of NCS chromophore. *MEM*, methoxyethoxymethyl

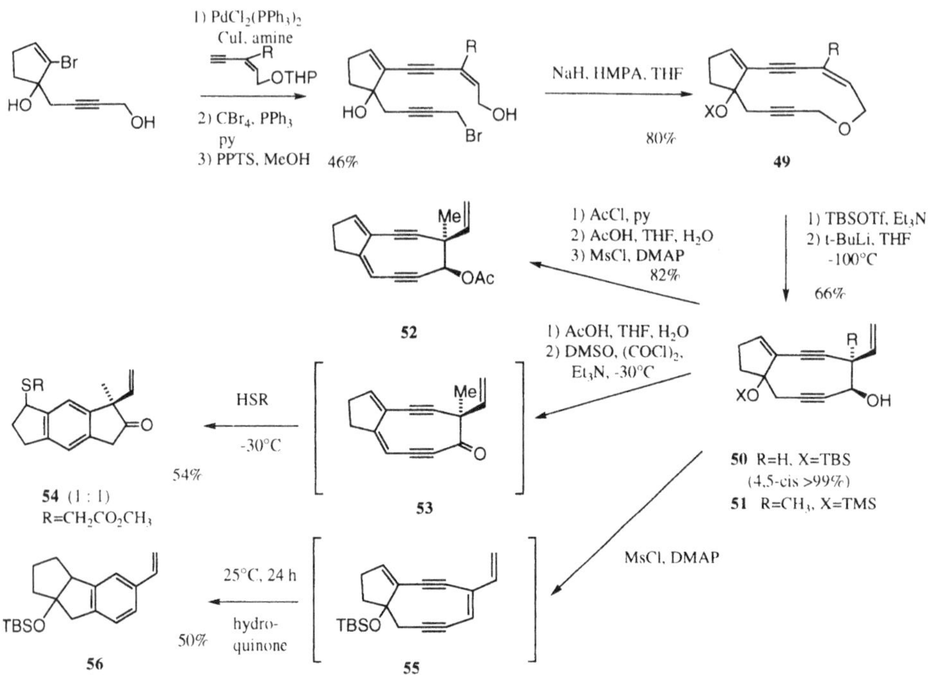

FIG. 16. Synthesis and characterization of NCS chromophore analogs. *OTHP*, tetrahydro-2-pyranyloxy

Moreover, Takahashi et al. accomplished the synthesis of the nine-membered enediyne system **55** from **50**, which easily decomposed under concentration even at −20°C. Labile **55** was demonstrated to undergo Masamune–Bergman cycloaromatization [26,27] at 25°C to give **56** in the presence of hydroquinone as the radical quencher.

Enantioselective Synthesis of the Epoxy Diyne Core

The first synthesis of the epoxy diyne core of the NCS chromophore **1** was reported by Myers et al. in 1991 [75], starting from the selectively silylated epoxydiyne **57** and the properly masked cyclopentenone derivative **58** (Fig. 17). 1,2-Addition of **57** to (2R,3R)-isomer **58** proceeded stereoselectively form the face *cis* to the acetal group. Regioselective sulfoxide elimination and mild acid hydrolysis of the acetal gave the enal **59** with the epoxy diyne intact. Cyclization of **59** to the highly strained nine-membered epoxy diyne **60** was remarkably achieved by the acetylide addition [excess $Li(TMS)_2$ at −78°C in the presence of $CeCl_3$] similar to that employed in the syntheses of ten-membered ring compounds related to the calicheamicin-esperamicins [23–25,39–42].

In 1992 Myers et al. [76] succeeded in the enantioselective synthesis of the complete epoxybicyclo[7.3.0]dodecadienediyne core structure **63** from **61**, prepared by a simple modification of a route described in Fig. 17 for the synthesis of

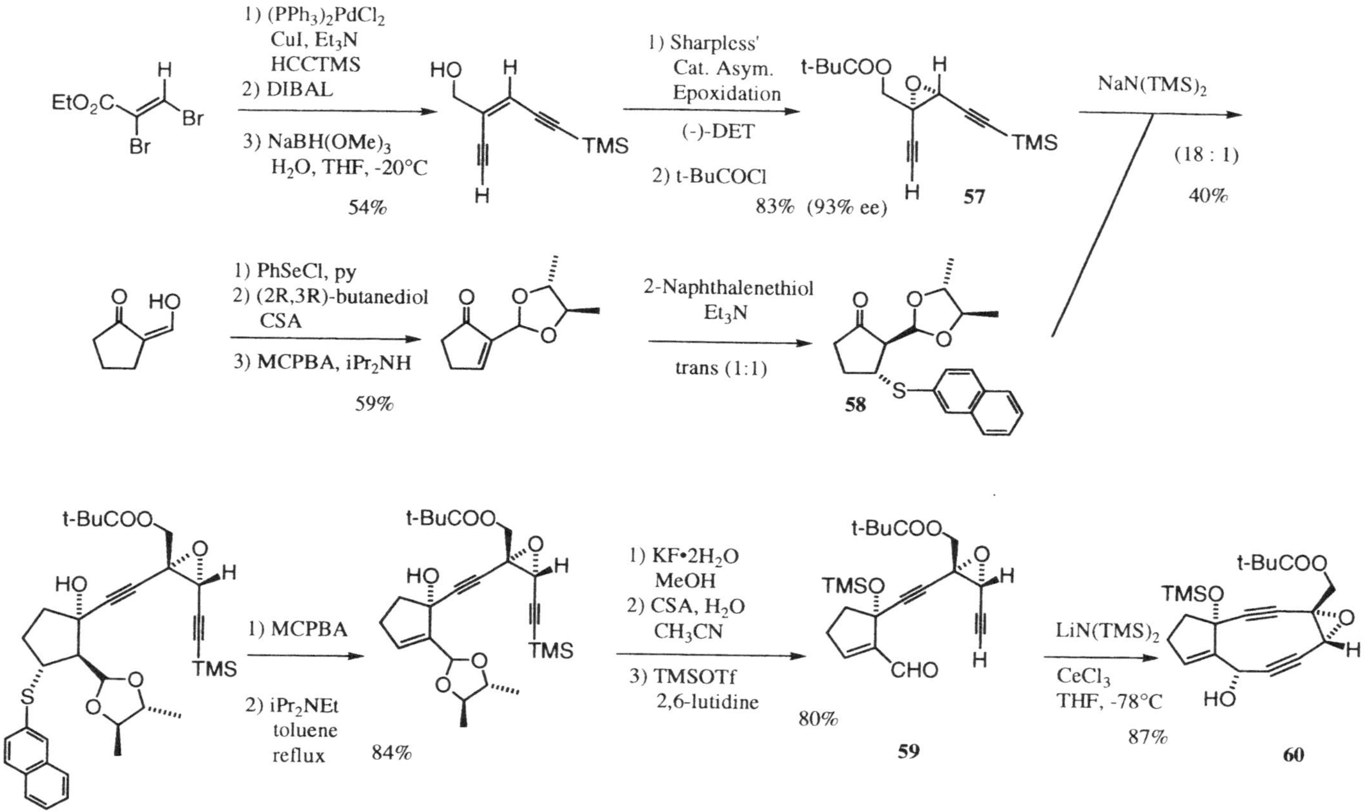

FIG. 17. Enantioselective synthesis of epoxydiyne core of NCS chromophore. *TMS*, trimethylsilyl

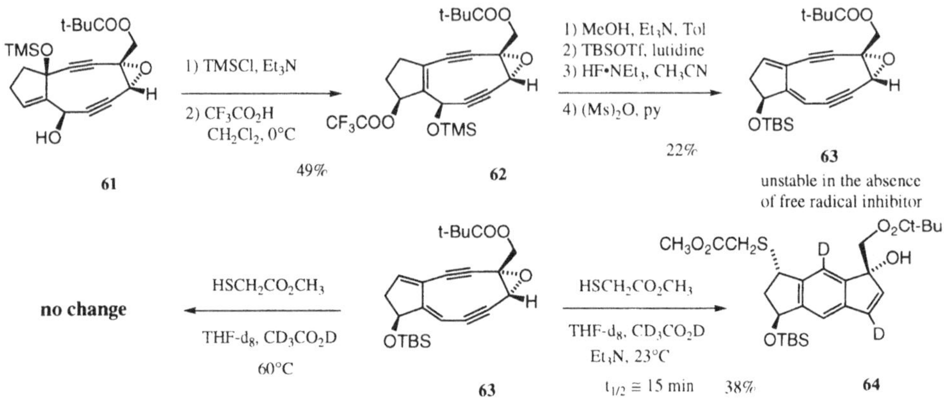

FIG. 18. Synthesis and reactivity of epoxydienediyne core of NCS chromophore. *TBS*, t-butyldimethylsilyl

diastereomer **60** (Fig. 18). A simple but skilled allylic *syn*-transposition of the trimethylsilylated **61** to the labile trifluoroacetate **62**, which has a correct absolute configuration of C-10, was developed by brief exposure to 0.2 *M* trifluoroacetic acid in dichloromethane. After protection of the C-10 hydroxy group as TBS ether, the TMS ether was selectively deprotected and subjected to dehydration with methanesulfonic acid anhydride-pyridine to afford **63**, which is less stable in neat form than chromophore **1**, decomposing within seconds upon concentration in the absence of free radical inhibitors. These labile molecules **62** and **63** were purified by flash chromatography (at 0°C), concentrated for brief periods in the presence of a free radical inhibitor, and fully characterized spectroscopically. These syntheses by Myers et al. are both intellectual and scrupulous.

Myers proposed that thiol activation of **1** is facilitated through participation of the carbohydrate amino group as an internal base even in vivo: nonbasic **63** is found to be completely inert to methyl thioglycolate in acetic acid-THF [76], while **1** reacts readily at −70°C (see Fig. 2) [28–31]. Addition of triethylamine caused a smooth formation of **64**. They also demonstrated that the nitroso derivative of **1** is inert to thiol at 0°C, and extended the conformational analysis of the aromatized chromophore to that of **1**, in which the methylamino group should be located over the β-face of the cyclopentane ring [7–13,62]. Thiol addition to **1** in water seems to be slow relative to addition **1** bound to DNA. There is a similar controversy on the aminoglycoside assistance to the internal thiolate addition in calicheamicin series. [77–80].

η^2-Hexacarbonyldicobalt-Mediated Aldol Approach to the Core Diyne Structure

In 1991, Magnus et al. [67] reported the successful synthesis of the η^2-Co(CO)$_6$-NCS chromophore core diyne **66**, adopting the novel intramolecular aldol reac-

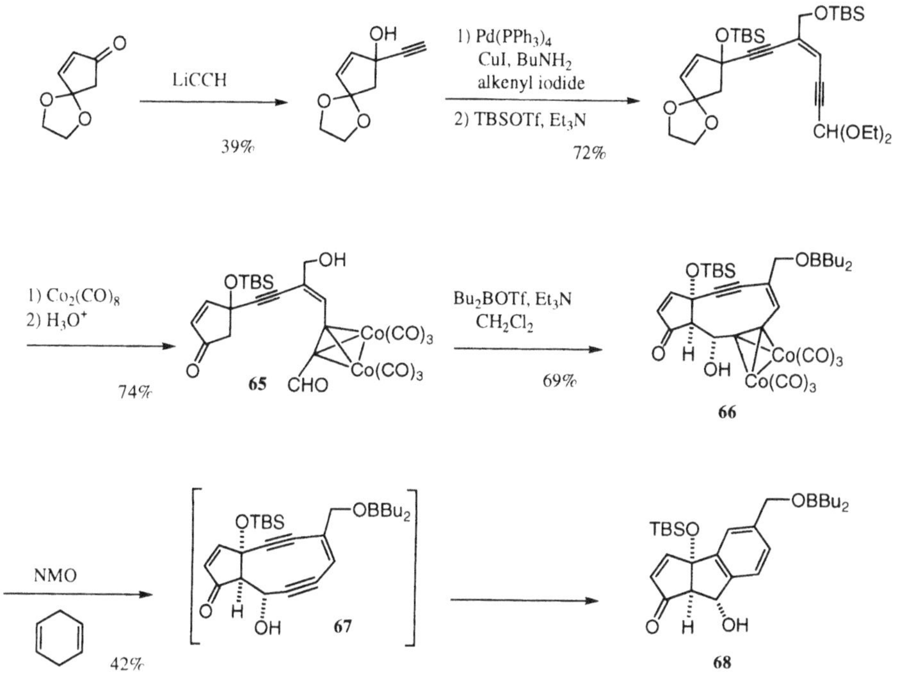

FIG. 19. Synthesis of nine-membered enediyne analog related to NCS chromophore

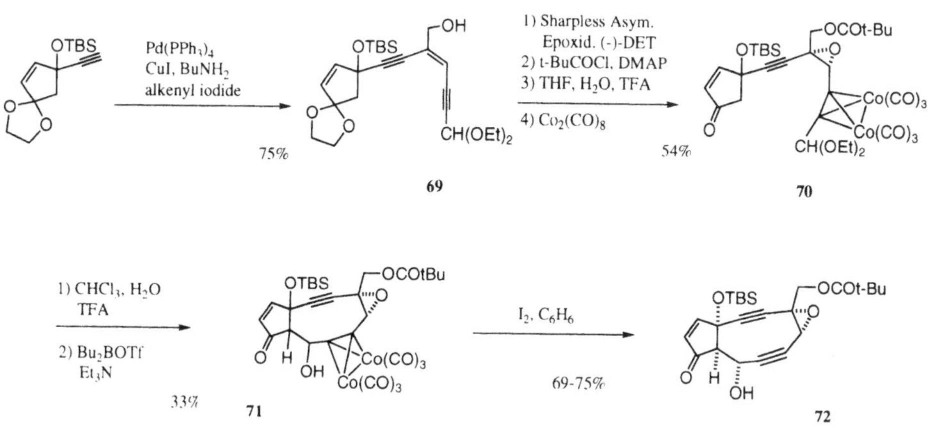

FIG. 20. Synthesis of nine-membered epoxydiyne core of NCS chromophore

tion of **65** to form the C-8–C-9 bond (Fig. 19). Oxidative decomposition of **66** in cyclohexa-1,4-diene gave the cycloaromatized **68**, which indicated the transient formation of the nine-membered enediyne **67**. Moreover, Magnus et al. extended their approach to the synthesis of the epoxydiyne core structure (Fig. 20) [81]. As removal of the BBu_2 group from **66** and attempted epoxidation of the allylic double bond were unsuccessful, the epoxide functionality was introduced at an earlier stage. Thus, **69** was asymmetrically epoxidized and the hydroxy group was protected as a stable pivaloyl ester that should retard epoxide opening at the stage of the dicobalt complex. While attempts to hydrolyze the two acetal groups simultaneously destroyed the molecule, the ethylene ketal could be hydrolyzed selectively.

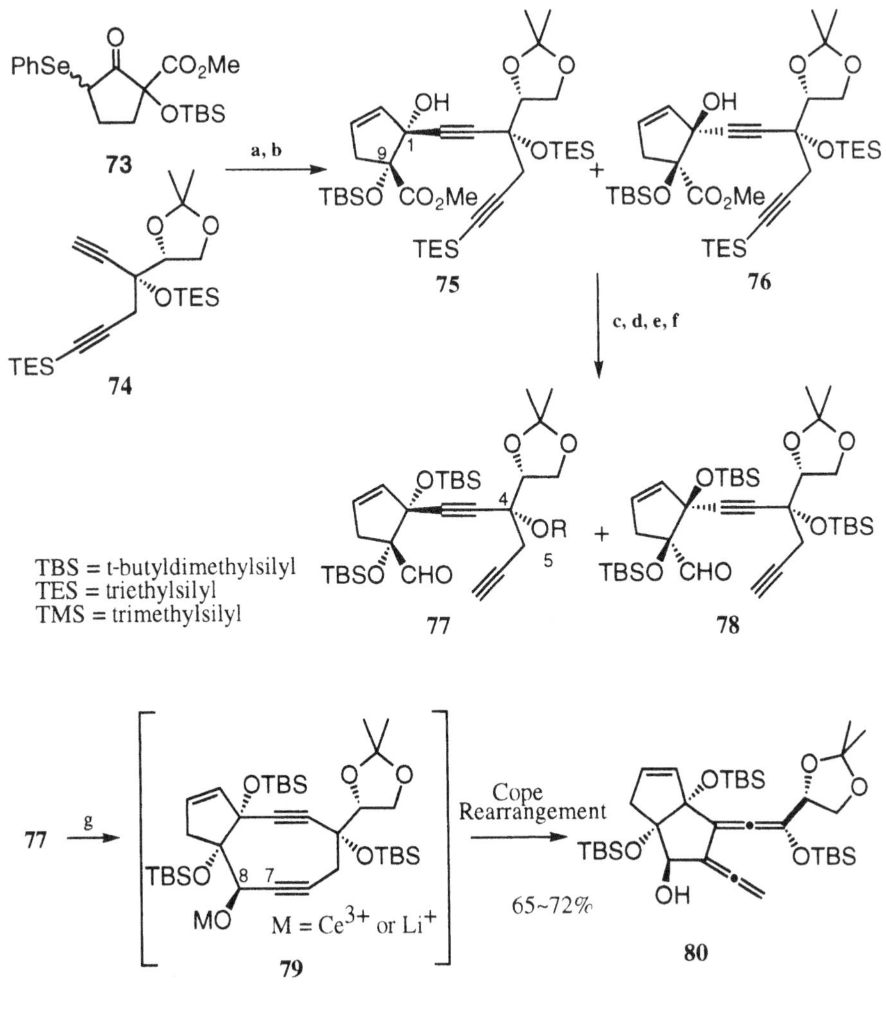

(a) EtMgBr, THF, -78°C~ rt, 79%; (b) H_2O_2, THF, 82%; (c) TBAF, THF, 93%; (d) $^tBuMe_2SiOTf$, 2,6-lutidine, CH_2Cl_2, 85%, or Me_3SiOTf (77%); (e) DIBAL-H, CH_2Cl_2, -60°C, 83-98%; (f) Dess-Martin periodinane, CH_2Cl_2, 87-97%; (g) $LiN(TMS)_2$ (6~23 mol eq), anhydrous $CeCl_3$ (7~20 mol eq), THF (3~6 mM), -30°C ~ rt, 30 min..

FIG. 21. Convergent synthesis and extremely facile Cope rearrangement of bicyclo-[7.3.0]dodecadiyne system related to C-1027 and kedarcidin chromophores

Complexation of the less hindered 6,7-alkyne bond with $Co_2(CO)_8$ gave **70**, which allowed the activated diethyl acetal to be hydrolyzed, and the subsequent Bu_2OTf-Et_3N-mediated aldol reaction afforded a diastereomeric mixture **71**. This illustrates the surprising compatibility of the 4,5-epoxide to the Lewis acid catalysis conditions. Oxidative decomposition of **71** gave the stable epoxydiyne core **72**.

General and Versatile Approach and Extremely Facile Cope Rearrangement

Thus, several groups have succeeded in synthesizing the relevant nine-membered diynes. They used techniques to minimize the high enthalpic and entropic barriers, such as ring contraction [68–71] assembly of either a *cis*-epoxide [75] or *cis*-olefin between the two acetylenic bonds, and bending of the acetylenic bond as a cobalt complex [81,82]. These results encouraged us to examine the straightforward construction of the bicyclo[7.3.0]dodecadiyne system through an intramolecular acetylide addition, from a precursor such as **77**, which possesses a conformationally nonrigid C-4–C-5 single bond.

Stereospecific addition of **74** to **73** gave the 1,9-*cis*-diol derivatives **75** and **76**, which were converted to the diastereomeric aldehydes **77** and **78**, respectively, as shown in Fig. 21 [32]. When cyclization of **77** was attempted by adding a large excess of lithium hexamethyldisilazide [$LiN(TMS)_2$, 20–30 eq] in THF at −50°C, only the C_2 symmetrical dimer was produced in 20%–40% yield, even under high dilution conditions (2 m*M*). In the presence of anhydrous cerium chloride [76,83,84]; however, a monomeric product was formed at a higher temperature (−30°C). Surprisingly, a bis-allene **80** was isolated as a single stereoisomer in a yield of up to 72%. Formation of the C-7–C-8 bond clearly indicated that the nine-membered cyclic diyne **79** acted as an intermediate and was followed by Cope rearrangement (Fig. 21) [32]. A similar dimerization and allene formation were also observed in the reactions of diastereomer **78**. Cope rearrangement of acyclic 1,5-diynes has been shown to occur above 200°C [85]. Thus, we found that the bicyclo[7.3.0]-dodecenediyne system, such as in **79**, undergoes an extremely facile Cope rearrangement [32].

Fine Tuning of the Extremely Facile Cope Rearrangement

We next examined how we could suppress the foregoing rearrangement. Review of the relevant isolated systems [68–71] suggested that a common structural feature among them is that they contain a cyclopentene double bond exo to the nine-membered ring. Therefore, we synthesized **81** and **83** from **75** and **76**, respectively. Addition of **81** to the $LiN(TMS)_2/CeCl_3$ mixture at −40°C followed by stirring at room temperature for 1 h yielded a cyclic diyne **82** as a

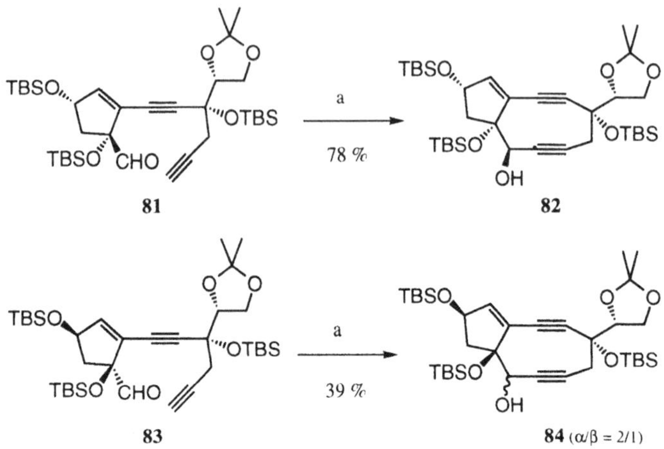

(a) $LiN(TMS)_2$ (10~23 eq), anhydrous $CeCl_3$ (11~25 mol eq), THF (1~2 mM), -40°C ~ rt, 1 h.

FIG. 22. Stereoselective construction of bicyclo[7.3.0]dodecadiyne system related to C-1027 and kedarcidin chromophores

single stereoisomer at a yield of 78% without contamination of the corresponding bis-allene (Fig. 22) [32]. Cyclization of the diastereomer **83** also occurred under the same reaction conditions, but nonstereoselectively, to give a 2:1 mixture **84** in a moderate yield. ^{13}C-NMR data for **82** and **84**, which are not stable at room temperature but can be stored in solution at –20°C without deterioration, are consistent with the nine-membered strained acetylenic bonds [3,32,75].

Cope rearrangement of **82** to a bis-allene took place in deoxygenated toluene-d_8 at a higher temperature. The half-life of this rearrangement at 50°C is 6.4 h (^{1}H-NMR analysis), which indicates that the rate of the Cope rearrangement can be modified by a small structural change, such as the shift of a double bond [32].

Previous syntheses of both cyclononadiyne and cyclodecadiyne rings related to enediyne antibiotics [18–25] via intramolecular acetylide additions were ensured by the presence of at least two structural elements that reduced the degree of conformational freedom of the substrate. i.e., *cis*-olefin (or *cis*-epoxide) and an additional five- or six-membered carbocyclic fused ring. In this study, we have demonstrated that $LiN(TMS)_2/CeCl_3$-mediated cyclization procedures [75,83] can be used to construct the highly strained cyclononadiyne system when the former principal element is absent. The diynes **82** and **84** possess the appropriate functionality to synthesize the chromophores of C-1027 (**2**) [14,15] and kedarcidin (**3**) [16].

Synthesis, Isolation, and Characterization of Nine-Membered Cyclic Enediynes

Synthesis, Isolation, and Reactions

Nine-membered diynes (**85** and **88**) were converted to enediynes (**86** and **90**, respectively) as shown in Fig. 23 [17]. The presence of an epoxide ring in the mesylate of **89** greatly facilitated the elimination reaction, which was completed within 30 min in the presence of 1,8-diazabicyclo[5.4.0]undec-7-ene (DBU) (~6 eq) in CH_2Cl_2 at 25°C; this is approximately ten times faster than that of the mesylate of **85**. The enediyne (**86**) was too labile to be isolated as anticipated [67,68] and rapidly underwent spontaneous cycloaromatization ($t_{1/2}$, ~11 min) in the presence of excess 1,4-cyclohexadiene at room temperature to afford **87** in a good yield (~87%). On the other hand, the epoxyenediyne (**90**) was more stable and could be purified by silica gel chromatography. Cycloaromatization of pure

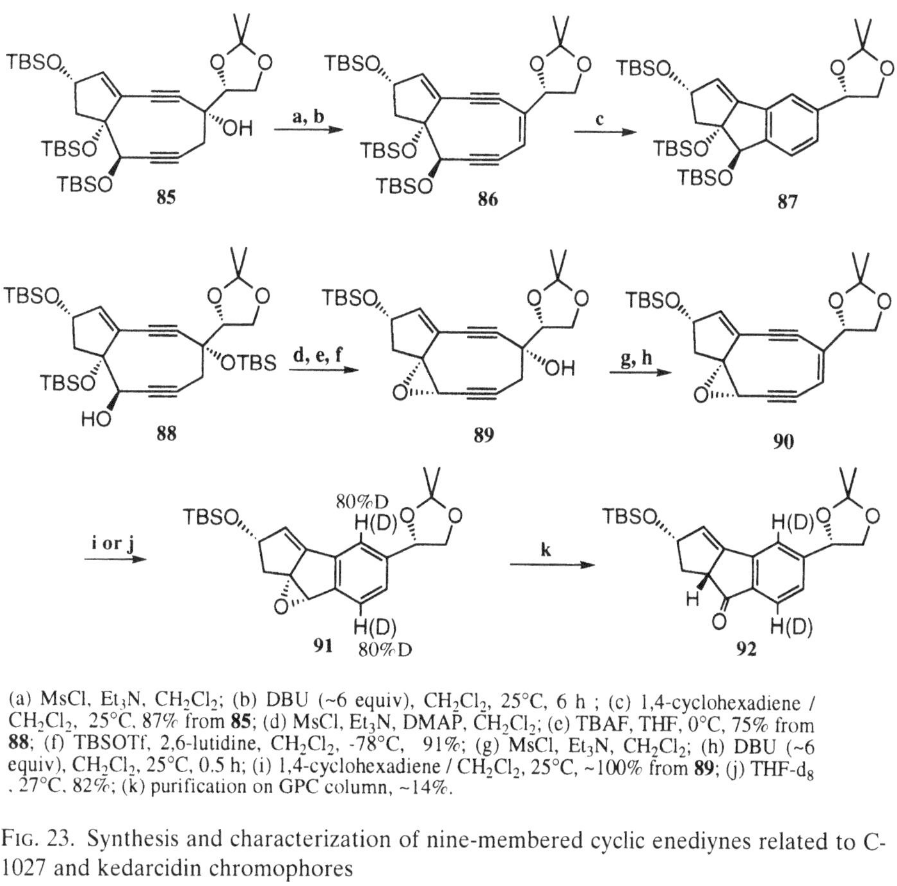

(a) MsCl, Et_3N, CH_2Cl_2; (b) DBU (~6 equiv), CH_2Cl_2, 25°C, 6 h ; (c) 1,4-cyclohexadiene / CH_2Cl_2, 25°C, 87% from **85**; (d) MsCl, Et_3N, DMAP, CH_2Cl_2; (e) TBAF, THF, 0°C, 75% from **88**; (f) TBSOTf, 2,6-lutidine, CH_2Cl_2, -78°C, 91%; (g) MsCl, Et_3N, CH_2Cl_2; (h) DBU (~6 equiv), CH_2Cl_2, 25°C, 0.5 h; (i) 1,4-cyclohexadiene / CH_2Cl_2, 25°C, ~100% from **89**; (j) THF-d_8, 27°C, 82%; (k) purification on GPC column, ~14%.

FIG. 23. Synthesis and characterization of nine-membered cyclic enediynes related to C-1027 and kedarcidin chromophores

TABLE 1. Remarkable solvent dependence of cycloaromatization rate of nine-membered enediyne (**90**) at 28°C[a]

Entry	Solvent	$t_{1/2}$ (min)[b]	k ($\times 10^{-5} s^{-1}$)[b]	Relative rate
1	CD_2Cl_2	680	1.7	0.035
2	CH_3CN	610	1.9	0.039
3	C_6H_6	330	3.5	0.071
4	1,4-Dioxane-d_8	310	3.7	0.076
5	1,4-Dioxane	110	11	0.22
6	THF-d_8[c]	220	5.4	0.11
7	THF	68	17	0.35
8	CD_3CD_2OD	130	8.8	0.18
9	CH_3CH_2OH	65	18	0.37
10	1,4-C_6D_8-CD_2Cl_2[d]	28	41	0.84
11	1,4-C_6H_8/CH_2Cl_2[e]	23	49	1.0

[a] Measured by HPLC except entry 6.
[b] Deviation: ±1% ± ±6%.
[c] Measured by ^{1}H-NMR.
[d] 1,4-Cyclohexadiene-d_8/CD_2Cl_2 = 1 : 1 (v/v).
[e] 1,4-Cyclohexadiene/CH_2Cl_2 = 1 : 1 (v/v).

90 was approximately four times slower than that of **86** in THF-d_8. Quantitative formation of unstable **91** in 1,4-cyclohexadiene-CH_2Cl_2 (1:1) was confirmed by NMR, but removal of the solvent or silica gel chromatography resulted in its complete decomposition. In this case, ketone **92** was isolated as a major product (~14%) by gel permeation chromatography (GPC) column filtration. Bis-deuterated **91**-d_2 and **92**-d_2 were produced in the perdeuterated solvents [17].

Remarkable Solvent Dependence of Rate of Cycloaromatization

Yoshida and co-workers recently observed the unexpected solvent dependence of the rate of cycloaromatization of **2** [86]. Because a pure nine-membered enediyne (**90**) that is soluble in most organic solvents is now available, we examined more precisely the cycloaromatization rate of **90** in various solvents[1] [87,88]. We were again surprised that the pseudo-first-order decay of **90** is highly dependent on the solvent as a hydrogen donor (Table 1) [17]. The data showed the relative rates of THF, benzene, and CH_3CN to be 1:0.2:0.1, which were only slightly lower than those reported for hydrogen abstraction by phenyl radical (1:0.1:0.02) [89,90]. A primary kinetic isotope effect was also noticed: its magnitude increased as the reaction became slow, as has been observed generally in hydrogen transfer reactions [91–93]. These results indicate that the hydrogen

[1] Dependence of $t_{1/2}$ of ten-membered enediynes on the concentration of 1,4-cyclohexadiene at 84°C was reported recently.

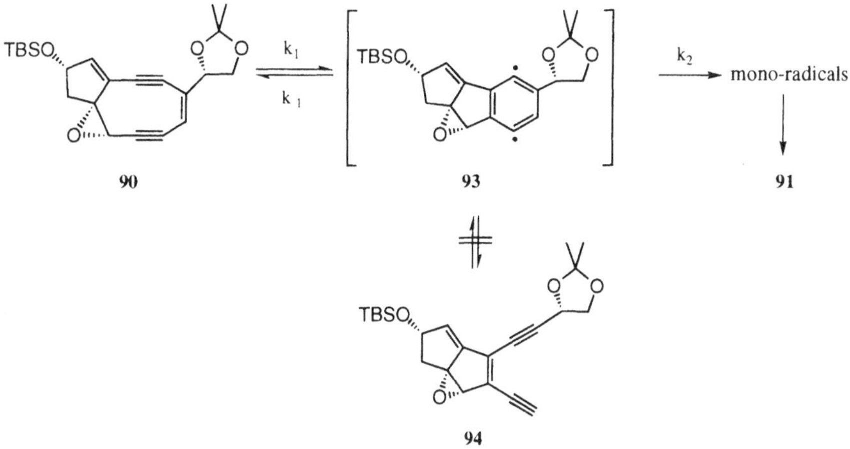

FIG. 24. Equilibration of synthetic nine-membered enediyne with *p*-benzyne biradical at ambient temperatures

abstraction step by a *p*-benzyne biradical intermediate (**93**) (for studies on *p*-benzene, see [94–104]) is kinetically significant, or rate limiting in the cycloaromatization of **90** (Fig. 24) [17], while the first cyclization step is known to be rate determining for acyclic enediynes [94,105].

Kinetics and Energetics of Cycloaromatization

Thus, **90** may be virtually in equilibrium with **93** in CH_3CN and CD_2Cl_2 in which cycloaromatization is substantially retarded (Fig. 24). If the hydrogen abstraction rate of **93** is slower than that of phenyl radical (CH_3CN, $k_H = 1.0 \times 10^5\,M^{-1}s^{-1}$) [89] because of steric hindrance or through-bond interaction [95–104] by a factor of about 100 [106–108][2], the equilibrium constant (K) in CH_3CN is estimated to be 2×10^{-9} ($\Delta G = \sim 12$ kcal/mol) by steady-state approximation to the cocentration of **93** ($K_{obs} = k_1k_2/k_{-1} = Kk_2$) based on the assumption of the pseudo-first-order kinetic constant $k_2 = k_H[CH_3CN] \times 10^{-2} \approx k_H \times 10^{-1}\,s^{-1}$. This ΔG seems not unreasonable because the sum of this value and an E_a for hydrogen abstraction by phenyl radical (4–7 kcal/mol) [90] and the increment from steric hindrance or through-bond interaction, ~3 kcal/mol, is in good agreement with an apparent activation energy for the decay of **90** [$E_a = 21.6$ kcal/mol (ln A = 28.5: 20°–32°C in 1,4-C_6H_8/CH_2Cl_2); 18.5 kcal/mol (ln A = 21.9: 40°–60°C in EtOH)] obtained by the Arrhenius plot of the rate constants [17].

[2] This value was arbitrarily selected, as the lifetimes of free radicals including phenyl are dramatically increased by surrounding the radical center with bulky substituents by a factor of up to $\sim 10^7$

Equilibration of Nine-Membered Enediyne with p-*Benzyne*

The ΔG (~12 kcal/mol) is similar to that reported for acyclic systems [86,94,95]. It suggests that energy of **93** may also be raised to such an extent that **90** is destabilized by nine-membered ring strain [17]. The destabilization should arise from the presence of a 1,8-double bond and an epoxide ring in the dehydrobenzopentalene core of **93** [32] as implied by MM2 calculations (MacroModel ver. 4.5) [17,109]. The barriers for cycloaromatization of acyclic (Z)-3,4-dipropylhex-3-ene-1,5-diyne to 2,3-dipropyl-1,4-dehydrobenzene (**95**), its back reaction, and an alternative ring opening of **95** to (Z)-dodec-6-ene-4,8-diyne were reported to be 27.4, ~16, and ~10 kcal/mol, respectively [94]. Therefore, the destabilization of both **90** and **93** should significantly decrease the barrier for their interconversion and consequently the process became reversible at ambient temperature. The foregoing Bergman's observation [94] and our failure to detect even trace amounts of an acyclic enediyne **94** strongly suggest that cleavage of the less substituted bond of **93** leading to **94** has a considerably higher barrier [86,17]. Nine-membered ring constraint may also contribute to lowering the barrier to **90** because the reaction of **93** to **90** requires less molecular deformation. Thus, the kinetic significance of hydrogen abstraction step in the present system may be attributable to the large energy difference between **90** and **93**, the very low barrier for back reaction the **90**, and the slower hydrogen abstraction by *p*-benzyne biradical **93** [17].

Conclusion

Kinetic Stabilization in Natural Chromoprotein Antibiotics

These observations for **90** suggest a hypothesis that the chromophores (**2** and **3**) may also be equilibrated with their *p*-benzyne form such as **95** and are stabilized kinetically by a specific apoprotein (Fig. 25). Thus, **2** and **3** may exist indefinitely if they remain free of hydrogen donor(s) in the holoprotein complex. Aromatic amino acid residues, such as Tyr32, as well as the internal tyrosine and benzoxazine rings of **2**, seem to play an important role in this kinetic stabilization [17]. An attempt to detect *p*-bezyne biradical intermediate (**95**) in equilibrium by electron spin resonance (ESR) spectroscopy are currently being investigated in our laboratory [17].

A Proposed Equilibration of the Cumulene Intermediate with the Dehydroindene Biradical Formed from the NCS Chromophore

By analogy, it could be proposed that there also exists a partial equilibrium between the cumulene intermediate (**5**) and the 1,4-*trans*-biradical (**6**) produced from the NCS chromophore (**1**) by thiol trigger (Fig. 25), although it has not yet been supported experimentally. This idea suggests that **5** and **6** should have an

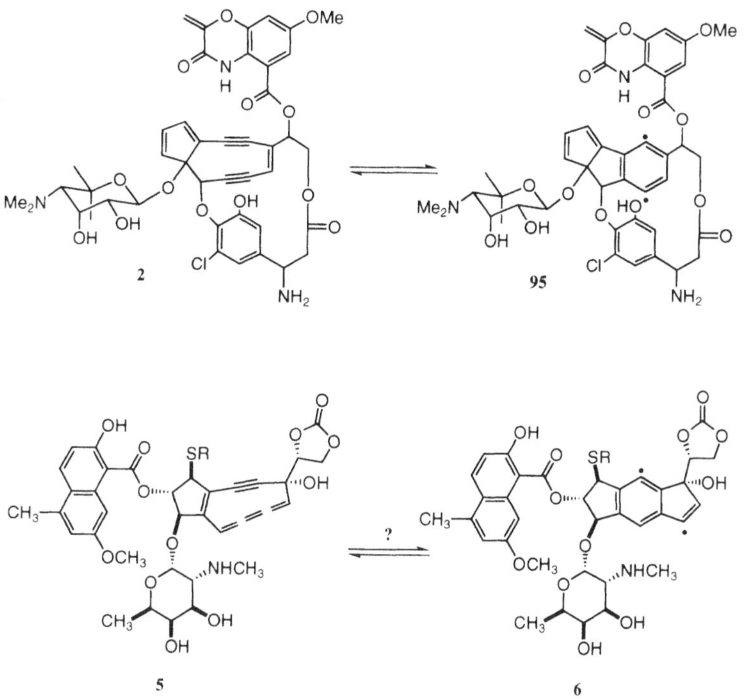

FIG. 25. Proposed equilibrations of natural chromophores with aromatized 1,4-biradicals

unusually long lifetime in the absence of a hydrogen donor, and that k_{obs} ($1.0 \times 10^{-4}s^{-1}$) and $\Delta G^{\ddagger}_{obs}$ (18.0 kcal/mol), reported for the reaction cascade of the neocarzinostatin chromophore, might apply to the rate-limiting hydrogen abstraction by **6**, but not to the cyclization step (**5** → **6**) as assigned previously [30,31]. Our proposition may provide a better understanding of the intermediacy of **5** with appreciably long life in the sequence-specific DNA cleavage [31,110].

References

1. Ishida N, Miyazaki K, kumagai K, Rikimaru M (1965) Neocarzinostatin, an antitumor antibiotic of high molecular weight. J Antibiot (Tokyo) Ser A 18:68–76
2. Goldberg IH (1991) Mechanism of neocarzinostatin action: role of DNA microstructure in determination of chemistry of bistranded oxidative damage. Acc Chem Res 24:191–198
3. Edo K, Mizugaki M, Koide Y, Seto H, Furihata K, Otake N, Ishida N (1985) The structure of neocarzinostatin chromophore possessing a novel bicyclo [7,3,0]dodecadiyne system. Tetrahedron Lett 26:331–334
4. Kuromizu K, Tsunasawa S, Maeda H, Abe O, Sakiyama F (1986) Reexamination of the primary structure of an antitumor protein, neocarzinostatin. Arch Biochem Biophys 246:199–205

5. Sakata N, Minamitani S, Kanbe T, Hori M, Hamada M, Edo K (1993) The amino acid sequence of neocarzinostatin apoprotein deduced from the base sequence of the gene. Biol Pharm Bull 16:26–28
6. Kappen LS, Goldberg IH (1980) Stabilization of neocazinostatin nonprotein chromophore activity by interaction with apoprotein and with HeLa cells. Biochemistry 19:4786–4790
7. Tanaka T, Hirama M, Fujita K, Imajo S, Ishiguro M (1993) Solution structure of antitumor antibiotic neocarzinostatin, a chromophore-protein complex. J Chem Soc Chem Commun: 1205–1207
8. Hirama M (1994) Neocarzinostatin complex: chromophore analogs, binding structure and stabilization mechanism (in English). J Synth Org Chem Jpn 52:980–991
9. Ishiguro M, Imajo S, Hirama M (1991) Modeling study of the structure of macromolecular antitumor antibiotics, neocarzinostatin. The origin of the stabilization of the chromophore. J Med Chem 34:2366–2373
10. Imajo S, Ishiguro M, Tanaka T, Hirama M, Teplyakov A (1995) On the conformation of Phe78 of chromoprotein antibiotic, neocarzinostatin. Bioorg Med Chem 3:429–436
11. Hirama M, Tanaka T (1994) Molecular recognition in neocarzinostatin complex: how does the apoprotein bind specifically and stabilize the chromophore? Pure Appl Chem 66:791–796
12. Gomibuchi T, Fujiwara K, Nehira T, Hirama M (1993) On the origin of photolability of neocarzinostatin chromophore. Tetrahedron Lett 34:5753–5756
13. Gomibuchi T, Hirama M (1995) Photoactivation of neocarzinostatin chromophore: photo-induced cycloaromatization via naphthoate participation. J Antibiot (Tokyo) 48:738–740
14. Minami Y, Yoshida K, Azuma R, Saeki M, Otani T (1993) Structure of an aromatization product C-1027 chromophore. Tetrahedron Lett 34:2633–2636
15. Yoshida K, Minami Y, Azuma R, Saeki M, Otani T (1993) Structure and cycloaromatization of a novel enediyne, C-1027 chromophore. Tetrahedron Lett 34:2637–2640
16. Lett JE, Schroeder DR, Langley DR, Colson KL, Huang S, Klohr SE, Lee MS, Golik J, Hofstead SJ, Doyle TW, Matson JA (1993) Chemistry and structure elucidation of the kedarcidin chromophore. J Am Chem Soc 115:8432–8443
17. Iida K, Hirama M (1995) Synthesis and characterization of nine-membered cyclic enediynes, models of C-1027 and kedarcidin chromophore: equilibration with a *p*-benzyne biradical and kinetic stabilization. J Am Chem Soc 117:8875–8876
18. Fujiwara K, Hirama M (1990) Chemistry of antitumor natural products and analogs generating biradicals. Gendai Kagaku (Chemistry Today) 7:14–21
19. Hirama M (1991) Design and synthesis of neocarzinostatin analogs. J Synth Org Chem Jpn 49:1032–1042
20. Hirama M (1992) Why do enediynes affect cancers? Kagaku (Chemistry) 47:162–166
21. Hirama M (1993) Synthesis and chemistry of neocarzinostatin analogs. In: Lukacs G (ed) Recent progress in the chemical synthesis of antibiotics and related microbial products, vol 2. Springer, Berlin Heidelberg New York, pp 293–329
22. Hirama M (1993) Molecular recognition in neocarzinostatin complex and synthesis of new DNA cleaving molecules. In: Krohn K, Kirst HA, Maag H (eds) Antibiotics and antiviral compounds: chemical synthesis and modification. VCH, Weinheim, pp 289–301

23. Nicolaou KC, Dai W-M (1991) Chemistry and biology of the enediyne anticancer antibiotics. Angew Chem Int Ed Engl 30:1387–1416
24. Skrydstrup T, Ulibarri G, Audrain H, Grierson DS (1993) The "enediyne" antibiotics calicheamicin γ_1^I, esperamicin A, and dynemycin A: physicochemical and biological properties—partial and total synthesis. In: Lukacs G (ed) Recent progress in the chemical synthesis of antibiotics and related microbial products, vol 2. Springer, Berlin Heidelberg New York, pp 213–291
25. Danishefsky SJ, Shair MD (1996) Observations in the chemistry and biology of cyclic enediyne antibiotics: total syntheses of calicheamicin γ_1^I and dynemicin. J Org Chem 61:16–44
26. Darby N, Kim CU, Salaün JA, Shelton KW, Takeda S, Masamune S (1971) Concerning the 1,5-didehydro[10]annulene system. J Chem Soc D Chem Commun: 1516–1517
27. Jones RR, Bergman RG (1972) *p*-Benzyne. Generation as an intermediate in a thermal isomerization reaction and trapping evidence for the 1,4-benzenediyl structure. J Am Chem Soc 94:660–661
28. Myers AG (1987) Proposed structure of the neocarzinostatin chromophore-methyl thioglycolate adduct: a mechanism for the nucleophilic activation of neocarzinostatin. Tetrahedron Lett 28:4493–4496
29. Myers AG, Proteau PJ, Handel TM (1988) Stereochemical assignment of neocarzinostatin chromophore. Structures of neocarzinostatin chromophore-methyl thioglycolate adducts. J Am Chem Soc 110:7212–7214
30. Myers AG, Proteau PJ (1989) Evidence for spontaneous, low-temperature biradical formation from a highly reactive neocarzinostatin chromophore-thiol conjugate. J Am Chem Soc 111:1146–1147
31. Myers AG, Cohen SB, Kwon B-M (1994) DNA cleavage by neocarzinostatin chromophore. Establishing the intermediacy of chromophore-derived cumulene and biradical species and their role in sequence-specific cleavage. J Am Chem Soc 116:1670–1682
32. Iida K, Hirama M (1994) Efficient route to the 9-membered cyclic diyne system: tuning of the extremely facile cope rearrangement of 1,5-diyne. J Am Chem Soc 116:10310–10311
33. Lee MD, Dunne TS, Siegel MM, Chang CC, Morton GO, Borders DB (1987) Calichemicins, a novel family of antitumor antibiotics. 1. Chemistry and partial structure of calichemicin γ_1^I. J Am Chem Soc 109:3464–3466
34. Lee MD, Dunne TS, Chang CC, Ellestad GA, Siegel MM, Morton GO, McGahren WJ, Borders DB (1987) Calichemicins, a novel family of antitumor antibiotics. 2. Chemistry and structure of calichemicin γ_1^I. J Am Chem Soc 109:3466–3468
35. Golik J, Clardy J, Dubay G, Groenewold G, Kawaguchi H, Konishi M, Krishnan B, Ohkuma H, Saitoh K, Doyle TW (1987) Esperamicins, a novel class of potent antitumor antibiotics. 2. Structure of esperamicin X. J Am Chem Soc 109:3461–3462
36. Golik J, Dubay G, Groenewold G, Kawaguchi H, Konishi M, Krishnan B, Ohkuma H, Saitoh K, Doyle TW (1987) Esperamicin, a novel class of potent antitumor antibiotics. 3. Structures of esperamicins A_1, A_2, and A_{1b}. J Am Chem Soc 109:3462–3464
37. Konishi M, Ohkuma H, Tsuno T, Oki T (1990) Crystal and molecular structure of dynemicin A: a novel 1,5-diyn-3-ene antitumor antibiotic. J Am Chem Soc 112:3715–3716

38. Schroeder DR, Colson KL, Klohr SE, Zein N, Langley DR, Lee MS, Matson JA, Doyle TW (1994) Isolation, structure determination, and proposed mechanism of action for artifacts of maduropeptin chromophore. J Am Chem Soc 116:9351–9352
39. Nicolaou KC, Hummel CW, Pitsinos EN, Nakada M, Smith AL, Shibayama K, Saimoto H (1992) Total synthesis of calicheamicin γ_1^I. J Am Chem Soc 114:10082–10084
40. Nicolaou KC, Hummel CW, Nakada M, Shibayama K, Pitsinos EN, Saimoto H, Mizuno Y, Baldenius K-U, Smith AL (1993) Total synthesis of calicheamicin γ_1^I. 3. The final stages. J Am Chem Soc 115:7625–7635
41. Hitchcock SA, Boyer SH, Chu-Moyer MY, Oson SH, Danishefsky SJ (1994) A convergent total synthesis of calicheamicin γ_1^I. Angew Chem Int Ed Engl 33:858–862
42. Hitchcock SA, Boyer SH, Chu-Moyer MY, Olson SH, Danishefsky SJ (1995) A remarkable glycosylation reaction: the total synthesis of calicheamicin γ_1^I. J Am Chem Soc 117:5750–5756
43. Myers AG, Fraley ME, Tom NJ, Cohen SB, Madar DJ (1995) Synthesis of (+)-dynemicin A and analogs. Chem Biol 2:33–43
44. Shair MD, Yood T-Y, Danishefsky SJ (1995) Total synthesis of (±)-dynemicin A. Angew Chem Int Ed Engl 34:1721–1723
45. Fujiwara K, Sakai H, Hirama M (1991) Enyne[3]cumulene. Synthesis and mode of aromatization. J Org Chem 56:1688–1689
46. Myers AG, Kuo EY, Finny NS (1989) Thermal generation of α,3-dehydrotoluene from (Z)-1,2,4-heptatrien-6-yne. J Am Chem Soc 111:8057–8059
47. Nagata R, Yamanaka H, Murahashi E, Saito I (1990) DNA cleavage by acyclic eneyne-allene systems related to neocarzinostatin and esperamicin-calicheamicin. Tetrahedron Lett 31:2907–2910
48. Scheuplein SW, Machinek R, Suffert J, Brückner R (1993) Activation of a dienediyne model of neocarzinostatin chromophore through an acid-mediated solvolysis. Evidence for a new cyclization mode of enyne[3]cumulenes. Tetrahedron Lett 34:6549–6552
49. Stille JK, Simpson JH (1987) Stereospecific palladium-catalyzed coupling reactions of vinyl iodides with acetylenic tin reagents. J Am Chem Soc 109:2138–2152
50. Hirama M, Fujiwara K, Shigematsu K, Fukazawa Y (1989) The 10-membered ring analogues of neocarzinostatin chromophore: design, synthesis and mode of decomposition. J Am Chem Soc 111:4120–4122
51. Fujiwara K, Kurisaki A, Hirama M (1990) Two diverse modes of aromatization of a new 10-membered ring analogue of neocarzinostatin chromophore. Tetrahedron Lett 31:4329–4332
52. Fujiwara K, Sakai H, Tanaka T, Hirama M (1994) Reactions of neocarzinostatin chromophore analogues with molecular oxygen. Chem Lett: 457–460
53. Tanaka T, Fujiwara K, Hirama M (1990) Oxidative triggering for aromatization of neocarzinostatin chromophore. Tetrahedron Lett 31:5947–5950
54. Sugiyama H, Yamashita K, Fujiwara T, Saito I (1994) Apoprotein-assisted unusual cyclization of neocarzinostatin chromophore. Tetrahedron 50:1311–1325
55. Chin D-H, Goldberg IH (1993) Identification of the active species responsible for holo-neocarzinostatin-induced DNA cleavage. J Am Chem Soc 115:9341–9342
56. Tokuda M, Fujiwara K, Hirama M (1991) Synthesis and cycloaromatization of neocarzinostatin chromophore analogue equipped with an intramolecular nucleophile. Synlett: 651–654

57. Hirama M, Gomibuchi T, Fujiwara K, Sugiura Y, Uesugi M (1991) Synthesis and DNA-cleaving abilities of functional neocarzinostatin chromophore analogues. Base discrimination by a simple alcohol. J Am Chem Soc 113:9851–9853
58. Tokuda M, Fujiwara K, Gomibuchi T, Hirama M, Uesugi M, Sugiura Y (1993) Synthesis of a hybrid molecule containing neocarzinostatin chromophore analogue and minor groove binder. Tetrahedron Lett 34:669–672
59. Galat A, Goldberg IH (1990) Molecular models of neocarzinostatin damage of DNA: analysis of sequence dependence in 5′GAGCG : 5′CGCTC. Nucleic Acids Res 18:2093–2099
60. Sugiyama H, Fujiwara T, Kawabata H, Yoda N, Hirayama N, Saito I (1992) Chemistry of neocarzinostatin-mediated cleavage of oligonucleotides. Competitive ribose C5′ and C4′ hydroxylation. J Am Chem Soc 114:5573–5578
61. Nishio M, Hirota M (1989) CH/p interaction: implication in organic chemistry. Tetrahedron 45:7201–7245
62. Roey PV, Beerman Ta (1989) Crystal structure analysis of auromomycin apoprotein (macromomycin) shows importance of protein side chains to chromophore binding selectivity. Proc Natl Acad Sci USA 86:6587–6591
63. Teplyakov A, Obmolova G, Wilson K, Kuromizu K (1993) Crystal structure of apo-neocarzinostatin at 0.15-μm resolution. Eur J Biochem 213:737–741
64. Roey PV, Beerman TA (1989) Crystal structure analysis of auromomycin apoprotein (macromomycin) shows importance of protein side chains to chromophore binding selectivity. Proc Natl Acad Sci USA 86:6587–6591
65. Khokhlov AS, Reshetov PD, Chupova LA, Cherches BZ, Zhigis LS, Stoyachenko IA (1976) Chemical studies on actinoxanthin. J Antibiot (Tokyo) 29:1026–1034
66. Takahashi K, Tanaka T, Suzuki T, Hirama M (1994) Synthesis and binding of simple neocarzinostatin chromophore analogues to the apoprotein. Tetrahedron 50:1327–1340
67. Magnus P, Pitterna T (1991) Synthesis of the neocarzinostatin chromophore a core diynene structure using an η^2-hexacarbonyldicobalt-mediated aldol reaction. J Chem Soc Chem Commun: 541–543
68. Doi T, Takahashi T (1991) Syntheses and transannular cyclizations of neocarzinostatin-chromophore and esperamicin-calichemicin analogues. J Org Chem 56:3465–3473
69. Wender PA, McKinney JA, Mukai C (1990) General methodology for the synthesis of neocarzinostatin chromophore analogues: intramolecular chromium-mediated closures for strained-ring synthesis. J Am Chem Soc 112:5369–5370
70. Takahashi T, Tanaka H, Hirai Y, Doi T, Yamada H, Shiraki T (1993) Synthesis of a nine-membered cyclic analogue of the neocarzinostatin chromophore and its DNA-cleaving activity. Angew Chem Int Ed Engl 32:1657–1659
71. Wender PA, Harmata M, Jeffrey D, Mukai C, Suffert J (1988) Studies on DNA-active agents: the synthesis of the parent carbocyclic subunit of neocarzinostatin chromophore A. Tetrahedron Lett 29:909–912
72. Wender PA, McKinney JA, Mukai C (1990) General methodology for the synthesis of neocarzinostatin chromophore analogues: intramolecular chromium-mediated closures for strained-ring synthesis. J Am Chem Soc 112:5369–5370
73. Wender PA, Grissom JW, Hoffmann U, Mah R (1990) Hetero-atom substituted chromium allyls: synthetic studies on neocarzinostatin chromophore analogues. Tetrahedron Lett 31:6605–6608

74. Krebs A, Wehlage T, Kramer C-P (1990) Synthesis of a bicyclic oxacycloalkenediyne system related to neocarzinostatin chromophore A. Tetrahedron Lett 31:3533–3536
75. Myers AG, Harrington PM, Kuo EY (1991) Enantioselective synthesis of the epoxy diyne core of neocarzinostatin chromophore. J Am Chem Soc 113:694–695
76. Myers AG, Harrington PM, Kwon B-M (1992) Evidence for aminoglycoside participation in thiol activation of neocarzinostatin chromophore. Synthesis and reactivity of the epoxy dienediyne core. J Am Chem Soc 114:1086–1087
77. Ellestad GA, Hamann PR, Zein N, Morton GO, Siegel MM, Pastel M, Borders DB, McGahren WJ (1989) Reactions of the trisulfide moiety in calicheamicin. Tetrahedron Lett 30:3033–3036
78. Zein N, Poncin M, Nilakantan R, Eellestad GA (1989) Calicheamicin γ_1^I and DNA: molecular recognition process responsible for site-specificity. Science 244: 697–699
79. Cramer KD, Townsend CA (1991) Kinetics of trisulfide cleavage in calicheamicin—assessing the role of the ethylamino group. Tetrahedron Lett 32:4635–4638
80. Chatterjee M, Smith PJ, Townsend CA (1996) The role of the amino sugar and helix binding in the thiol-induced activation of calicheamicin for DNA cleavage. J Am Chem Soc 118:1938–1948
81. Magnus P, Davies M (1991) Synthesis of the 4,5-epoxybicyclo[7.3.0]dodecadiyne neocarzinostatin core structure. Surprising compatibility of the 4,5-epoxide with a η^2-hexacarbonyldicobalt-mediated aldol reaction. J Chem Soc Chem Commun: 1522–1524
82. Magnus P (1994) A general strategy using $\eta^2Co_2(CO)_6$ acetylene complexes for the synthesis of the enediyne antitumor agents esperamicin, calicheamicin, dynemicinan and neocarzinostatin. Tetrahedron Symp 50:1397–1418
83. Nishikawa T, Isobe M, Goto T (1991) Synthesis studies on the bicyclo [7.3.1]tridecenediyne system in an antitumor antibiotic, dynemicin A. Synlett: 393–395
84. Imamoto T, Takiyama N, Nakamura K, Hatajima T, Kamiya Y (1989) Reactions of carbonyl compounds with Grignard reagents in the presence of cerium chloride. J Am Chem Soc 111:4392–4398
85. Viola A, Collins JJ, Filipp N (1981) Intramolecular pericyclic reactions of acetylenic compounds. Tetrahedron 37:3765–3811
86. Yoshida K, Minami Y, Otani T, Tada Y, Hirama M (1994) Remarkable kinetic solvent isotope effect on the cyclolaromatization of C-1027 chromophore, a 9-membered enediyne, and the thermochemistry. Tetrahedron Lett 35:5253–5256
87. Semmelhack MF, Neu T, Foubelo F (1992) Arene 1,4-diradical formation from *o*-dialkynylarenes. Tetrahedron Lett 33:3277–3280
88. Semmelhack MF, Neu T, Foubelo F (1994) Arene 1,4-diradical formation from *o*-dialkynylarenes. J Org Chem 59:5038–5047
89. Scaiano JC, Stewart LC (1983) Phenyl radical kinetics. J Am Chem Soc 105:3609–3614
90. Kryger RG, Lorand JP, Stevens NR, Herron NR (1977) Radicals and scavengers. 7. Diffusion-controlled scavenging of phenyl radicals and absolute rate constants of several phenyl radical reactions. J Am Chem Soc 99:7589–7600
91. Wiberg KB, Motell EL (1963) The kinetic isotope effect in the photo-chlorination of methane. Tetrahedron 19:2009–2023
92. Westheimer FH (1961) The magnitude of the primary kinetic isotope effect for compounds of hydrogen and deuterium. Chem Rev 61:265–273

93. Poutsma ML (1973) Halogenation. In: Kochi JK (ed) Free radicals, vol 2. Wiley, New York, p 174
94. Lockhart TP, Comita PB, Bergman RG (1981) Kinetic evidence for the formation of discrete 1,4-dehydrobenzene intermediates. Trapping by inter- and intramolecular hydrogen atom transfer and observation of high-temperature CIDNP. J Am Chem Soc 103:4082–4090
95. Wenthold PG, Squires RR (1994) Biradical thermochemistry from collision-induced dissociation threshold energy measurements. Absolute heats of formation of ortho-, meta-, and para-benzyne. J Am Chem Soc 116:6401–6412
96. Hoffmann R, Imamura A, Hehre WJ (1968) Benzynes, dehydroconjugated molecules, and the interaction of orbitals separated by a number of intervening *s* bonds. J Am Chem Soc 90:1499–1509
97. Lockhart TP, Bergman RG (1981) Evidence for the reactive spin state of 1,4-dehydrobenzenes. J Am Chem Soc 103:4091–4096
98. Snyder JP (1990) Monocyclic enediyne collapse to 1,4-diyl biradicals: a pathway under strain control. J Am Chem Soc 112:5367–5369
99. Koga N, Morokuma K (1991) Comparison of biradical formation between enediyne and enyne-allene. Ab initio CASSCF and MRSDCI study. J Am Chem Soc 113:1907–1911
100. Nicolaides A, Borden WT (1993) CI calculcation on didehydrobenzenes predict heats of formation for the meta- and para isomers that are substantially higher than previous experimental values. J Am Chem Soc 115:11951–11957
101. Wierschke SG, Nash JJ, Squires RR (1993) A multiconfigurational SCF and correlation-consistent CI study of the structures, stabilities, and singlet-triplet splittings of *o*-, m-, and *p*-benzyne. J Am Chem Soc 115:11958–11967
102. Kraka E, Cremer D (1994) CCSD(T) investigation of the Bergman cyclization of enediyne. Relative stability of *o*-, *m*-, and *p*-didehydrobenzene. J Am Chem Soc 116:4929–4936
103. Lindh R, Persson BJ (1994) Ab initio study of Bergman reaction: the autoaromatization of hex-3-ene-1,5-diyne. J Am Chem Soc 116:4963–4969
104. Davico GE, Bierbaum VM, DePuy CH, Ellison GB, Squires RR (1995) The C-H bond energy of benzene. J Am Chem Soc 117:2590–2599
105. Grissom JW, Calkins TL, McMillen HA, Jiang Y (1994) Determination of the activation parameters for the Bergman cyclization of aromatic enediynes. J Org Chem 59:5833–5835
106. Griller D, Ingold KU (1976) Persistent carbon-centered radicals. Acc Chem Res 9:13–19
107. Brunton G, Griller D, Barclay LRC, Ingold KU (1976) Kinetic applications of electron paramagnetic resonance spectroscopy. 26. Quantum-mechanical tunneling in the isomerization of sterically hindered aryl radicals. J Am Chem Soc 98:6803–6811
108. Brunton G, Gray JA, Griller D, Barclay LRC, Ingold KU (1978) Kinetic applications of electron paramagnetic resonance spectroscopy. 32. Further studies of quantum-mechanical tunneling in the isomerization of sterically hindered aryl radicals. J Am Chem Soc 100:4197–4200
109. Mohamadi F, Richards NGJ, Guida WC, Liskamp R, Lipton M, Canfield C, Chang G, Hendrickson T, Still WC (1990) MacroModel—an integrated software system for modeling organic and bio-organic molecules using molecular mechanics J Comput Chem 11:440

110. Frank BL, Worth L Jr, Christner DF, Kozarich JW, Stubbe J, Kappen L, Goldberg IH (1991) Isotope effects on the sequence-specific cleavage of DNA by neocarzinostatin: kinetic partitioning between 4′- and 5′-hydrogen abstraction at unique thymidine sites. J Am Chem Soc 113:2271–2275

Solution Structure of Neocarzinostatin Determined by Homonuclear Two-Dimensional Nuclear Magnetic Resonance

HIROYUKI TAKASHIMA[1] and YUJI KOBAYASHI[2]

Summary. The secondary and tertiary structures of apo- and holo-neocarzinostatin have been investigated by x-ray crystallography and spectroscopic methods. This chapter reviews the determination process of the three-dimensional structure of neocarzinostatin by the combined use of homonuclear two-dimensional nuclear magnetic resonance (NMR) and distance geometry calculations, which has been dramatically improved in the past decade. The protein moiety consists mainly of antiparallel β-sheets forming a β-barrel structure and a backbone sheet. The chromophore molecule is tightly enwrapped by the barrel and thus shielded from exposure to the solvent. The protein–chromophore interaction mechanism is also discussed in the context of release of the active chromophore and the activation mechanism of holo-neocarzinostatin.

Key words: Holo-neocarzinostatin—Apo-neocarzinostatin—Homonuclear two-dimensional NMR—NOE—NOESY—Distance geometry calculation—Restrained molecular dynamics calculation

Introduction

An understanding of protein structure and its role in the mechanism of action and function related to biological activity can only be accomplished with a knowledge of its three-dimensional structure at atomic resolution. Two primary methods of analyzing protein structure have been established, single-crystal X-ray diffraction of the solid and nuclear magnetic resonance (NMR) spectroscopy of protein solutions. However, the X-ray method has several intrinsic limitations that arise from the prerequisite crystallization step. Many proteins do not crystallize easily, and even when they do crystallize, heavy atom incorporation is often required to

[1] International Research Laboratories Ciba-Geigy Japan, Ltd. 10-66 Miyuki-cho, Takarazuka, Hyogo, 655 Japan
[2] Faculty of Pharmaceutical Sciences Osaka University, 3-3 Yamadaoka, Suita, Osaka, 565 Japan

help solve the structure. Furthermore, most proteins function in solution but may crystallize in several conformations or a conformation other than the active solution conformation with substrate.

Multidimensional NMR [1–3] of protein solutions can avoid these crystallization difficulties and still provide structural information at atomic resolution. Although NMR can be limited by protein solubility and size or molecular weight, the techniques and instrumentation have improved dramatically in the last decade to overcome these limitations for a large number of proteins [4–6]. Furthermore, the NMR method not only provides information about the static or bound structure but also about its dynamics, internal motions, conformational changes, and fluctuations. In particular, motions or mobility related to protein–ligand binding, cooperative effects, or enzymatic activity may also be probed by heteronuclear NMR. The NMR technique is a unique method for probing molecular dynamics and structure and their implications to biological activity.

Neocarzinostatin (NCS or holo-NCS) is a complexed protein containing the noncovalently bound enediyne-containing antitumor antibiotic chromophore (NCS-chrom) that intercalates into DNA, binds to specific DNA sequences (preferentially thymidine and adenosine), and nicks the double-stranded DNA by abstracting hydrogen atoms [7–12]. The protein was first isolated from *Streptomyces carzinostaticus* by Ishida et al. [13], and later NCS-chrom was separated by Koide et al. and Goldberg et al. [14,15]. NCS-chrom possesses the scission activity, but the strained enediyne ring is chemically unstable in the free state and forms a cumulene intermediate that spontaneously rearranges to an active biradical species. When NCS-chrom complexes or binds with the protein moiety, it is stabilized in solution, even in vivo.

Because NCS-chrom has attractive cytotoxic and mutagenic properties as an anticancer agent, the conformational features that stabilize it in its complexes have attracted much attention for years. Early NMR and X-ray studies investigated only apo-NCS [16–20]. The three-dimensional structure of holo-NCS, determined by two-dimensional NMR in 1991 by two groups [21,22], revealed the protein–chromophore interaction mechanism through its bound structure. Later the X-ray diffraction conformation was determined (at 1.8 Å) for holo-NCS in 1994 [23], and when the results are compared with the earlier NMR structure, the holo-NCS crystal structure and solution structure are similar.

Recent improvements in both heteronuclear and multidimensional NMR have extended conformational analysis beyond the limitations of conventional homonuclear two-dimensional NMR methods. For example, apo-NCS relaxation was studied by ^{13}C-NMR at natural abundance [24,25]. In addition, heteronuclear multidimensional NMR methods have been applied to a related antitumor antibiotic chromoprotein, ^{13}C-, ^{15}N-isotopically enriched apokedarcidin [26]. Despite the improvements, natural abundance NMR is still difficult, and producing isotopically labeled protein in quantities needed for NMR is labor intensive and expensive, thereby defining a rate-limiting step in the analytical procedure. Although multidimensional triple-resonance NMR does improve the ease of spin system assignments and sequence connectivities for larger proteins (100–200 residues), a considerable amount of information can be obtained from conventional two-dimensional methods before resorting to isotopic enrichment. That is, homonuclear two-dimensional NMR is still the starting point to the first stage of structural determination. In this context, we demonstrate here the power of this technique to make assignments to sequential amino acids in holo-NCS, to interpret NOEs and coupling constants in terms of the complex, and to demonstrate the contact points of NCS-chrom.

NMR Spectra and Resonance Assignments of Holo-NCS

Basic to all forms of NMR structural analyses is the identification of the resonances associated with each amino acid in the protein sequence and their connectivities [1]. Often this process is complicated by overlapping resonances, degeneracies, incomplete spin systems caused by exchange, and even artifacts in the complex spectra. Nonetheless there are two key experiments in homonuclear two-dimensional NMR from which we can extract information concerning the chemical structure as well as the three-dimensional structure of a protein and its ligand: correlational spectroscopy (COSY) and nuclear Overhauser enhancement spectroscopy (NOESY). These two experiments have several derivative or associated experiments. For example. COSY-type experiments include double quantum filtered COSY (DQF-COSY) and total correlation spectroscopy (TOCSY) experiments, which are all widely used to investigate the intra-residue connectivity within a residue [27–29]. In this case, connectivity means through covalent bond interaction by an intra-residue J-coupling that is an index of coupling strength between pairwise atoms or protons in the homonuclear NMR. On the other hand, NOESY experiments are used to independently investigate the inter-residue coupling through space connectivities (nuclear Overhauser effects, NOE). NOEs provide distance information between pairwise atoms or protons in the case of homonuclear NMR [30–32].

The 20 amino acid residues that are the main components of the protein show characteristic NMR peak (resonance) patters. Individual resonances associated with an amino acid spin system are representative of the proton type (e.g., NH, CαH, CβH, etc.). COSY experiments help identify individual resonances that belong to each spin system and classify it with respect to the amino acid residue type (for example, a glycine spin system is readily distinguishable from a lysine spin system because of its characteristic couplings and chemical shifts). The power of the NMR technique lies in the fact that each proton gives rise to a specific resonance in the spectrum, and even for identical amino acid residues, such as two alanines, conformationally dependent chemical shift perturbations normally cause sufficient dispersal of resonances to permit them to be resolved. A key aspect of NMR structure determination is the initial requirement to assign resonances to corresponding atoms and then to uniquely elucidate the connectivities through bonds or space [1].

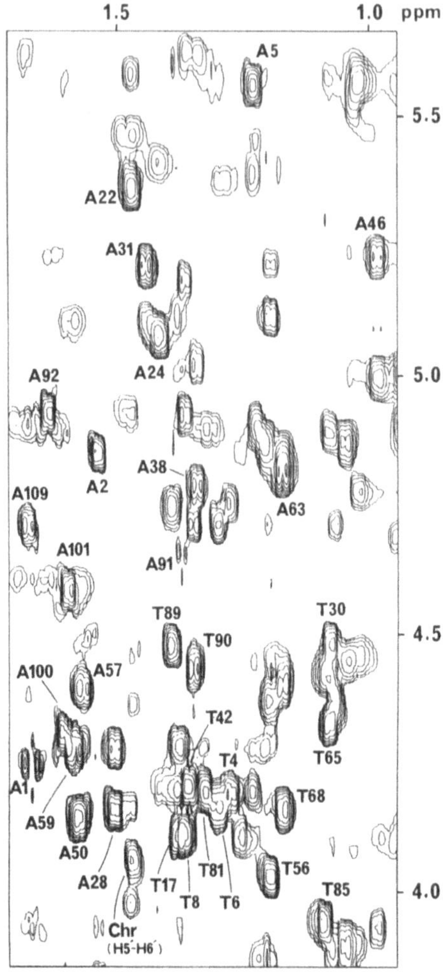

FIG. 1. A portion of the nuclear Overhauser enhancement spectroscopy (NOESY) spectrum of holo-NCS (holo-neocarzinostatin) showing resonance assignments for alanine and threonine. *Horizontal and vertical axes* indicate nuclear magnetic resonance (NMR) chemical shifts in parts per million (*ppm*). In combination with a double quantum filtered-correlation spectroscopy (DQF-COSY) spectrum, the intra-residue NOE cross-peaks between a side-chain methyl group and methine protons were readily distinguished from other NOEs and were specifically assigned to alanine or threonine. Using TOCSY cross-peaks raised from backbone amide protons and Cα protons, some of these peaks were uniquely assigned to alanine (represented by *A* in this figure), or to threonine (represented by *T*). Subsequently, after residue-specific assignment, sequence-specific assignments were made (see text) and are presented here by numbers of the residue in the sequence, e.g., *T89*. By the two-step assignment process, all protons of the 18 alanines and 13 threonines were assigned. The cross-peak between H5′ and H6′ of the methyl group in NCS-chrom (see text and Fig. 6) was also identified (*Chr*)

Residue-Specific Assignment of Holo-NCS

Initially the resonance assignment process is divided into two steps. The first step is to assign chemical shift patterns to residue spin types. Figure 1 shows an example of alanine and threonine assignments. Because methyl protons in these two amino acids normally show strong doublets around 1.5 ppm, with connectivity to a distinctively coupled methine proton at 4.5 ppm, they are readily identified in DQF-COSY spectra. Because NCS contains 18 alanines and 13 threonines, 31 cross-peaks around the intersection of 1.5 and 4.5 ppm are expected as a result of the connectivity between the side-chain methyl group and the methine protons. A cross-peak designates pairwise correlated resonance peaks in the two-dimensional NMR spectrum. These 31 cross-peaks were then connected to backbone amide protons and Cα protons by DQF-COSY, TOCSY, and NOESY. Thus, these 31 Thr and Ala spin systems were identified.

Glycines are similarly distinctive in DQF-COSY and NOESY spectra. NCS has 13 glycines that show characteristic cross-peaks between the strongly coupled geminal backbone Cα protons (see Fig. 2). Likewise, cross-peaks between the amide protons and the Cα protons of glycine show unique patterns and are readily assigned as glycine spin systems.

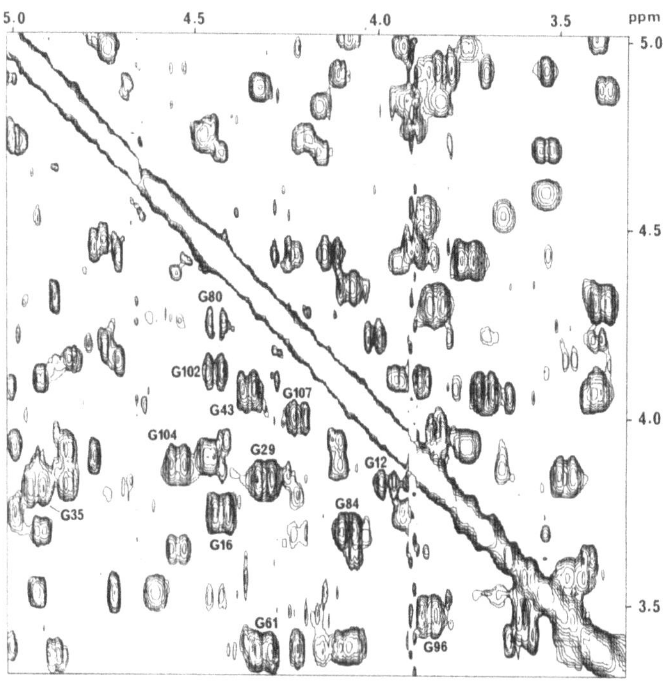

FIG. 2. Part of the NOESY spectrum of holo-NCS and resonance assignments of the geminal Cα protons of glycine. The assignments are represented here by the one-letter code and sequence number of the glycine residue

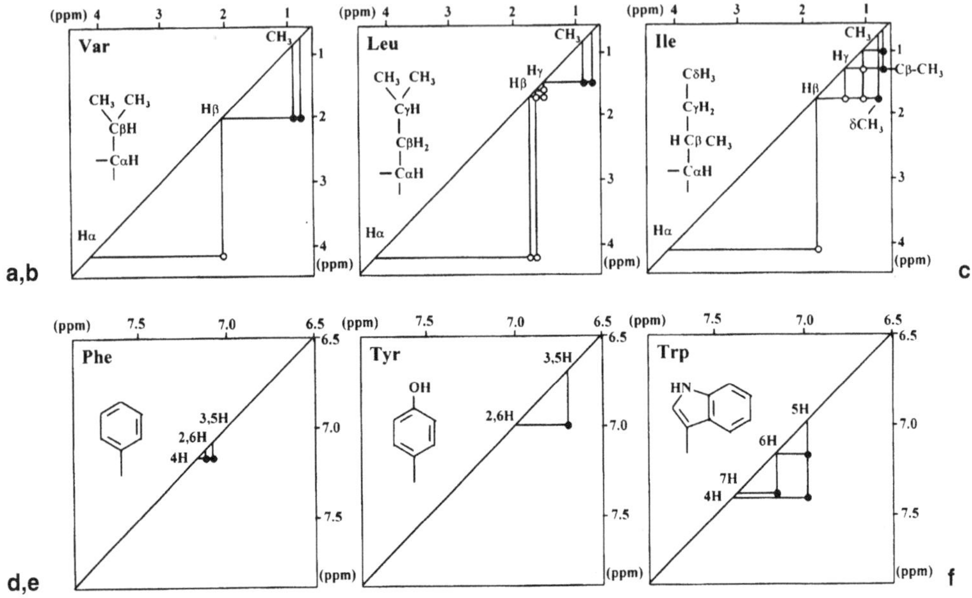

FIG. 3a–f. Schematic representation of the cross-peak patterns in a COSY spectrum for the aliphatic protons of (**a**) valine, (**b**) leucine, and (**c**) isoleucine and the unexchangeable aromatic protons of (**d**) phenylalanine, (**e**) tyrosine, and (**f**) tryptophan. *Horizontal* and *vertical axes* indicate NMR chemical shifts. The cross-peaks are represented by *open circles* and are connected by *lines* denoting spin system connectivities. *Filled circles* indicate cross-peaks derived from methyl groups and aromatic rings, which are sharp and readily distinguishable from other peaks in the spectra. Conformation-dependent chemical shift perturbations cause dispersal of resonances from average values. It therefore becomes possible to resolve individual resonances for amino acid residues of the identical type. On the other hand, it may cause ambiguity in resonance assignments. It is therefore necessary to recognize cross-peak patterns in two-dimensional NMR to make complete residue-specific assignments

The side-chain protons of the aromatic residues, e.g., phenylalanine, tyrosine, and tryptophan, and the methyl groups of valine, leucine, and isoleucine also show unique cross-peak patterns that permit relatively easy residue-specific assignments. The COSY cross-peak patterns of these residues are summarized in Fig. 3. Owing to the peak pattern similarity of aspartate, asparagine, cysteine, and aromatic residues in the aliphatic region, these residues are classified into one group type, that is, the AMX spin system. Cross-peaks between Cα protons and amide protons or Cβ protons of AMX spin systems are usually well separated and are useful to simplify the analysis. NCS has a total of 24 AMX spin systems.

The remainder of the residues, Lys, Arg, Pro, etc., are usually more complicated to identify at this stage in the analysis because not all cross-peaks appear strongly or are seen unobstructed by other resonances. Nonetheless, at this point a large number of residues (about 80%) have been either identified uniquely or identified as belonging to a specific group such as an AMX spin system. On the basis of this partial assignment, the results of the residue-specific assignments, or spin group assignments, will be corroborated by NOE data and the position of residues in the protein sequence established.

Sequence-Specific Assignment of Holo-NCS

This NOE-based approach involves identification of relatively strong NOESY cross-peaks that arise from protons on adjacent residues interacting though space (in homonuclear NMR, unlike heteronuclear NMR, the proton coupling is weak through the amide bond, and hence COSY-type interactions are discontinuous and cannot directly provide information about inter-residue connectivity). This key procedure of sequence-specific resonance assignment was introduced by Wüthrich [1,5]. The assignment proceeds by tracing the connections between inter-residue cross-peaks and intra-residue cross-peaks in a sequential manner (Fig. 4). In this context, sequential manner means making the connection from one residue to its next adjacent neighbor. By comparing NOESY and DQF-COSY spectra, sequential NOEs (inter-residue NOEs) are readily distinguished from intra-residue NOEs.

Holo-NCS (113 residues) has 108 backbone amide protons, excluding the N-terminus and the prolines. Therefore, at least 108 intra-residue cross-peaks are expected between the amide proton and the Cα proton in the DQF-COSY and NOESY spectra (glycine has two Cα protons.). In addition, in the NOESY spectrum, 108 inter-residue cross-peaks are expected between the amide proton and the Cα proton of the preceding residue. This region of the NOESY spectra is called the "fingerprint" region. Because the backbone protons, especially the amide protons, are so sensitive to their sequence, conformation, and environment, the cross-peak pattern of the fingerprint region can also be useful for protein identification.

From these 216 cross-peaks of the fingerprint region, the sequential relationships among the spin systems are identified. To facilitate the complete identification and make the sequential connections between spin groups, the reported primary sequence of NCS [33–35] was used for the assignment process. If there is some unique amino acid residue in the primary sequence, the residue-specific assignment of the residue must be identical with the absolute sequence-specific assignment of the residue. In NCS, Lys 20 and Ile 110 are unique. Unfortunately, lysine has a cross-peak pattern difficult to distinguish from the cross-peaks of arginine. Nonetheless, if we identify Ile 110, we can also number the preceding Ala as 109 and the subsequent Ser as 111. Because the NOE of an amide proton to the preceding Cα proton is much stronger than that to the Cα of the following residue, when an Ile spin system (which is attached to both an Ala and a Ser) is identified it allows the assignment of Ala 109 and Ser 111. In principle, this

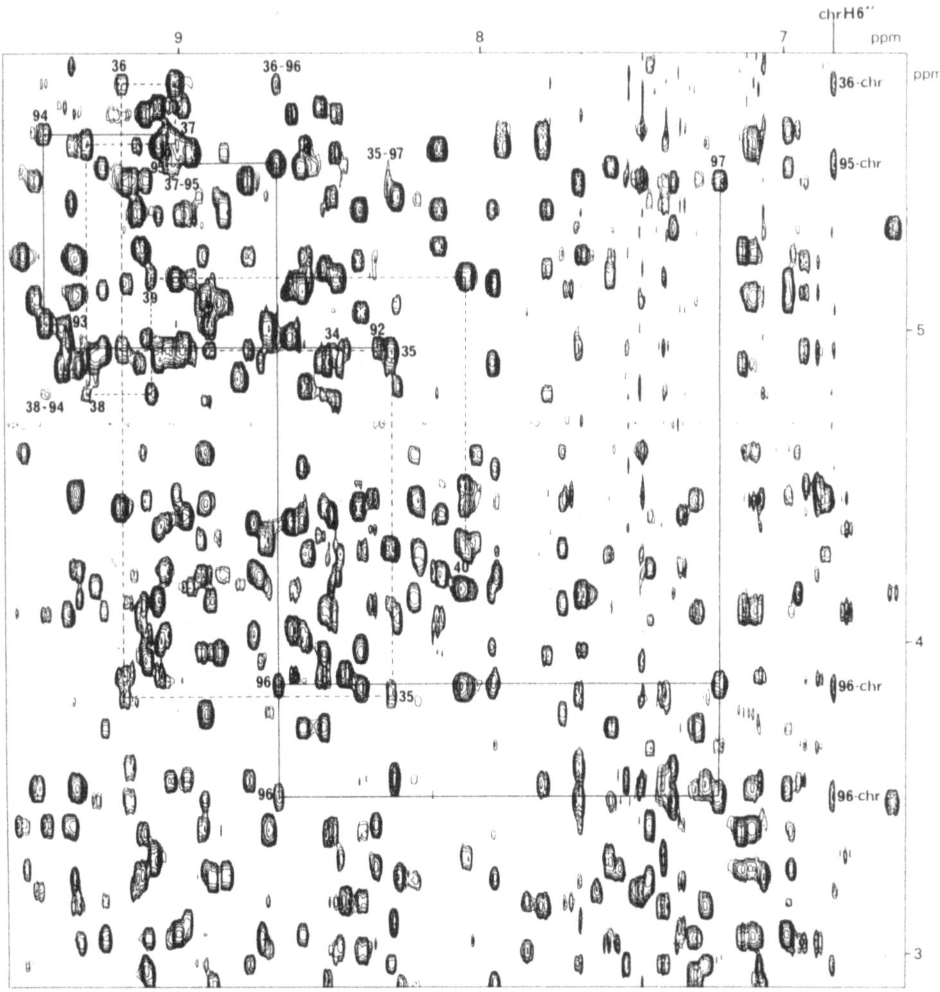

FIG. 4. A portion of the NOESY "fingerprint region" for holo-NCS. Examples of the procedure in the sequence-specific assignment of two segments, e.g., residues from 34 to 40 and residues from 92 to 97, are represented by a *broken line* and *solid line*, respectively. In the procedure, the intra-residue cross-peaks between amide protons and coupled Cα protons (represented by a residue number in this figure) and inter-residue cross-peaks between an amide proton and an Cα proton of its preceding residue (sequential NOE, "dαN"; see text and Fig. 4) have to be alternately connected. The sequence-specific assignments were performed in conjunction with other sequential NOE, dNN (an NOE between two successive amide protons), and dβN (an NOE between an amide proton and a Cβ proton of the preceding residue), and the results of residue-specific assignments. Successive strong dαN (cross-peaks between amide protons and Cα protons) connectivities indicate that the two segments consist of β-sheet strands, and interstrand NOEs (represented by pairwise residue numbers) reveal that they are paired components of an antiparallel β-sheet. Some of the intermolecular NOEs between the NCS-chrom H6" proton and the protein moiety are also included in this figure

procedure can be repeated from Ala 109 to the N-terminus and from Ser 111 to the carboxyl end of the protein.

However, the repeated sequence-specific assignment procedure is broken at proline positions because of the absence of the backbone amide proton in proline residues. Consequently, a minimum of five starting points is required to complete the residue-specific assignments because there are four prolines. In reality many starting points are examined until a self insistent set of assignments is completed. We used "unique-shot sequences" for starting points. Even though there are a number of alanines in the sequence, connecting two or three neighboring residues often is enough to identify them. For example, an Ala-Ala-Gly sequence can be uniquely assigned as Ala 100-Ala 101-Gly 102 because that particular sequence occurs only once. Figure 4 shows an example of such a sequence-specific assignment. Thus, by repeatedly mapping a chain of sequentially connected spin systems to a portion of the amino acid sequence, the sequence-specific assignment is made.

During this repetitive procedure, residue assignments are simultaneously completed as well, and 98% (106 of the 108 residues) sequential NOE cross-peaks between the amide protons and the Cα protons were finally observed in the NOESY spectrum and summarized (Fig. 5). The cross-peaks, represented as "dαN," are shown as filled squares in Fig. 5. The intensity of the individual NOE peaks is also indicated. A strong NOE, which indicates a close contact between a pair of protons in space, is represented by a big square and a weak NOE by a small square.

The method can be extended to side-chain protons. In Fig. 5, "dβN" represents an NOE between an amide proton and a Cβ proton of the preceding residue. An NOE between two successive amide protons is designated by "dNN" in Fig. 5.

FIG. 5. Amino acid sequence of the protein moiety of holo-NCS and sequential NOE connectivities observed in NOESY spectra. The amino acid sequence is shown by its one-letter code [35]. Each *filled square* shows the magnitude of the NOE interaction of the backbone amide protons with amide, Cα, or Cβ protons, respectively. The residues in *circles* displayed a slowly exchanging amide proton

Secondary Structure in Holo-NCS

In addition to sequence information, sequential NOEs contain conformational information that can be interpreted in terms of protein folding, e.g., secondary structure, as summarized in Fig. 6. For instance, many successive strong sequential dαN NOEs suggest that holo-NCS is mainly composed of β-sheet structures. Figure 7 is a schematic representation of β-sheet structure with its successive strong sequential dαNs and interstrand NOEs. The few strong sequential NOEs between an amide proton and the amide proton of its preceding residue suggest the existence of turn structures.

The secondary structure of holo-NCS was deduced to satisfy the structural information obtained from both the sequential NOEs and the long-range NOE interactions (Fig. 8). It consists mainly of antiparallel β-sheets, which can be divided into two sections or domains, i.e., sheet-a and sheet-b. How this information was deduced is explained next.

In the NOESY spectrum of an deuterium aqueous (D_2O) solution, some slowly exchanging amide protons were observed. (These are represented by circles about the residue numbers in Figs. 5 and 8.) In general, amide proton exchange is too fast to be observed for amide protons accessible to the solvent, D_2O. Deuterium-exchange experiments indicate slowly exchanging amide protons that are assumed to be buried within the molecule or are involved in hydrogen bonds resulting from structural features such as α-helix and β-sheet. The positions of the nonexchanging protons were consistent with the β-sheet indicated by NOE data (Fig. 8). In this case, the amide protons are protected from the solvent by interstrand hydrogen bonds.

In general, the side chains in a β-sheet structure point in an alternating fashion to the plane of the sheet, with the Cα–Cβ bonds pointing out, nearly perpendicular to the plane of the sheet. Consequently, in sheet-b of Fig. 8, the side chains of

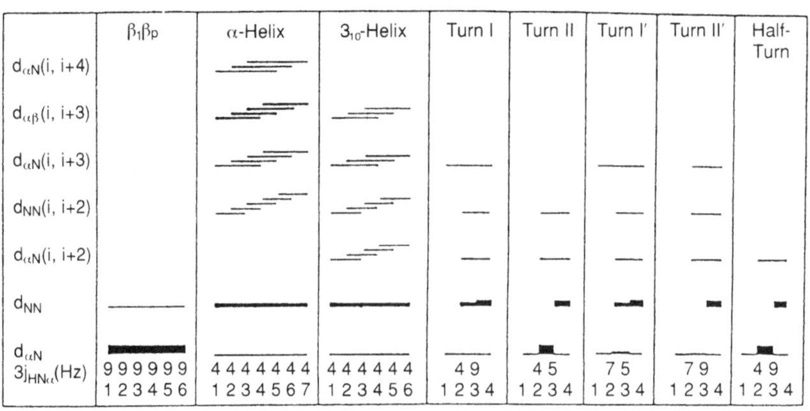

FIG. 6. Summary of the sequential NOEs and medium-range NOEs that are indicative of protein secondary structure

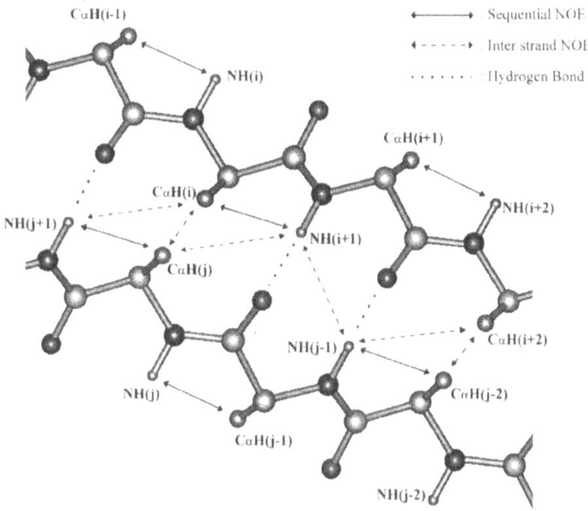

FIG. 7. Schematic representation of two-stranded β-sheet. Protein backbones of the two pairwise strands are represented by a ball-and-stick model. CαH and NH are indicted by letters with their sequences. Successive strong sequential daN NOEs, e.g., NH(i+1) to CαH(i), NH(i+2) to CαH(i+1), and NH(i) to CαH(i−1), are represented by *solid arrows*. The sequential NOEs are typical features of the β-sheet structure. On the other hand, sequential dNN distances, e.g., NH(i+1) to NH(i), are long and the dNN NOEs are small in the structure. In addition to the feature of sequential NOEs, interstrand medium- and long-range NOEs *(broken arrows)* are also useful to elucidate the β-sheet structure. *Dotted lines* are H bonds

residues 33, 35, 37, 45, 47, 52, 54, 76, 94, 98, and 109 face in the same direction. These residues are designated by their one-letter codes and a filled triangle; the remaining residues face the opposite side of the sheet. Because residues 35, 75, 96, and 107 are glycyl residues, these residues do not actually have side chains.

NMR Peak Assignments of NCS-Chrom

As previously mentioned, NCS-chrom has a unique molecular strcuture. Some of its NMR resonances are distinct and resolved from the protein in DQF-COSY and TOCSY spectra and thus were unambiguously assigned. For instance, protons of the naphthalene moiety and the protons around the enediyne structure were readily assigned because of their unique chemical shifts and cross-peak patterns. H5, H8, H6″, and H8″ protons (Fig. 9) give rise to singlets that have no COSY connectivities to any other protons and were assigned by intramolecular NOE connectivities.

Assignments of the other NCS-chrom peaks that had chemical shifts and cross-peak patterns similar to those of the protein were ultimately confirmed by the

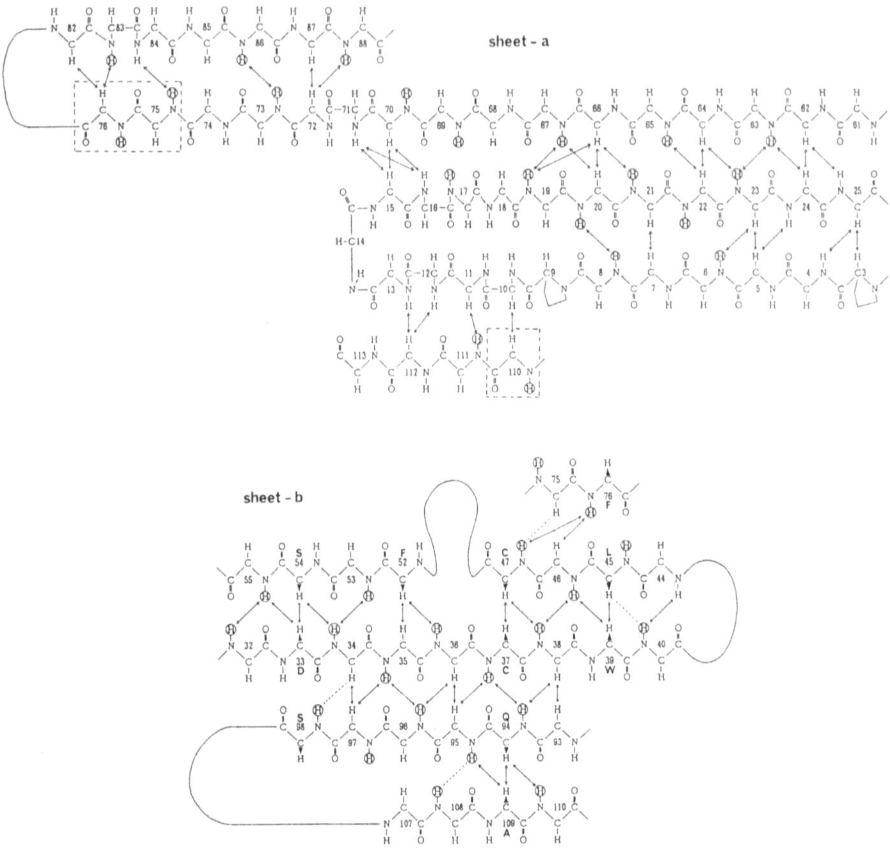

FIG. 8. β-sheet (sheet-a, *top*, and sheet-b, *bottom*) diagram of holo-NCS. The long-range interstrand NOEs are indicated by *arrows*. Using these NOEs and sequential NOEs (summarized in Fig. 5), the β-sheet structures were determined. The slowly exchanging amide protons are *circles* (as in Fig. 4). The residues shared by sheets a and b are indicated by *squares with broken lines* in *sheet-a*. The *filled triangles* in *sheet-b* indicate the chirality of Cα protons, which determines the orientation of side chains as upward or downward on the surface of sheet-b

elimination of all resonances derived from the protein. For instance, the H5′ proton and the H6′ methyl group showed a cross-peak pattern similar to alanine and threonine (see Fig. 1). The assignment for NCS-chrom H5′ and H6′ methyl was made after complete residue-specific assignments of alanines and threonines. It then became evident that the NCS-chrom methyl group has peak widths and chemical shifts similar to those of the protein moiety of NCS. Because peak width is strongly affected by molecular mobility, it indicates that the binding between the protein moiety and NCS-chrom is strong. The tumbling rate for NCS-chrom (a small molecule) is the same as that of the protein on the NMR time scale (nanosecond to microsecond range) at physiological conditions. After complete

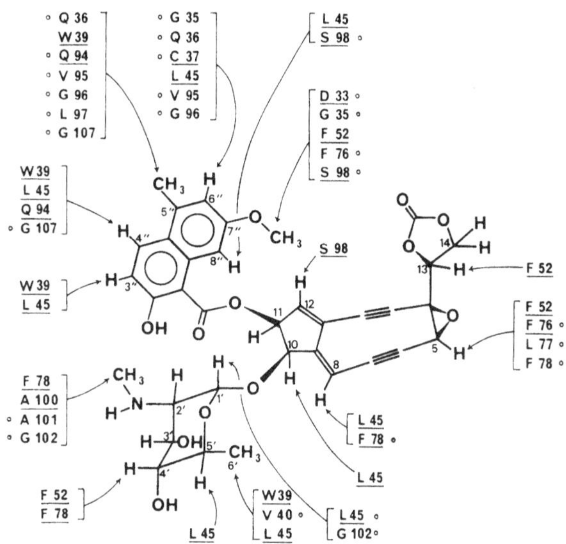

FIG. 9. A summary of the intermolecular NOEs between NCS-chrom and the protein moiety. The amino acid residues responsible for the intermolecular NOE and the proton in the structural formula of NCS-chrom are represented and connected by *arrows*. These NOEs are classified as side-chain-mediated NOEs or backbone-mediated NOEs, designated by *underlines* and *small circles*, respectively

assignments of holo-NCS, both the protein moiety and the chromophore moiety, intermolecular NOE connectivities between them were detected.

Intermolecular NOEs Between Holo-NCS Complex

The 73 intermolecular NOEs detected for NCS complex were sufficient to draw the structure of holo-NCS. The residues responsible for NOE interactions with NCS-chrom are summarized in Fig. 9, labeled with their residue numbers. NCS-chrom protons give rise to NOEs almost exclusively with protons of sheet-b (see Fig. 8), indicating that sheet-b is the binding site for NCS-chrom.

A number of NOEs were seen between NCS-chrom and protein moiety side chains, primarily on those of 76, 78, 94, 98, and 100. These residues are coincidentally on the same side of the sheet, as mentioned earlier. This indicates that NCS-chrom interacts with the protein on only one surface of sheet-b. The remaining NOEs were observed between NCS-chrom and protons of the protein backbone. Residues 35, 36, 95, 96, 97, 102, and 107 (but not 40 and 77) are residues along a diagonal line from the lower left to upper right of sheet-b (Fig. 10a).

There are some pairs of residues whose side chains interact with the same protons on NCS-chrom but that show no direct NOE interactions to each other; Leu45-Ser98 and Asp33-Phe76 are such examples. They may not be close

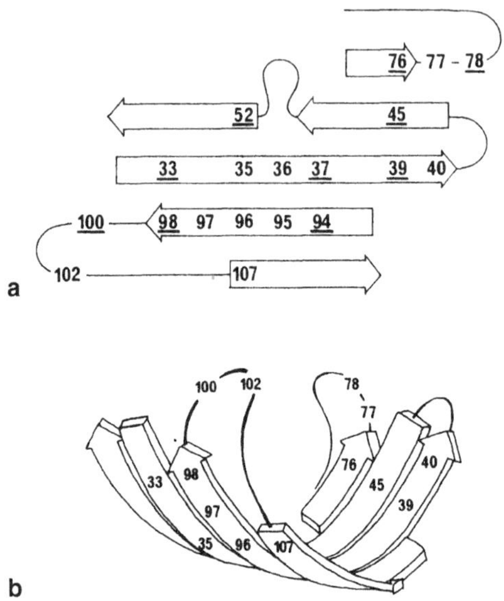

FIG. 10. **a** Arrangement of β-strands in sheet-b. The numbered residues show intermolecular NOEs with NCS-chrom. **b** A profile of the global arrangement of β-strands forming a barrel structure

enough, i.e., within a distance of about 5 Å, to show NOE cross-peaks. This situation is interpreted as arising from the complex formation sandwiching NCS-chrom between the two parts of sheet-b, designated by the diagonal line. Such a spatial arrangement results from a double bend in the protein along the line forming the β-barrel, as shown in Fig. 10b. Support for this model is that the inside surface of the barrel, which forms a groove exposing only backbone protons, causes no steric hindrance from side chains along the fold.

These backbone protons give rise to NOEs only with protons of the naphthalene moiety in NCS-chrom. Consequently, NCS-chrom is deeply anchored at the bottom of the groove. In this model, the protein moiety forms a hydrophobic cluster consisting of the side chains of Trp39, Leu45, Phe52, Phe76, and Phe78, which are again inside the barrel. Some NOEs seen between protons of these side chains are indicative of the hydrophobic interaction between the cluster and the labile and hydrophobic parts of NCS-chrom, i.e., the acetylenic and the epoxide groups. These observations led to the conclusion that the stability of the complex, as well as protection of the reactive NCS-chrom, result from the occlusion of NCS-chrom in a hydrophobic environment.

Figure 11 shows the intermolecular NOEs that designate the folding characteristics of NCS-chrom and summarize the NOEs mentioned earlier. The arrangement of the enediyne moiety (A) and naphthalene moiety (D) within the binding pocket are schematically represented in Fig. 11b. More structural features are discussed next.

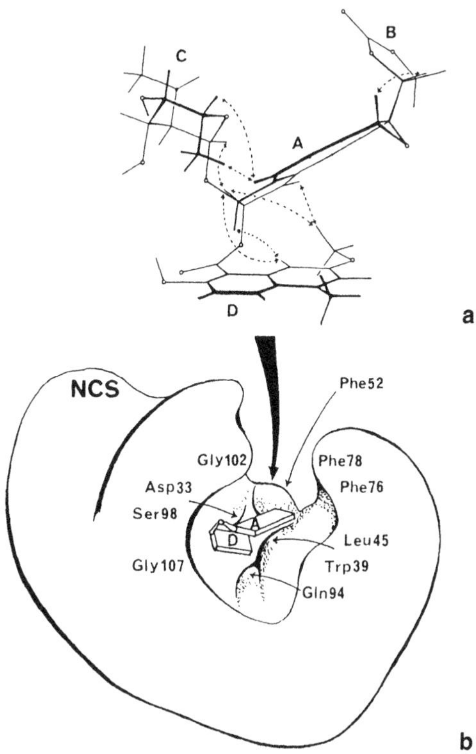

FIG. 11a,b. Summary of the intramolecular NOEs of NCS-chrom and a schematic representation of the arrangement of NCS-chrom. The NOEs are indicated by *arrows with broken lines* (**a**). The NCS-chrom enediyne moiety (*A*), carbonate moiety (*B*), amino sugar moiety (*C*), and naphthalene moiety (*D*) and the protein residues responsible for intermolecular NOEs are schematically represented to comprise the binding pocket of holo-NCS (**b**)

Structural Comparison Between Holo-NCS and Apo-NCS

Because the secondary structure of apo-NCS in solution had been reported [16–18], it was possible to compare the structures of protein moiety in the holo and apo states. For each state the secondary structure and global folds resemble each other. Chemical shift differences between the two states are indicated in Fig. 12. Almost all the significant differences arise from residues that display intermolecular NOEs with NCS-chrom, except for several residues around Arg 82.

The most striking difference between the holo and apo states was observed in the spatial arrangement of the Arg70-Cys88 portion of the protein. Arg70-Cys88 forms a two-stranded antiparallel β-sheet with significantly different alignments in holo- and apo-NCS. In holo-NCS, the Gly75-Phe76 portion is located at the junction of sheets a and b and is involved in the dense hydrogen-bonded net-

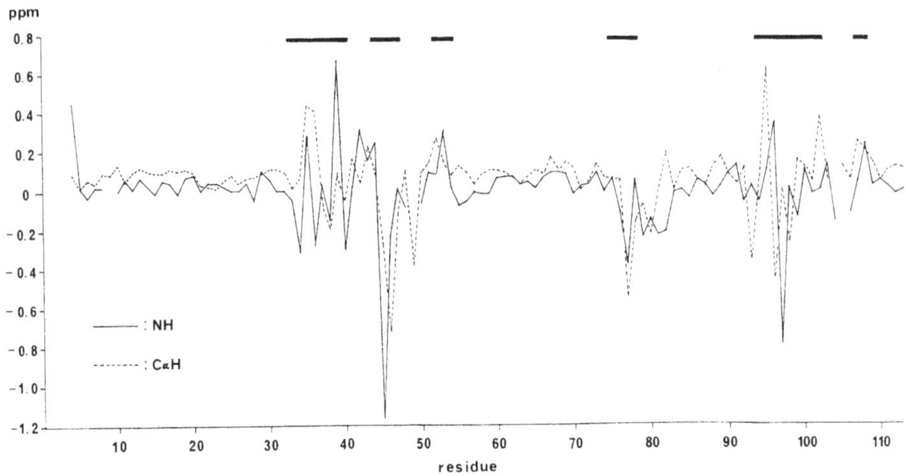

FIG. 12. Chemical shift value differences of main chain amide protons (*solid lines*) and Cα protons (*broken lines*) between apo-NCS and holo-NCS. Residues associated with the intermolecular NOEs are indicated by *thick solid bars*

works distributed over both sheets. This indicates that the spatial arrangements of these two sheets are fixed such that no significant internal movement is expected in this part of the molecule. In this context, it is noteworthy that the β-strands in sheet-a are twisted at three points located around the site of the connection of this two-stranded β-sheet with the main part of sheet-a, as judged from the sequential NH–NH NOEs found at Arg71-Ser72, Gly16-Thr17, and Gly12-Leu13. The stabilization of this conformation with such a distortion can be attributed to the dense hydrogen-bonded network.

On the contrary, in apo-NCS, the two-stranded β-sheet, consisting of the peptide segment Arg70-Cys88, is isolated from other hydrogen-bonded networks of the molecule. The loss of hydrogen bonds at Gly75 and Phe76 allows the movement of this two-stranded β-sheet. The flexibility of this moiety, which might be obtained in the apo state, was suggested by the X-ray analysis of macromomycin [36]. In this member of the superfamily of antibiotic proteins from *Streptomyces*, the strands, corresponding to the two-stranded β-sheet in apo-NCS, are characterized by higher thermal parameters than the rest of the molecule, indicating the more pronounced flexibility of this portion of the molecule. The sequential NH–NH NOES from NH–NH interactions, which act as a clamp holding the twisted portions of sheet-a, are not found in apo-NCS.

The differences in conformation between the holo and apo states indicate that the two-stranded β-sheet covers, NCS-chrom, protecting it from exposure to the solvent in the holo state. Then, after a certain conformational change, this part becomes more flexible so that it releases NCS-chrom. Many NOEs detected from NCS-chrom to Phe76, Leu77, and Phe78 support the representation of the two-stranded β-sheet covering NCS-chrom.

The following hypothetical pathway for the release of NCS-chrom is suggested on the basis of these conformational differences. An environmental change, such as a pH change, occurs and affects the conformation around the site of the connection between the two-stranded β-sheet and sheet-a, where a dense hydrogen-bonded network and the twisted-sheet moieties are localized. For example, Asp15 might detect the pH change at the center of this area. The twists may then relax to trigger a larger conformational change that can be transferred to the region around Gly75 and Phe76. Ultimately the segment covering NCS-chrom could then be removed.

The foregoing hypothetical mechanism for the exposure or release of the active principle of NCS is supported by the refinement of the three-dimensional structure of NCS (holo-NCS) using both distance geometry calculations and energy minimization approaches to further explain the experimental NMR data.

Distance Geometry Calculations of the Protein Moiety of Holo-NCS

The interproton distance constraints derived from the NOESY experiments, when combined with distance geometry calculations, allow the three-dimensional structure of the protein to be further refined. Since distance geometry calculations were first applied by Braun and Go [37–39], many studies have successfully applied these methods to protein conformation determinations. Details of this approach can be found in reviews of Wüthrich and collaborators [1,5]. Briefly, from the intensities of the NOEs, distance limits are imposed upon the distance separating the two protons. Then energy calculations are performed until the lowest energy conformation of the protein is found that satisfies all the distance constraints simultaneously. This serves as the starting conformation to the further refined in holo-NCS, as outlined next.

The distance geometry calculation consists of minimization of the square sum of the differences between atomic distances in the calculated structure and the corresponding distances obtained from NOE information. First, the error function, T, for a given set of constraints is determined as the target function to be minimized:

$$\mathrm{T} = \sum{}' \frac{\left(U_{ij}^2 - r_{ij}^2\right)^2}{U_{ij}^2} + \sum{}' \frac{\left(L_{ij}^2 - r_{ij}^2\right)^2}{L_{ij}^2} + \sum{}' w\left[\left(s_i + s_j\right)^2 - r_{ij}^2\right]^2$$

where r_{ij}, U_{ij}, and L_{ij} are the interatomic distance between atoms i and j and the upper and lower constraints of the corresponding r_{ij}, respectively. s_i and s_j are the repulsive core radii of the i and j atoms. Σ' means summation only over the terms that violate distance constraints, and w is the weight of the third term relative to that of the first and second terms.

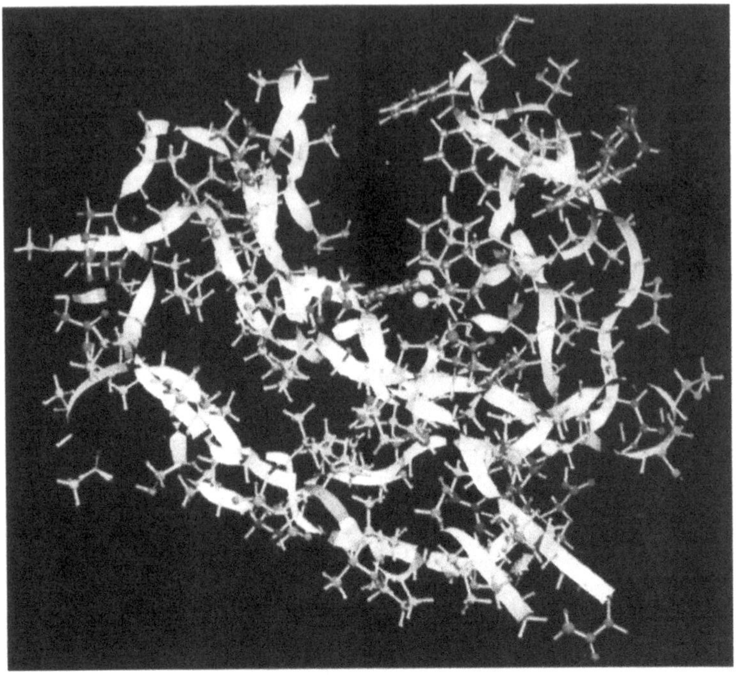

FIG. 13. Three-dimensional structure of protein moiety of holo-NCS determined by distance geometry calculations, with a ribbon drawing of the main chain and a ball-and-stick model drawing of the side chains. (*For color reproduction see color insert in front matter*)

Next, an initial conformation is generated by using random values of dihedral angles, and r_{ij} are calculated. The target function then is minimized by changing the values for the dihedral angles, with fixed values for bond angles and bond lengths of peptide residues. To avoid the multiple minimum problem during the minimization, the distance constraints are applied stepwise according to the difference between residue numbers to which each of the paired protons under consideration belongs. Starting from the conformation obtained in the previous stage, a series of minimizations is then carried out with constraints between further residues along the primary structure.

Using 840 inter-residue NOEs, as well as the two disulfide bond constraints, 100 starting random structures were subjected to the calculation. The calculations were carried out with the DIANA program [40,41], and the result is shown in Fig. 13.

Three-Dimensional Structure of Holo-NCS

Based on the structure of the protein moiety determined by the distance geometry calculations and the intermolecular NOEs between the protein and NCS-chrom, the structure of the holo-NCS complex was determined by restrained molecular dynamics calculations [42,43]. The method contains an energy minimization step and a dynamic structure simulation step. In combination with the experimental distance constraints from NOE experiments, this method is used to determine the three-dimensional structure of the complex and assess the accuracy of the distance geometry calculations. Because holo-NCS has a noncovalently bound ligand, the three-dimensional arrangement of the binding pocket had to be confirmed by the restrained energy minimization process. The result is indicated in Fig. 14.

Holo-NCS is an acidic molecule containing few basic residues, primarily Lys20, Arg71, and Arg82. All these basic residues, as well as many acidic residues, point out toward the protein surface and the solvent. On the other hand, a majority of hydrophobic residues are distributed on the interface of the two β-sheet structures, namely sheet-a and sheet-b (in Fig. 8) and the binding pocket of NCS-chrom. The complex β-sheet assembly and the interface between them are stabilized primarily by hydrophobic interactions between residues.

NCS has two disulfide bonds, one of which consists of Cys37 and Cys47, which is close to the enediyne moiety of NCS-chrom in the three-dimensional structure. As shown in Fig. 8, the region surrounding the two cysteines consists of an antiparallel β-sheet and a turn structure that contains both Trp39 and Leu45, which comprise part of the hydrophobic cluster stabilizing the enediyne moiety.

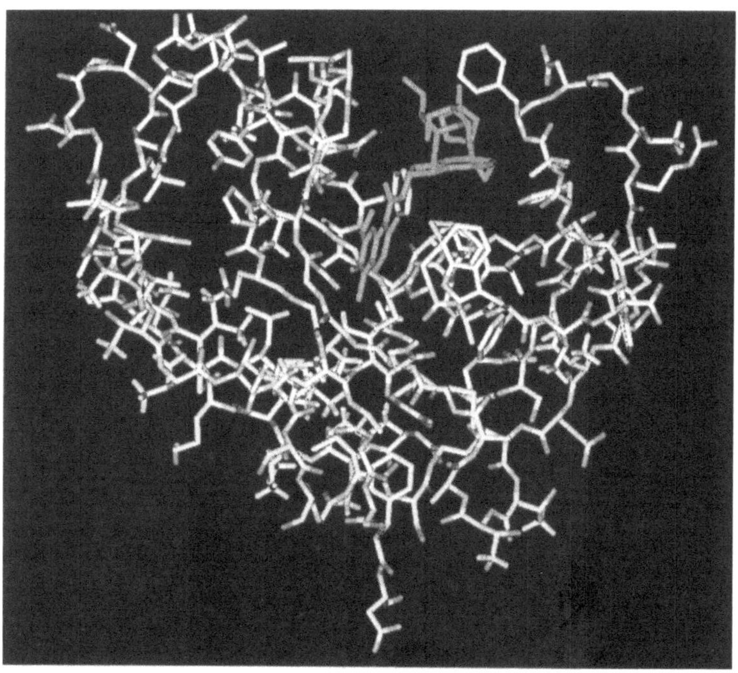

FIG. 14. Three-dimensional structure of holo-NCS determined by NMR data and restrained molecular dynamics calculations. The NCS-chrom molecule is displayed in *purple*. (*For color reproduction see color insert in front matter*)

Residues Trp39 and Leu45 have bulky side chains that come face to face with each other in the β-sheet assembly of NCS. This stabilization by the disulfide bond is deemed essential to maintain the hydrophobic cluster. The peptide chain from Asn48 to Asp51 contains Pro49 in a loop protruding form the β-sheet assembly of sheet-b. The side-chain amino protons of Asn48 have some unique spectral features in both their chemical shift values and hydrogen exchange rates. Both the side-chain amino protons are observed for Asn48. The amino protons show a large low-field shift (9.55 ppm) and a high-field shift (6.21 ppm); furthermore, these two amino protons never exchange at physiological conditions.

Although the assignments were confirmed by intra-residue NOEs between the Cβ and amino protons, the assignment for these amino protons was not independently confirmed until DNA sequencing placed Asn at position 48 [35]. Considering the three-dimensional structure of holo-NCS, it is seen that the amino protons of Asn48 are protected by strong hydrogen bonding to backbone carbonyl protons of Arg70 or Arg71. Thus it is interesting that the basic amino acid domain, which comprises the hinge structure mentioned previously, is connected by a hydrogen-bonded network to the binding pocket of NCS-chrom. Because a proline residue often introduces conformational variation into a peptide chain through *cis-trans* isomerism, the loop region containing Pro49 (connected to Cys47 and Asn48) is assumed to have a role in maintaining the activity of holo-NCS. In the three-dimensional structure, this region is covered by a phenyl ring of Phe52 that is in close contact with the NCS-chrom molecule.

Esperamicin and calicheamicin, which are enediyne-type antibiotics, have quite different structural features when compared to NCS. These two antibiotics have no carrier protein but have trisulfide bonds within the molecules that are considered to play an important role in stabilization as well as activation of the labile molecules. Similarly the disulfide bond between Cys37 and Cys47 is assumed to mediate a stabilizing mechanism for the NCS chromophore. Hirama et al. [44] have postulated an activation mechanism for NCS-chrom through a radical indication of the disulfide bond. Radicals tend to be oxidative toward disulfides.

In addition to the entropic effects of a disulfide bridge, hydrophobic forces and hydrophobic residues play an important role in stabilizing NCS-chrom. Trp39, Leu45, Phe76, and Phe78 all display intermolecular NOEs with NCS-chrom (see Fig. 9). As mentioned previously, the two phenylalanine residues protect the enediyne moiety from solvent. They are located in the loop region elongated from the hinge region around Arg70 and Arg71 (part of sheet-a; see Fig. 8) and appear to be flexible in apo-NCS [20,25]. Thus, dynamics pertaining the loop region may have interesting implications for the activation and release mechanisms of NCS-chrom.

In general, the binding between NCS-chrom and apo-NCS is relatively specific. In addition to NCS-chrom, ethidium bromide and daunomycin bind to apo-NCS. Because ethidium bromide also intercalates with DNA, it is interesting to note that many of the residues important in the NCS-chrom complex are also present in ethidium bromide and the apo-NCS complex [45]. The binding site and many of the binding interactions are comparable in the two systems.

The complex formation and binding features between NCS-chrom and apo-NCS were first studied by computer modeling [46]. The binding orientation was simulated using presolved X-ray crystallographic data of macromomaycin [36], one of the superfamily containing NCS and molecular dynamics calculations. In this simulation, the NCS-chrom molecule bound to the protein in a different manner than shown by the results presented here. The most striking difference was the orientation of the enediyne moiety.

After the publication of NMR results by two groups [21,22], X-ray crystallographic data of holo-NCS were published [23] that presented results similar to those from the NMR groups. Although the X-ray crystallographic group mentioned that their results regarding the binding feature were different from NMR data as well as modeling results, the essential formation and orientation were quite consistent between NMR data and X-ray data. Close contact of the benzene ring of Phe78, Phe76, Phe52, and polar residues of the binding pocket and NCS-chrom molecule, which were reported in their work, have been observed by many intermolecular NOEs, as already reported. Some detailed differences were likely caused by a crystal packing effect or crucial instability of NCS-chrom against X-ray irradiation. In the NMR experiments, as mentioned earlier, we could determine NCS-chrom and the protein structure independently at physiological conditions without crystallization and light irradiation.

The X-ray crystallographic group have reported π-interactions between the side chains of Phe52 and Phe78 and the NCS-chrom enediyne and amino sugar

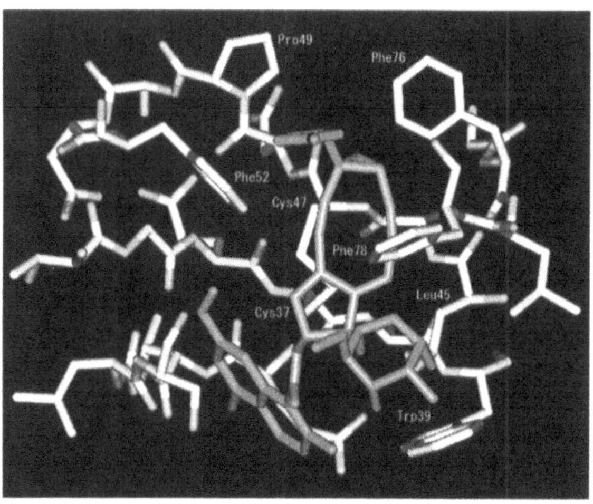

FIG. 15. NCS-chrom binding site of holo-NCS; the point of view is from top of Fig. 13. NCS-chrom (*purple*) and the residues from Val34 to Trp39, Leu45 to Ser53, Phe76 to Phe78, and Gln94 to Leu97 are indicated. The disulfide bond bridge between Cys 37 and Cys47 is displayed in *yellow*. (*For color reproduction see color insert in front matter*)

groups. The existence of the π-interaction from the Phe78 phenyl ring to the amino sugar group was also assumed in our restrained molecular dynamics result and was supported by many intermolecular NOEs among them (Fig. 15) and the result of chemical shift changes of NCS-chrom [47]. However, no π-interaction between Phe52 and the enediyne ring was detected. The two faces of the rings are almost perpendicular to each other and have no overlaps in our results.

In the continuation of study of the stabilization and activation mechanism of holo-NCS, the interactions are interesting because Phe78 is located on the top of the loop region based on two Arg residues (see above), and Phe52 is located on the edge of the loop-out region, including Pro49, from sheet-b. Although we cannot determine the flip angle and overall tumbling rate of the phenyl ring in Phe78 by NOE data, the mobility of the amino sugar (which is a hydrophilic moiety and is exposed on the surface of the complex) is considered to resemble that of the protein moiety by its line shape of NMR peak (see above). The tumbling rate and mobility of the phenyl rings will be investigated by multinuclear NMR relaxation studies using ^{15}N-^{13}C-labeled NCS.

Based on the statistical conformational analysis described here, it is an important next step to study the dynamical structure of holo-NCS. The dynamic features of the loop region and binding pocket in holo-NCS, and their implications regarding the mechanism of activation and release of NCS-chrom by environmental variables and interaction with other biological macromolecules, are expected to be elucidated by this method. The results will provide useful information to design new anticancer drugs.

References

1. Wüthrich K (1986) NMR of proteins and nucleic acids. Wiley, New York
2. Kessler H, Gehrke M, Griesinger C (1988) Two-dimensional NMR spectroscopy: background and overview of the experiments. Angew Chem Int Ed Engl 27: 490–536
3. Ernst RR, Bodenhausen G, Wokaun A (1986) Principles of nuclear magnetic resonance in one and two dimensions. Clarendon, Oxford
4. Clore GM, Gronenborn AM (1991) Structures of larger proteins in solution: three- and four-dimensional heteronuclear NMR spectroscopy. Science 252:1390–1399
5. Wüthrich K (1989) Protein structure determination in solution by nuclear magnetic resonance spectroscopy. Science 243:45
6. Edison AS, Abildgaard F, Westler WM, Mooberry ES, Markley JL (1994) Practical introduction to theory and implementation of multinuclear, multidimensional nuclear magnetic resonance experiments. Methods in Enzymology, vol 239. Academic, New York, pp 3–79
7. Takeshita J, Maeda H, Koike K (1980) Subcellular action of neocarzinostatin. J Biochem (Tokyo) 88:1071–1080
8. Edo K, Mizugaki M, Koide Y, Seto H, Furihata K, Otake N, Ishida N (1985) The structure of neocarzinostatin chromophore possessing a novel bicyclo[7,3,0] dodecadiyne system. Tetrahedron Lett 26:331–334

9. Myers AG (1987) Proposed structure of the neocarzinostatin chromophore-methyl thioglycolate adduct; a mechanism for the nucleophilic activation of neocarzinostatin. Tetrahedron Lett 28:4493–4496
10. Kappen LS, Goldberg IH (1989) Identification of 2-deoxyribonolactone at the site of neocarzinostatin-induced cytosine release in the sequence d(AGC). Biochemistry 28:1027–1032
11. Kappen LS, Goldberg IH (1992) Mismatch-induced switch of neocarzinostatin attack sites in the DNA minor groove, Biochemistry 31:9081–9089
12. Meshchwitz SM, Schultz RG, Ashley GW, Goldberg IH (1992) Selective abstraction of 2H from C-1′ of the C residue in AGC. ICT by the radical center at C-2 of activated neocarzinostatin chromophore: structure of the drug/DNA complex responsible for bistranded lesion formation. Biochemistry 31:9117–9121
13. Ishida N, Miyazaki K, Kumagai K, Rikimaru M (1965) Neocarzinostatin, and antitumor antibiotic of high molecular weight. Isolation, physicochemical properties and biological activities. J Antibiot (Tokyo) 18:68–76
14. Koide Y, Ishii F, Hasuda K, Koyama Y, Edo K, Katamine K, Kitame F, Isida N (1980) Isolation of a nonprotein component and a protein component from neocarzinostatin (NCS) and their biological activities. J Antibiot 33:342–346
15. Goldberg IH, Hatayama T, Kappen LS, Napier MA, Povirk LF (1981) Molecular actions and targets for cancer chemotherapeutic agents. Academic, New York, pp 163–191
16. Adjadj E, Mispelter J, Quiniou E, Dimicoli JL, Favaudon V, Lhoste JM (1990) Proton NMR studies of apo-neocarzinostatin from *Streptomyces carzinostaticus*, sequence-specific assignment and secondary structure. Eur J Biochem 190:263–271
17. Remerowski ML, Glaser SJ, Sieker LC, Samy TSA, Drobny GP (1990) Sequential ^{1}H-NMR assignments and secondary structure of aponeocarzinostatin in solution. Biochemisty 29:8401–8409
18. Gao X, Burkhart W (1991) Two- and three-dimensional proton NMR studies of apo-neocarzinostatin. Biochemistry 30:7730–7739
19. Sieker LC, Jensen LH, Samy TSA (1976) Neocarzinostatin: an antitumor protein. A preliminary x-ray diffraction study. Biochem Biophys Res Commun 68:358–362
20. Teplyakov A, Obmolova G, Wilson K, Kuromizu K (1993) Crystal structure of apo-neocarzinostatin at 0.15-nm resolution. Eur J Biochem 213:737–741
21. Takashima H, Amiya S, Kobayashi Y (1991) Neocarzinostatin: interaction between the antitumor-active chromophore and the carrier protein. J Biochem (Tokyo) 109:807–810
22. Tanaka T, Hirama M, Ueno M, Imajo M, Ishiguro M, Mizugaki M, Edo K, Komatsu H (1991) Proton NMR studies on NCS-chrom binding structure in neocarzinostatin complex. Tetrahedron Lett 32:3175–3178
23. Kim KH, Kwon BM, Myers AG, Rees DC (1993) Crystal structure of neocarzinostatin, an antitumor protein-chromophore complex. Science 262:1042–1046
24. Lefevre C, Adjadj E, Quiniou E, Mispelter J (1994) Assignment of the protonated ^{13}C resonances of apo-neocarzinostatin by 2D heteronuclear NMR spectroscopy at natural abundance. J Biomolecular NMR 4:689–702
25. Mispelter J, Lefevre C, Adjadj E, Quiniou E, Favaudon V (1995) Internal motions of apo-neocarzinostatin as studied by ^{13}C-NMR methine relaxation at natural abundance. J Biomol NMR 5:233–244
26. Constantine KL, Colson KL, Wittekind M, Friedrichs MS, Zein N, Tuttle J, Langley DR, Leet JE, Schroeder DR, Lam KS, Tarmer II BT, Metzler WJ, Bruccoleri RE,

Mueller L (1994) Sequential ^{1}H, ^{13}C, and ^{15}N NMR assignments and solution conformation of apokedarcidin. Biochemistry 33:11438–11452
27. Nagayama K, Kumar A, Wüthrich K, Ernst RR (1980) Experimental techniques of two-dimensional correlated spectroscopy. J Magn Reson 40:321–334
28. Braunschweiler L, Ernst RR (1983) Coherence transfer by isotropic mixing: application to proton correlation spectroscopy. J Magn Reson 53:521–528
29. Bax A, Davis DG (1985) MLEV-17-based two-dimensional homonuclear magnetization transfer spectroscopy. J Magn Reson 65:355–360
30. Kumar A, Ernst RR, Wüthrich K (1980) A two-dimensional nuclear Overhauser enhancement (2D NOE) experiment for the elucidation of complete proton-proton cross-relaxation networks in biological macromolecules. Biochem Biophys Res Commun 95:1–6
31. Macura S, Wüthrich K, Ernst RR (1982) Separation and suppression of coherent transfer effects in two-dimensional NOE and chemical exchange spectroscopy. J Magn Reson 46:269–282
32. Keepers JW, James TL (1984) A theoretical study of distance determinations from NMR. Two-dimensional nuclear Overhauser effect spectra. J Magn Reson 57:404–426
33. Hirayama K, Ando T, Takahashi R, Murai A (1986) A revised primary structure of neocarzinostatin using the combination of endopeptidase, carboxypeptidase Y and fast-atom-bombardment mass spectrometry. Bull Chem Soc Jpn 59:1371–1378
34. Kuromizu K, Tsunasawa S, Maeda H, Abe O, Sakiyama F (1986) Reexamination of the primary structure of an antitumor protein, neocarzinostatin. Arch Biochem Biophys 246:199–205
35. Sakata N, Minamitani S, Kanebe T, Hori M, Hamada M, Edo K (1993) The amino acid sequence of neocarzinostatin apoprotein deduced from the base sequence of the gene. Biol Pharm Bull 16:26–28
36. Roey PV, Beerman TA (1989) Crystal structure analysis of auromomycin apoprotein (macromomycin) shows importance of protein side chains to chromophore binding selectivity. Proc Natl Acad Sci USA 86:6587–6591
37. Braun W, Go N (1985) Calculation of protein conformations by proton-proton distance constraints. A new efficient algorithm. J Mol Biol 186:611–626
38. Braun W (1987) Distance geometry and related methods for protein structure determination from NMR data. Q Rev Biophys 19:115–127
39. Kobayashi Y, Ohkubo T, Kyogoku Y, Nishiuchi Y, Sakakibara S, Braun W, Go N (1989) Solution conformation of conotoxin GI determined by ^{1}H nuclear magnetic resonance spectroscopy and distance geometry calculations. Biochemistry 28:4853–4860
40. Guntert P, Braun W, Wüthrich K (1991) Efficient computation of three-dimensional protein structures in solution from nuclear magnetic resonance data using the program DIANA and the supporting programs CALIBA, HABAS and GLOMSA. J Mol Biol 217:517–526
41. Guntert P, Wüthrich K (1991) Improved efficiency of protein structure calculations from NMR data using the program DIANA with redundant dihedral angle constraints [REDAC]. J Biomol NMR 1:447–460
42. Borgias BA, James TL (1988) COMATOSE, a method for constrained refinement of macromolecular structure based on two-dimensional nuclear Overhauser effect spectra. J Magn Reson 79:493–512
43. Andersen NH, Lai XN, Hammen PK, Marschner TM (1990) Computer-aided conformational analysis based on NOESY signal intensities. Basic Life Sci 56:95–134

44. Tanaka T, Hirama M, Fujita K, Imajo S, Ishiguro M (1993) Solution structure of the antitumor antibiotic neocarzinostatin, a chromophore-protein complex. J Chem Soc Chem Commun 15:1205–1207
45. Mohanty S, Sieker LC, Drobny GP (1994) Sequential ^{1}H-NMR assignment of the complex of aponeocarzinostatin with ethidium bromide and investigation of protein-drug interactions in NCS-chrom binding site. Biochemistry 33:10579–10590
46. Ishiguro M, Imajo S, Hirama M (1991) Modeling study of the structure of the macromolecular antitumor antibiotic neocarzinostatin. Origin of the stabilization of NCS-chrom. J Med Chem 34:2366–2373
47. Imajo S, Ishiguro M, Tanaka T, Hirama M, Teplyakov A (1995) On the conformation of Phe78 of a chromoprotein antibiotic, neocarzinostatin. Bioorg Med Chem 3:429–436

Three-Dimensional Structure of Neocarzinostatin

Kyoung-Hee Kim[1], Andrew D. Myers[2], and Douglas C. Rees[1]

Summary. Neocarzinostatin (NCS) is a member of a class of natural antitumor antibiotics each composed of a labile chromophore component with DNA-cleaving activity and a protein component that binds and stabilizes the chromophore. Crystal structures of the protein–chromophore complex (holo-NCS) and the apoprotein form (apo-NCS) have been determined at 1.8-Å resolution. The protein fold of holo-NCS and apo-NCS consists of two domains composed of antiparallel β-sheets; the chromophore is bound noncovalently in a pocket between the two domains, primarily through van der Waals contacts with nonpolar residues. The chromophore displays marked nonlinearity in the C—C≡C bonds, and the two π-faces of the enediyne ring interact with the benzene ring edges of Phe^{78} and Phe^{52}. Phe^{78} additionally contacts the amino group of the amino sugar (presumed protonated) through its π-face. Within the chromophore, the amino sugar and carbonate residues are solvent exposed; however, the epoxide, acetylene groups, and carbon C-12, the site of nucleophilic thiol addition during chromophore activation, are sequestered from solution. The latter features likely contribute to the stability of the chromophore when bound to apo-NCS. The position of the amino group of the amino sugar relative to C-12 supports the idea that the amino group plays a role in thiol activation. The most significant difference between the holo-NCS and apo-NCS is the positioning of Phe^{78}, which rotates to a more solvent-exposed position in apo-NCS. Although the basic protein structure is conserved among at least four chromoprotein antibiotics, Phe^{78} is unique to NCS, perhaps accounting for the binding specificity of NCS for its chromophore.

Key words: Neocarzinostatin—X-ray structure—Enediyne antibiotics—Chromoprotein

---

[1] Division of Chemistry and Chemical Engineering (147-75CH)
[2] Division of Chemistry and Chemical Engineering (164-30Cr), California Institute of Technology, Pasadena, CA 91125, USA

Introduction

Neocarzinostatin (NCS) is a natural chromoprotein antibiotic isolated from the culture filtrate of *Streptomyces carzinostaticus* [1] and is composed of a binding protein (apo-NCS) and a labile nonprotein chromophore component (NCS-chrom) [2,3]. As a potent cytotoxic agent, NCS has undergone clinical evaluation in the treatment of various cancers [4–6]. In the presence of a thiol cofactor, NCS induces single- and double-strand cleavage of DNA both in vivo and in vitro. This activity is found to reside exclusively within the chromophore component [7] which, through a series of spectroscopic, degradative, and synthetic studies, has been proposed to possess the structure containing the unique epoxybicyclo [7.3.0]dodecadienediyne core (Fig. 1) [8–10]. It has been shown that NCS-chrom undergoes efficient thiol addition to form a highly reactive, carbon-centered biradical, thus presenting a potential molecular basis for the antitumor and DNA-cleaving properties of NCS [11,12]. The latter feature places NCS among the broader class of biradical-forming acetylenic (enediyne) antibiotics such as calichemicin [13,14], esperamicin [15,16], and dynemicin [17]. NCS-chrom also cleaves a bulged DNA at a specific site without the presence of thiol [18,19].

Apo-NCS is an acidic protein (pI = 3.2) with 113 amino acid residues [20–23]. It possesses two disulfide bonds and a large portion of alanine, glycine, serine,

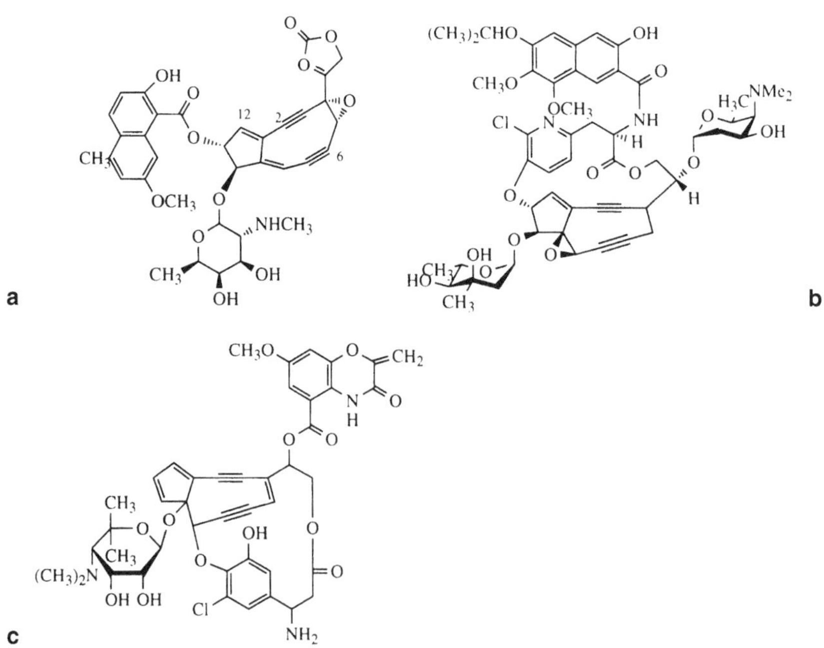

FIG. 1a–c. Structures of the chromophores of (**a**) neocarzinostatin (NCS), (**b**) kedarcidin (KDC), and (**c**) C-1027

and threonine residues (61% of the total residues). Apo-NCS binds the chromophore with high affinity and is found to greatly stabilize the labile chromophore; in the absence of its binding protein, the chromophore degrades rapidly in aqueous solution, particularly at elevated pH [24]. The reactivity of NCS-chrom toward a thiol changes when complexed with apo-NCS [25,26]. It has been suggested that the protein may function in binding to a cell-surface receptor, or to transport the chromophore across the cell membrane into the cytoplasm, possibly for release at the level of the cell nucleus [27–30]. The NCS protein may also function as a protease to degrade histones bound to DNA in the cell nucleus [31].

Other chromoprotein antibiotics have been isolated from diverse microbial sources including actinoxanthin (ACX) [32], auromomycin (also known as macromomycin, MCM) [33], kedarcidin (KDC) [34], and C-1027 [35]. Their antibacterial activity is believed to result from DNA cleavage, and each consists of an apoprotein and an active chromophore. The protein component of a sixth antitumor chromoprotein, maduropeptin (~31 kDa), is not related to the NCS class proteins, although its chromophore has been proposed to contain the nine-membered enediyne ring system [36,37]. The apoproteins show a high degree of sequence identity (>36%) (Table 1). Evidence has been presented that C-1027 and KDC, as well as holo-NCS, possess specific proteolytic activity [31,38,39]. Structures of the KDC chromophore and the C-1027 chromophore [40,41] have been established, and the activation mechanism for both chromophores is proposed to be similar to that of NCS-chrom, although the specific activation pathways may vary. Each chromophore is distinct from NCS-chrom, suggesting that each apoprotein maintains specificity for its respective chromophore.

The crystal structures of the apoproteins ACX [42] and MCM [43], the nuclear magnetic resonance (NMR) structure of the KDC apoprotein [44], and the NMR and crystal structures of apo-NCS [45–48] have been reported, which demonstrate that these apoproteins have similar three-dimensional structures (Table 2). The instability of the isolated chromophores has complicated direct structural characterization of these species. We have determined the three-dimensional structure of the NCS-chrom complexed to the NCS protein by X-ray crystallography at 1.8-Å resolution. The details of the NCS-chrom structure have been established to be in agreement with the proposed structure on the basis of

Table 1. Percent amino acid sequence identities between apoproteins

	MCM	ACX	C-1027	KDC
NCS	46.8	48.1	48.2	39.6
MCM		38.0	38.2	36.0
ACX			93.5	41.1
C-1027				44.0

NCS, Neocarzinostatin; MCM, macromomycin; ACX, actinoxanthin, KDC, kedarcidin.

TABLE 2. Status of the structural characterizations of antibiotic chromoproteins

Name	Source	Residue (*n*)	Three-dimensional structure determination		
			Chromophore	Apo-protein	Holo-protein
NCS	*Streptomyces carzinostaticus*	113	Yes	Yes	Yes
ACX	*Actinomyces globisporus*	108	No	Yes	No
MCM	*Streptomyces macromomycetius*	112	No	Yes	No
C-1027	*Streptomyces globisporus*	110	Yes	No	No
KDC	Actinomycete strain L-586-6	114	Yes	Yes	No

methods of chemical analysis and NMR spectroscopy. The apoprotein has two domains composed of antiparallel β-sheets, and the chromophore is bound in a hydrophobic pocket between the two domains and stabilized primarily through van der Waals contacts with many nonpolar residues.

Crystal Structure Determination of NCS

NCS crystals were grown by vapor diffusion against 70% 2-methyl-2,4-pentanediol (MPD) and 50 m*M* sodium acetate buffer (pH 4.6). The crystals grow in space group $P6_5$ with cell dimensions a = b = 90.4 Å and c = 52.2 Å, with two protein molecules per asymmetric unit. This crystal form is distinct from that previously reported by Sieker et al. [49]. The crystal structure analysis of NCS encompassed a combination of multiple isomorphous replacement, molecular replacement, and noncrystallographic symmetry averaging methods [50].

Somewhat surprisingly, although the initial crystallization solution contained a 1:1 complex of apoprotein:chromophore, only one of the NCS protein molecules in the crystallographic asymmetric unit (designated by residue numbers with the prefix "B") contained the chromophore; the other NCS protein molecule (designated by residue numbers with the prefix "A") has an MPD in the place of the chromophore. The apparent dissociation of chromophore from about half the NCS molecules in crystals grown from the low pH, aqueous MPD solution is consistent with the known dissociation behavior of NCS in an acidic, organic medium.

The current model contains 2 protein molecules, 1 chromophore, 1 MPD, and 161 water molecules in the asymmetric unit. This model has an R factor of 19.9% between 5.0- and 1.8-Å resolution with root mean square (rms) deviations from ideal bond distances and angles of 0.014 Å and 2.84°, respectively. The average B factor of the protein and the chromophore is about 22 $Å^2$ and that of solvents is about 27 Å^2. The Ramachandran plot shows no residues in unfavorable regions. The maximum coordinate error is estimated as about 0.22 Å by the Luzzati plot [51]. The structure coordinates have been deposited in the Protein Data Bank (entry 1NCO).

Structure of the NCS protein

The structure of the NCS protein in both the holo- and apo-forms is L shaped and has overall dimensions of 20 Å by 25 Å by 40 Å (Fig. 2). It consists primarily of β-sheets organized into two domains, separated by a cleft for the chromophore binding. The large domain forms a seven-stranded antiparallel β-barrel that contains two sheets: an external sheet consisting of strands 1, 2, and 5 and an internal sheet formed by strands 4, 3, 6, and 7 (Table 3). The small domain consists of two twisted, two-stranded antiparallel β-sheets: strands 3a and 3b and strands 5a and 5b. These two β-sheets are almost perpendicular to each other. Strand 3a is continuous from the β-barrel to the small β-sheet and is one of the most conserved regions among different apoproteins. There is a short segment of 3_{10} helix between residues 49 and 52, which stacks the proline ring of the residue 49 to the benzene ring of residue 52. In contrast, there are no helical segments in

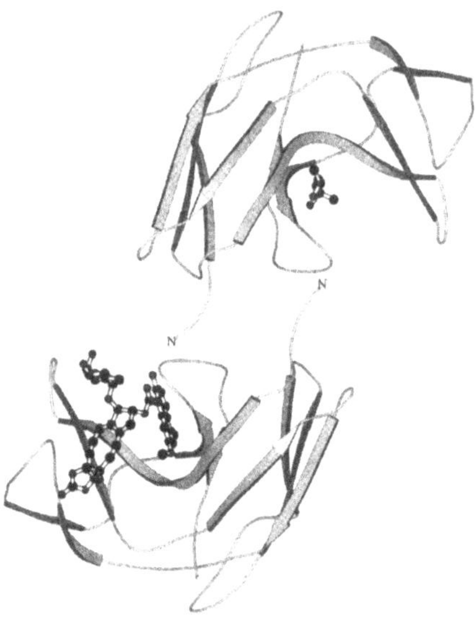

Fig. 2. Structures of two protein molecules in the asymmetric unit prepared with the program MOLSCRIPT [70]. The polypeptide chain is represented by a ribbons diagram; NCS-chrom and the 2-methyl-2,4-pentanediol (MPD) molecule are represented by ball-and-stick models. Holo-NCS (the *lower* molecule) contains NCS-chrom, while apo-NCS (the *upper* molecule) contains an MPD in the binding pocket. (*For color reproduction see color insert in the frontmatter*)

TABLE 3. Elements of secondary structure in the NCS apoprotein

Types of structure	Number	Residues
External β-sheet	1	Thr^{4}–Thr^{8}
	2	Val^{19}–Ala^{24}
	5	Ser^{62}–Ser^{66}
Internal β-sheet	4	Ser^{53}–Ala^{57}
	3	Thr^{30}–Gly^{35}
	6	Gln^{94}–Ser^{98}
	7	Val^{108}–Ile^{110}
Two-stranded β-sheet	3a	Gln^{36}–Val^{40}
	3b	Val^{44}–Asn^{48}
Twp-stranded β-sheet	5a	Ser^{72}–Gly^{75}
	5b	Gly^{84}–Asp^{87}
3_{10} Helix		Pro^{49}–Phe^{52}

Secondary structure assignments according to [69].

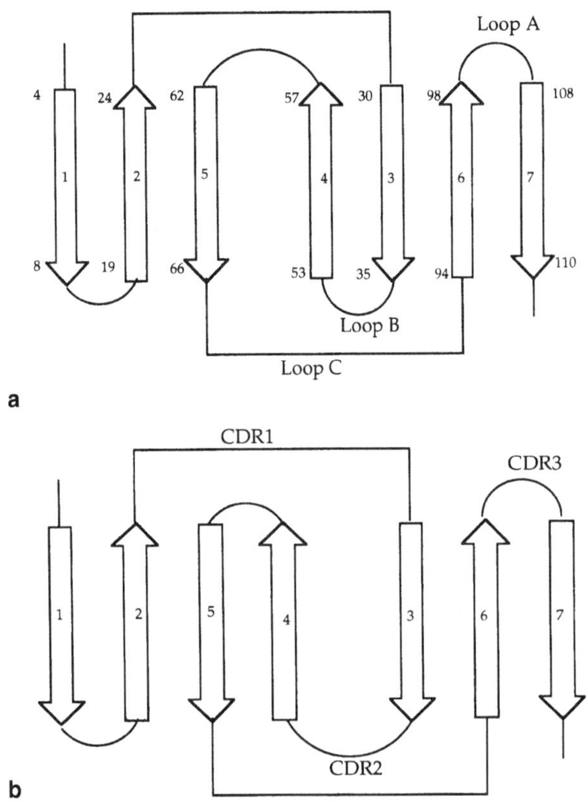

FIG. 3a,b. Topological diagrams of (**a**) NCS and (**b**) immunoglobulin (Ig) variable domains. For NCS, the strands are defined by the indicated residue numbers. *Loops A–C* of NCS and *CDR1–CDR3* of the Ig variable domain form the "active site" of these proteins

the structures of MCM and ACX. With respect to the hydrogen bonding network in the β-structure, the crystal structures are in agreement with NMR structures [45–47].

The β-barrel folding of NCS is similar to that of the immunoglobulin (Ig) variable domain (Fig. 3) [52]. The most noticeable difference is that the strand 4 belongs to the second sheet in NCS, and not to the first sheet as in the Ig variable domain. The topology with the strand 4 "sheet switching" is also observed in the structures of the bacterial chaperonin PapD [53], the D2 domain of CD4 [54], a fibronectin type III domain [55], and the extracellular domains of human growth hormone receptor [56]. Despite the topological similarity of the large domain of NCS with the Ig family, this similarity was not detected by a profile-based sequence analysis [57].

NCS is a highly acidic protein, containing eight aspartates and two glutamates but only three arginines and one lysine. The charged residues are fully exposed to the solvent (Fig. 4). Arg^{70} makes a salt bridge with the carboxylate group of Asp^{51} and has a close contact with the carboxylate group of Asp^{15}. The conformations of side chains of these three residues are well conserved both in apo-NCS and in holo-NCS, in spite of their solvent accessibility. The acidic groups of the protein appear to play an important role in its biological activity [58], because the modification of carboxyl groups into isopropylamide groups reduced the affinity of the protein for the chromophore. Lys^{20} is isolated from other charged residues, in agreement with the report that acylation of this residue results in no change in the protein conformation [59]. The amino-terminal group and the Nζ amino group of Lys^{20} have been used for attachment to poly(styrene-co-maleic acid) to form the conjugate SMANCS, for the purpose of stabilizing the NCS complex in the bloodstream and reducing its toxicity (see chapter by H. Maeda, this volume).

In contrast to the hydrophilic residues, many of the hydrophobic residues of NCS are buried in the protein interior of the large domain (Fig. 5). Half the aromatic residues—Trp^{39}, Phe^{52}, Phe^{76}, and Phe^{78}—are involved in the chromophore binding. Although Trp^{83} is exposed to the solvent, as predicted on the

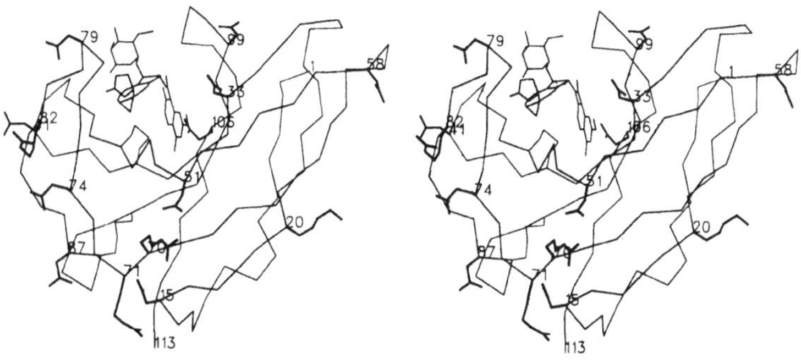

FIG. 4. Stereo view of charged residues (Asp, Glu, Lys, Arg) in holo-NCS. Cα positions and the chromophore are shown by *thin lines*; side chains are shown by *thick lines* and *residue numbers*

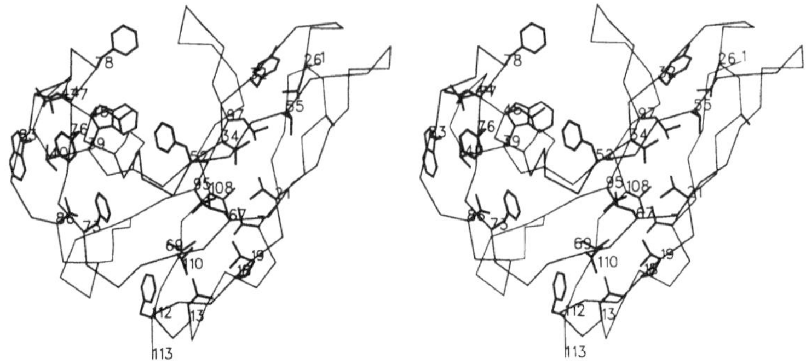

FIG. 5. Stereo view of hydrophobic residues (Phe, Tyr, Trp, Leu, Ile, Val) in holo-NCS. Cα positions are shown by *thin lines*; side chains are shown by *thick lines* and *residue numbers*

basis of an oxidation experiment with *N*-bromosuccinimide [60]. Tyr^{32} and Phe^{112} are buried in the protein and are well conserved among different apoproteins, except for replacement with Ala^{32} in KDC. Aromatic residues appear to be important in the side-chain packing. Stacking interactions are observed between the methyl groups and the aromatic rings for residues Val^{86}/Phe^{73}, Leu^{13}/Phe^{112}, Val^{45}/Trp^{39}, and Val^{40}/Trp^{83}.

Structure of the NCS Chromophore

The resolution of the refined holo-NCS structure is sufficiently high to establish the details of the chromophore structure (Fig. 6). The NCS-chrom structure is in complete agreement with the structure that was proposed on the basis of synthesis data and ^{1}H-NMR spectroscopy [10], including stereochemical assignments. The epoxy nonadiyne ring of the chromophore is essentially planar, with an rms deviation of the nine carbon atoms from the least-squares plane of only 0.06 Å. The four C—C≡C bond angles of the two triple bonds in the enediyne ring, which were not restrained during refinement, display marked nonlinearity with an average value of 161.5° ± 1.2°. The amino sugar group of the chromophore adopts a chair conformation, with the amino group oriented above C-12 at a distance (4.97 Å) that would accommodate the sulfur atom of a thiol between these two groups. This relationship supports the proposal that the amino group plays a role in the thiol activation mechanism [61].

NCS-chrom is bound in the hydrophobic pocket formed by the internal β-sheet of the large domain and the small domain (Fig. 7). The binding pocket can be divided into three parts. Residues 34–52 and 94–98 form the bottom of the pocket for the naphthoate group, residues 76–80 (loop C) interact with the carbonate group, and residues 100–102 (loop A) have contacts with the amino sugar group. The cyclononadiyne ring is placed above the Cys^{B37}–Cys^{B47} disulfide bond, and the naphthoate group is placed at the bottom of the pocket. The amino sugar and the

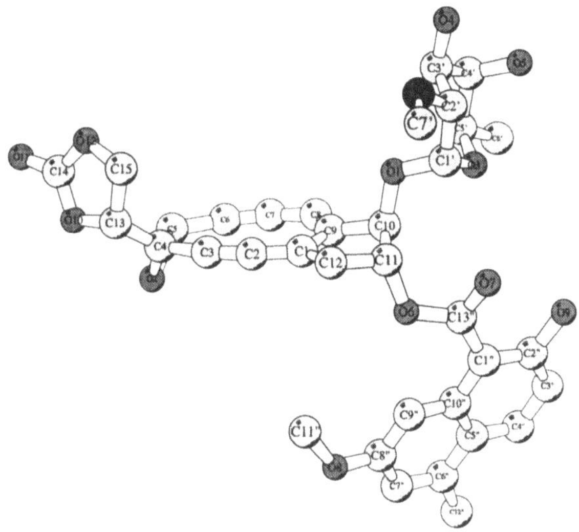

FIG. 6. Ball-and-stick model of NCS-chrom in the protein. Carbon atoms, *light gray*; oxygen atoms, *medium gray*, nitrogen atom, *dark gray*

TABLE 4. Van der Waals interactions defined as interatomic distances of less than 4.0 Å between the chromophore and the apoprotein

Residue	No.	Residue	No.	Residue	No.	Residue	No.
Val^{B34}	1	**Gly**^{B35}	8	**Cys**^{B37}	4	Trp^{B39}	2
Gly^{B43}	1	Leu^{B45}	9	**Cys**^{B47}	8	**Pro**^{B49}	7
Phe^{B52}	8	**Phe**^{B76}	5	**Leu**^{B77}	9	**Phe**^{B78}	30
Gly^{B80}	7	Gln^{B94}	3	Val^{B95}	2	**Gly**^{B96}	10
Ser^{B98}	10	Ala^{B100}	1	Ala^{B101}	2	**Gly**^{B102}	6

Residues closer than 3.5 Å are shown in **boldface**.

carbonate groups are facing outward and the epoxide is facing down toward the bottom of the hydrophobic pocket. The binding of NCS-chrom is achieved through extensive van der Waals contacts with many apolar residues and several aromatic residues in the binding pocket (Table 4). These same features of binding are believed to account for the stabilization of NCS-chrom on binding to apo-NCS. Gln^{B94} and Ser^{B98} are the only polar side chains around the chromophore. Phe^{B78} interacts with the chromophore most extensively.

The two π-faces of the nine-membered ring are sandwiched between the Phe^{B78} benzene ring on one side and the Phe^{B52} benzene ring and the Cys^{B37}–Cys^{B47} disulfide bond on the other. The benzene rings of Phe^{B52} and Phe^{B78} are perpendicular to the cyclononadiyne ring, which is a favorable orientation often exhibited by aromatic systems [62,63]. The distance between the C-12 atom and the Cδ1 atom of Phe^{B78} is 3.85 Å, and the solvent-accessible area of C-12 decreased from 24.2 Å^2 in the free form to 2.5 Å in holo-NCS. The positioning of the Phe^{B78} side

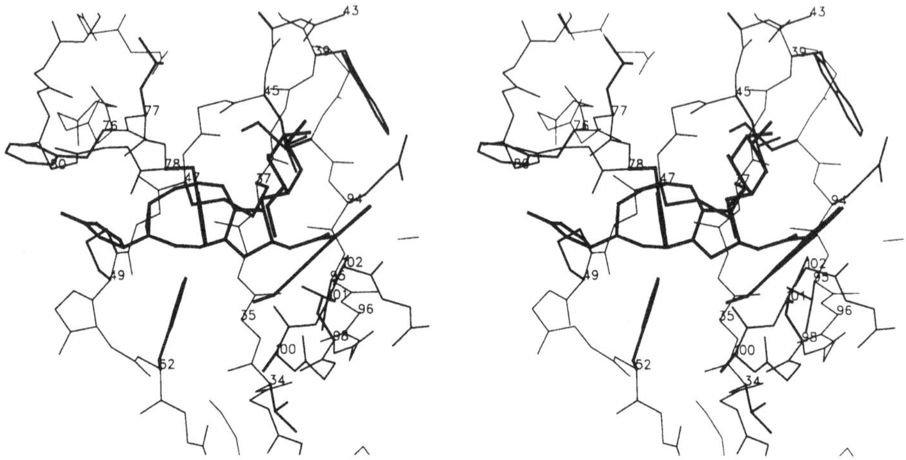

a

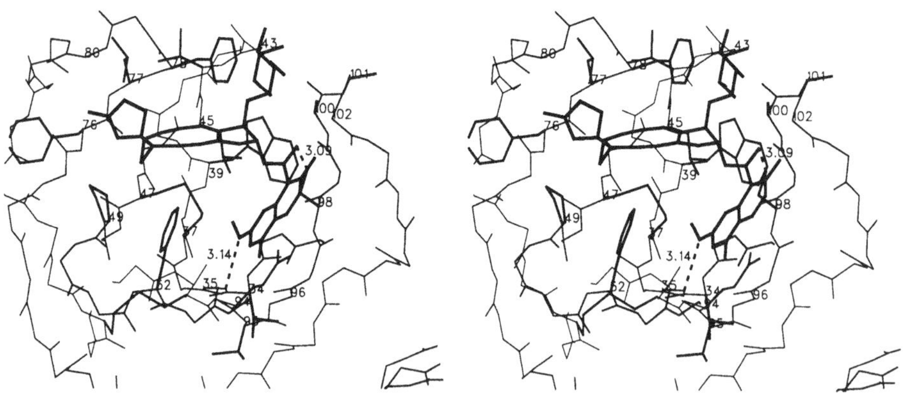

b

FIG. 7a,b. Stereo views of the protein environment of NCS-chrom. The main chain atoms of the protein are shown by *thin lines*; NCS-chrom and the side chains of residues within 4 Å from the chromophore (*numbered*) are represented by *thick lines*. (**a**) View down the cyclononadiyne ring shows the nonlinearity of two triple bonds. (**b**) View at the side of the cyclononadiyne ring; two hydrogen bonds between the NCS-chrom and the protein are shown by *dotted lines*

chain, in particular, should stabilize the chromophore by precluding attack of nucleophiles at C-12. The C2–C3 triple bond has contacts with the Cδ1 atom of Phe^{B78} and the Cε2 atom of Phe^{B52}. The C6–C7 triple bond is located just above the sulfur atom of Cys^{B47}, and is covered with the main-chain atoms of Phe^{B76}, Leu^{B77}, and Phe^{B78}. The epoxide is completely buried by the side chains of Cys^{B47}, Pro^{B49}, and Phe^{B52}, a feature that may preclude epoxide opening (occurring during thiol activation) by preventing solvation of the resulting alkoxide ion.

The naphthoate group is in proximity to the bottom of the pocket where strands 3 and 6 are exposed. The base of the binding pocket is very flat because of the absence of side chains at residues 35 and 96. The naphthoate group forms two distorted hydrogen bonds to the protein, one between the carbonyl oxygen atom (O-7) and the Oγ atom of Ser^{B98} (3.09 Å), and the other between the methoxy oxygen atom (C-11″) and the nitrogen atom of Gly^{B35} (3.14 Å). There is another hydrogen bond occurring at a lattice contact between the naphthoate hydroxyl oxygen atom (O-9) and the nitrogen atom of Ala^{A1}, the amino terminus of the other protein molecule in the asymmetric unit (2.64 Å). The methyl group (C-12″) of the naphthoate has contacts with Gln^{B94}, Val^{B95}, and Gly^{B96}, and the methoxy carbon (C-11′) is close to Gly^{B35} and Phe^{B52}. The sulfur atom of Cys^{B37} has contacts with the C-6″–C-9″ site of the naphthoate, and Leu^{B45} has contacts with the C-1″–C-4″ site. The hydroxyl oxygen (O-9) is close to the carbonyl oxygen of Gly^{B102} (3.71 Å). Trp^{B39} makes two contacts with C-3″ in the naphthoate.

The methyl amino group of the amino sugar stacks against the Phe^{B78} benzene ring in holo-NCS, which suggests that the amino group in the crystal is protonated. The distance between the nitrogen atom of the methyl amino group and the Phe^{B78} benzene ring is about 3.5 Å. The favorable interaction between a (presumed) positively charged group and the π-face of an aromatic system has been observed in a variety of protein and model systems [64–66]. The methyl group of the amino sugar is close to loop A. The carbonate group has contacts with the side chain of Pro^{B49} and the carbonyl oxygen atoms of Leu^{B77} and Phe^{B78}. The carbonyl oxygen of the carbonate has contacts with Gly^{B80}.

The solvent-accessible areas for the chromophore and the protein were calculated as defined by Lee and Richards [67]. The bound chromophore exposes about 171 $Å^2$ of surface area to the solvent, or about 18% of the total accessible area of the free NCS-chrom. The solvent-exposed regions of NCS-chrom are provided primarily by the amino sugar, the carbonate, and the naphthoate hydroxyl groups. The complete burial of the active epoxide in the protein suggests that the protein stabilizes the chromophore by sequestration of this group from the solvent, as suggested earlier. The changes in solvent-accessible areas of the protein before and after the chromophore binding are compared in Table 5. The biggest change occurs in Phe^{78}. Cys^{B37}, Cys^{B47}, Leu^{B45}, Gly^{B96}, and Ser^{B98} are completely buried by the chromophore binding. The surface areas of Gly^{B102}, Trp^{B39}, Phe^{B52}, and Pro^{B49} are also decreased significantly in the complex.

Comparison Between Apo-NCS and Holo-NCS

The structures of apo-NCS and holo-NCS are similar, with rms deviations of 0.59 Å for 113 Cα atoms and 0.31 Å for 35 Cα atoms in the seven-stranded β-barrel of the large domain. Differences between apo-NCS and holo-NCS are found in strands 5a and 5b and in three loops around the binding pocket (Fig. 8). The most significant change is the position of the Phe^{78} benzene ring. This side

TABLE 5. Solvent-accessible areas before and after chromophore binding in holo-NCS

Residue		Solvent-accessible area (Å^2)		
		Before	After	Difference
78	Phe	199.0	88.5	110.5
45	Leu	50.6	0.0	50.6
102	Gly	57.2	12.6	44.6
98	Ser	31.5	0.0	31.5
39	Trp	129.9	99.1	30.8
52	Phe	68.0	37.2	20.7
49	Pro	54.6	29.3	25.3
43	Gly	74.2	54.1	20.1
80	Gly	57.8	38.5	19.3
37	Cys	12.7	0.0	12.7
101	Ala	97.6	85.3	12.3
96	Gly	10.4	0.0	10.4
47	Cys	10.0	0.0	10.0
107	Gly	18.8	10.5	8.3
105	Pro	9.3	2.8	6.5
33	Asp	59.3	53.2	6.1
106	Glu	151.8	147.9	3.9
97	Leu	3.7	0.0	3.7
77	Leu	35.5	32.0	3.5
76	Phe	55.0	52.2	2.8
99	Asp	18.3	16.4	1.9
35	Gly	1.7	0.0	1.7
44	Val	45.9	44.6	1.3
94	Gln	27.6	26.1	1.5
100	Ala	102.3	101.3	1.0
34	Val	0.4	0.0	0.4
104	Gly	20.6	20.3	0.3

Residues are listed in the order of decreasing difference in contact area with the chromophore.

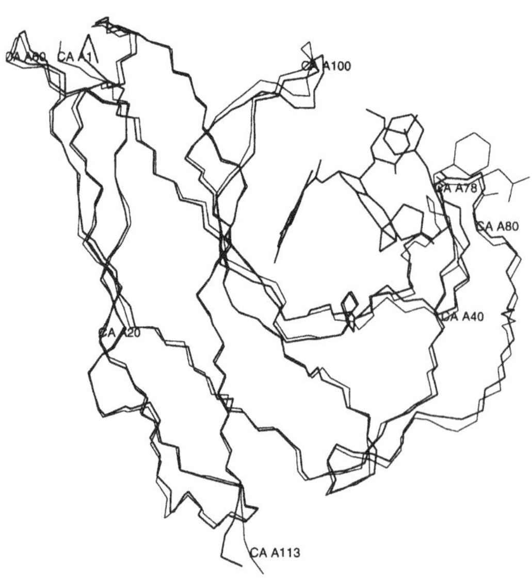

FIG. 8. Superposition of N, C, and Cα atoms of apo-NCS (*thin line*) and holo-NCS (*thick line*) also shows NCS-chrom and the side chains of Phe^{78} and Asp^{A79}

chain rotates from a stacking position over the cyclononadiyne ring in holo-NCS to a solvent-exposed position in apo-NCS, where it abuts the side chain of Asp^{A79}. The distance between the Phe^{A78} benzene ring and the Asp^{A79} carboxylate is about 4.2 Å. This amino acid interaction may underlie the pH-sensitive dissociation of the chromophore from NCS. Differences between apo-NCS and holo-NCS increase the width of the binding pocket about 5 Å to accommodate the chromophore. The distance between the Gly^{B102} $C\alpha$ atom and the Phe^{B78} $C\zeta$ atom is 4.54 Å in the holoprotein, while the distance between the Gly^{A102} $C\alpha$ atom and the Phe^{A78} $C\beta$ atom in apo-NCS increases to 9.54 Å. The opening of the binding pocket is mainly accomplished by the repositioning of the Phe^{78} benzene ring. In the absence of NCS-chrom, a hydrogen bond in the crystal between MPD and the $O\gamma$ atom of Ser^{A98} of apo-NCS (2.57 Å) replaces the distorted hydrogen bond between the chromophore and the $O\gamma$ atom of Ser^{B98} in holo-NCS.

The mean B factors of apoproteins in apo-NCS and holo-NCS are 23.3 Å^2 and 21.9 Å^2, respectively. Residues in the loop between strands 3a and 3b have the highest B factors in the structure. The largest differences in the mean B factors between apo-NCS and holo-NCS are found for residues in the loop between strands 6 and 7, which participate in the chromophore binding. Strands 4 and 5 and the loop between them have significantly higher B factors in apo-NCS than in holo-NCS.

A crystal structure of apo-NCS was determined at 1.5 Å resolution by Teplyakov et al. [48], and their coordinates are available from the Brookhaven

NCS	1	AAPTAT**V**T**P**SS**GL**SD**G**TV**V**K**V**AGAGLQAGTAYDVG**QCA**WV	40
MCM	1	.APGVT**V**T**P**AT**GL**SN**G**QT**V**T**V**SATGLTPGTVYHVG**QCA**VV	39
ACX	1	.APAFS**V**S**P**AS**GL**SD**G**QS**V**S**V**SV..AAAGETYYIA**QCA**PV	37
KDC	1	ASAAVS**V**S**P**AT**GL**AD**G**AT**V**T**V**SASGFATSTSATAL**QCA**IL	40
NCS	41	DT**G**VLA**C**NPADFSSVTADAN**G**SASTSLT**V**RR**S**FEGFL.FD**G**	80
MCM	40	EP**G**VIG**C**DATTSTDVTADAA**G**KITAQLK**V**HS**S**FQAVVGAD**G**	80
ACX	38	G.**G**QDA**C**NPATATSFTTDAS**G**AASFSFT**V**RK**S**YAGQT.PS**G**	76
KDC	41	AD**G**RGA**C**NVAEFHDFSLSG.**G**EGTTSVV**V**RR**S**FTGYVMPD**G**	80
NCS	81	TRW**G**T**V**D**C**TTA..A**C**QVGLSDAAGNGPEGV**A**IS**F**N	113
MCM	81	TPW**G**T**V**N**C**KVV..S**C**SAGLGSDSGEGA.AQ**A**IT**F**A	112
ACX	77	TPV**G**S**V**D**C**ATD..A**C**NLGAGNSGLNLGH.V**A**LT**F**G	108
KDC	81	PEV**G**A**V**D**C**DTAPGG**C**QIVVGGNTGEYGNA.**A**IS**F**G	114

Fig. 9. Amino acid sequence alignment of NCS [23], MCM (macromomycin) [71], ACX (actinoxanthin) [21], and KDC [72]. Residues that are absolutely conserved in all four proteins are shown in *boldface type*

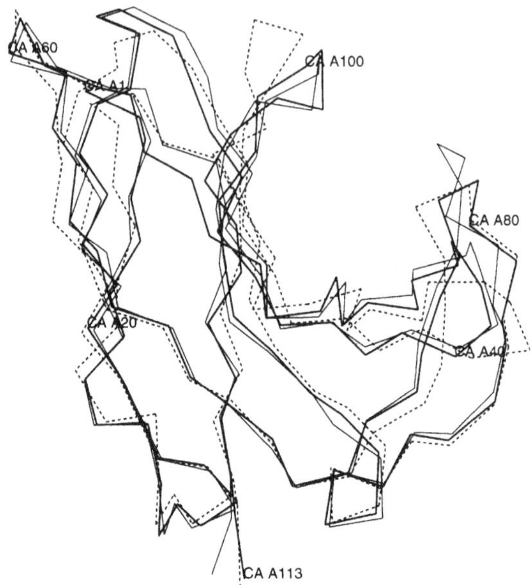

FIG. 10. Superposition of Cα atoms of NCS (*thick line*), MCM (*thin line*), and ACX (*dashed line*) apoproteins

Protein Data Bank as 1NOA.PDB. The rms deviations of the main chain atoms from apo-NCS and holo-NCS are 0.54 Å and 0.73 Å, respectively. The conformations of the side chains are in agreement with those of apo-NCS and holo-NCS, except for the Phe^{78} benzene ring and the Trp^{39} ring. The Phe^{78} benzene ring intrudes into the binding pocket in this structure, which suggests this residue is flexible. The side-chain conformations of Asp^{15}, Asp^{51}, and Arg^{70} are also well conserved in this structure.

Comparison with Other Apoproteins

The NCS, MCM, ACX, and KDC apoproteins share the same basic structural architecture, which consists of a large domain with a seven-stranded β-barrel and a small domain of two antiparallel β-sheets. Crystal structures are available for MCM and ACX apoproteins [42,43]. Twenty-two residues are conserved among five apoproteins in the family of antitumor antibiotic chromoproteins, including two disulfide bonds (Fig. 9). Strand 3a is the most conserved sequence region, having three conserved residues out of five, which indicates the importance of this sequence to either the function or the architecture of the protein. The Nε2 and Oε1 atoms of Gln^{36} form two hydrogen bonds with the carbonyl oxygen and amide nitrogen atoms of Arg^{70}, which is also a conserved residue, except in MCM. The region of residues 80–88 is also highly conserved, although they are exposed to the solvent. Strand 6 and the loop with residues 99–107 are poorly conserved and are important in chromophore recognition, as indicated in the solvent-accessible area changes on chromophore binding in NCS.

The MCM structure can be superimposed into the apo-NCS and holo-NCS structures with rms deviations of 1.02 Å and 1.04 Å, respectively, for 111 common Cα atoms. The rms deviations of the ACX structure from apo-NCS and holo-NCS are 1.84 Å and 1.97 Å, respectively, for 107 common Cα atoms. As shown in Fig. 10, the external β-sheet of NCS superimposes well on the other two structures, while the internal β-sheet is shifted significantly across the strands. The major differences occur in the loops around the binding cleft. Differences in Cα positions of equivalent residues of the loops range up to 6.9 Å for Gly^{102} of NCS and Leu^{98} of ACX.

The common core of these proteins, as defined by Chothia and Lesk [68], contains 90 residues (80% of the total). The rms deviations for the common core are 0.76 Å between holo-NCS and MCM, 1.27 Å between holo-NCS and ACX, and 1.17 Å between MCM and ACX. The regions outside the common core (residues 26–30, 41–44, 78–79, and 99–107, respectively) are less conserved and, except for Phe^{78} and Asp^{79}, correspond to the complementarity-determining regions in the Ig variable domains.

The binding pockets of NCS, ACX, and MCM have similar overall shapes, but the locations of side chains around the binding pockets are quite different. Among 20 residues close to NCS-chrom, only 3 residues—Cys^{37}, Cys^{47}, and Gly^{80}—are conserved. The size of the chromophore binding pocket is biggest in KDC, followed by ACX, and smallest in NCS and MCM. The distance between the Cα atoms of Met^{77} and Gly^{104} is 18.86 Å in KDC [44], and that between the Cα atoms of Pro^{74} and Asn^{95} is 12.69 Å in ACX. The distance between the Cα atoms of Ala^{78} and Asp^{100} is 9.61 Å in MCM. The sizes of the binding pocket are consistent with the relative size of the chromophores.

In NCS and MCM, the bottom of the binding pocket is flat because of the residues Gly^{35} and Gly^{96}. Ala^{35} in ACX replaces Gly^{35}, while Leu^{35} and Val^{97} are the corresponding residues in KDC. KDC also has the side chains of Val^{49} and Phe^{52}, as well as the disulfide bond, in the bottom of the pocket, which retains the hydrophobic nature. Considering the large functional groups of the KDC chromophore and large size of the binding pocket, it is likely that the enediyne ring of the KDC chromophore is located above the disulfide bond, with all other functional groups facing outward. The wall of the binding pocket in NCS is formed by strands 3a and 3b on one side and by strands 3 and 6 on the other, and contains the disulfide bond and Leu^{45}, Gln^{94}, and Ser^{98}. The two exits are closed by Trp^{39}, Pro^{49}, and Phe^{52} to limit the solvent accessibility of the naphthoate group. The hydrophobic nature of the wall is conserved in both MCM and ACX, but the two exits are open to solvent. The functional groups at the bottom of the binding pocket in MCM and ACX are expected to have hydrophilic groups.

Loop A and loop C in NCS are involved in chromophore recognition by interacting with the functional groups of the chromophore. Asp^{100} in MCM and Asp^{99} and His^{102} in ACX might interact with the corresponding chromophore through a hydrogen bond or a salt bridge. His[32] in MCM and Tyr[30] in ACX are likely to participate in chromophore recognition, although Asp[33] in NCS is positioned away from the chromophore. Tyr[106] and Met[78] in KDC are likely to participate in the chromophore recognition, as they displayed a high degree of disorder in the NMR structure of the kedarcidin apoprotein.

The positions and orientations of the diacetylene ring systems above the disulfide bond are expected to be different among the different apoproteins. NCS-chrom nicely fits into the binding pocket of MCM, but not into the binding pocket of ACX. The binding selectivity of a chromophore and its partner protein must be the result of interactions between the chromophore and specific side chains, such as Phe[78] of NCS, that represent variable elements within the protein structures. Detailed comparisons of chromophore recognition by different apoproteins will only be possible when the three-dimensional structures of other chromoproteins are determined.

References

1. Ishida N, Miyazaki K, Kumagi K, Rikimaru M (1965) Neocarzinostatin, an antitumor antibiotic of high molecular weight—isolation, physicochemical properties and biological activities. J Antibiot (Tokyo) A18:68–76
2. Napier MA, Holmquist B, Strydom DJ, Goldberg IH (1979) Neocarzinostatin: spectral characterization and separation of a non-protein chromophore. Biochem Biophys Res Commun 89:635–642
3. Koide W, Ishii F, Hasuda K, Koyama Y, Edo K, Katamine S, Kitame F, Ishida N (1980) Isolation of a non-protein component and a protein component from neocarzinostatin and their biological activities. J Antibiot (Tokyo) 33:342–346
4. Maeda H (1981) Neocarzinostatin in cancer chemotherapy (review). Anticancer Res 1:175–181
5. Kimoto A, Konno T, Kawaguchi T, Miyauchi Y, Maeda H (1992) Antitumor effects of SMANCS on rat mammary tumor induced by 7,12-dimethyl[a]anthracene. Cancer Res 52:1013–1017
6. Konno T (1992) Targeting chemotherapy for hepatoma—arterial administration of anticancer drugs dissolved in lipidol. Eur J Cancer 28:403–409
7. Kappen LS, Napier MA, Goldberg IH (1980) Roles of chromophore and apo-protein in neocarzinostatin action. Proc Natl Acad Sci USA 77:1970–1974
8. Edo K, Mizugaki M, Koide Y, Seto H, Furihata K, Otake N, Ishida N (1985) The structure of neocarzinostatin-chromophore possessing a novel bicyclo [7,3,0]dodecadiyne system. Tetrahedron Lett 26:331–334
9. Edo K, Akiyama Y, Saito K, Mizugaki M, Koide Y, Ishida N (1986) Absolute-configuration of the amino sugar moiety of the neocarzinostatin chromophore. J Antibiot (Tokyo) 39:1615–1619
10. Myers AG, Proteau PJ, Handel TM (1988) Stereochemical assignment of neocarzinostatin chromophore structures of neocarzinostatin-chromophore-methyl thioglycolate adducts. J Am Chem Soc 110:7212–7214
11. Myers AG (1987) Proposed structure of the neocarzinostatin-chromophore-methyl thioglycolate adduct; a mechanism for the nucleophilic activation of neocarzinostatin. Tetrahedron Lett 28:4493–4496

12. Myers AG, Proteau PJ (1989) Evidence for spontaneous, low-temperature biradical formation from a highly reactive neocarzinostatin-chromophore-thiol conjugate. J Am Chem Soc 111:1146–1147
13. Lee MD, Dunne TS, Siegel MM, Chang CC, Morton GO, Borders DB (1987) Calichemicins, a novel family of antitumor antibiotics. 1. Chemistry and partial structure of calichemicin-γ-1. J Am Chem Soc 109:3464–3466
14. Lee MD, Dunne TS, Chang CC, Ellestad GA, Siegel MM, Morton GO, McGahren WJ, Borders DB (1987) Calichemicins, a novel family of antitumor antibiotics. 2. Chemistry and structure of calichemicin-γ-1. J Am Chem Soc 109:3466–3468
15. Golik J, Clardy J, Dubay G, Groenewold G, Kawaguchi H, Konishi M, Krishnan B, Ohkuma H, Saitoh K, Doyle TW (1987) Esperamicins, a novel family of antitumor antibiotics. 2. Structure of esperamicin-X. J Am Chem Soc 109:3461–3462
16. Golik J, Dubay G, Groenewold G, Kawaguchi H, Konishi M, Krishnan B, Ohkuma H, Saitoh K, Doyle TW (1987) Esperamicins, a novel family of antitumor antibiotics. 3. Structure of esperamicin-A_1, esperamicin-A_2 and esperamicin-A_{1B}. J Am Chem Soc 109:3462–3464
17. Konishi M, Ohkuma H, Tsuno T, Oki T, VanDuyne GD, Clardy J (1990) Crystal and molecular structure of dynemicin-A, a novel 1,5-diyn-3-ene antitumor antibiotic. J Am Chem Soc 112:3715–3716
18. Kappen LS, Goldberg IH (1993) DNA conformation-induced activation of an enediyne for site-specific cleavage. Science 261:1319–1321
19. Kappen LS, Goldberg IH (1993) Site-specific cleavage at a DNA bulge by neocarzinostatin chromophore via a novel mechanism. Biochemistry 32:13138–13145
20. Maeda H, Glaster CB, Czombos J, Meienhofer J (1974) Structure of the antitumor protein neocarzinostatin: amino acid sequence. Arch Biochem Biophys 164:379–385
21. Gibson BW, Herlihg WC, Anantha Samy TS, Hahm K-S, Maeda H, Meienhofer J, Biemann K (1984) A revised primary structure for neocarzinostatin based on fast atom bombardment and gas chromatographic-mass spectrometry. J Biol Chem 259:10801–10806
22. Kuromizu K, Tsunasawa S, Maeda H, Abe O, Sakiyama F (1986) Reexamination of the primary structure of an antitumor protein neocarzinostatin. Arch Biochem Biophys 246:199–205
23. Sakata N, Minamitani S, Kanbe T, Hori M, Hamada M, Edo K (1993) The amino acid sequence of neocarzinostatin apoprotein deduced from the base sequence of the gene. Biol Pharm Bull 16:26–28
24. Povirk LF, Goldberg IH (1980) Binding of the nonprotein chromophore of neocarzinostatin to DNA. Biochemistry 19:4773–4780
25. Sugiyama H, Yamashita K, Fugiwara T, Saito I (1994) Apoprotein-assisted unusual cyclization of neocarzinostatin chromophore. Tetrahedron 50:1311–1325
26. Chin D-H, Goldberg IH (1993) Identification of the active species responsible for holo-neocarzinostatin-induced DNA cleavage. J Am Chem Soc 115:9341–9342
27. Maeda H, Aikawa S, Yamashita A (1975) Subcellular fate of protein antibiotic neocarzinostatin in culture of a lymphoid cell line from Burkitt's lymphoma. Cancer Res 35:554–559
28. Lazarus H, Razo V, Anantha Samy TS (1977) In vitro inhibition of human leukemic cells by agarose-immobilized neocarzinostatin. Cancer Res 37:3731–3736
29. Takeshita J, Maeda H, Koike K (1980) Subcellular action of neocarzinostatin. J Biochem (Tokyo) 88:1071–1080

30. Sakamoto S, Maeda H, Ogata J (1979) An uptake of fluorescein isothiocyanate-labeled neocarzinostatin into the cancer and normal cells. Experientia (Basel) 35:1233–1235
31. Zein N, Reiss P, Bernatowicz M, Bolgar M (1995) The proteolytic specificity of the natural enediyne-containing chromoproteins is unique to each chromoprotein. Chem & Biol 2:451–455
32. Khokholov AS, Cherches BZ, Reshetov PD, Smirnova GM, Sorokina IB, Prokoptzeva TA, Koloditskaya TA, Smirnov VV, Navashin SM, Formina IP (1969) Physico-chemical and biological studies on actinoxanthin, an antibiotic from *Actinomyces globisporus* 1131. J Antibiot (Tokyo) 22:541–544
33. Chimura H, Ishizuka M, Hamada M, Hori S, Kimura K, Iwanga J, Takeuchi T, Umezawa H (1968) A new antibiotic, macromomycin, exhibiting antitumor and antimicrobial activity. J Antibiot (Tokyo) 21:44–49
34. Lam KS, Hesler GA, Gustavson DR, Crosswell AR, Veitch JM, Forenza A, Tomita K (1992) Kedarcidin, a new chromoprotein antitumor antibiotic. 1. Taxonomy of producing organism, fermentation and biological activity. J Antibiot (Tokyo) 44:472–478
35. Hu J, Xue Y-C, Xie M-Y, Zhang R, Otani T, Minami Y, Yamada Y, Marunaka T (1988) A new macromolecular antitumor antibiotic C-1027. I. Discovery, taxonomy of the producing organism, fermentation and biological activity. J Antibiot (Tokyo) 41:1575–1585
36. Hanada M, Ohkuma H, Yonemoto T, Tomita K, Ohbayashi N, Kaemi H, Miyaki T, Konishi M, Kawaguchi H (1991) Maduropeptin, a complex of new macromolecular antitumor antibiotics. J Antibiot (Tokyo) 44:403–414
37. Schroeder DR, Colson KL, Khlor SE, Zein N, Langley DR, Lee MS, Matson JA, Doyle TW (1994) Isolation, structure determination, and proposed mechanism of action for artifacts of maduropeptin chromophore. J Am Chem Soc 116:9351–9352
38. Sakata N, Tsuchiya KS, Moriya Y, Hayashi H, Mori M, Otani T, Nagai M, Aoyagi T (1992) Aminopeptidase activity of an antitumor antibiotic C-1027. J Antibiot (Tokyo) 45:113–117
39. Zein N, Casazza AM, Doyle TW, Leet JE, Schroeder DR, Solomon W, Nadler SG (1993) Selective proteolytic activity of the antitumor agent kedarcidin. Proc Natl Acad Sci USA 90:8009–8012
40. Leet JL, Schroeder DR, Langley DR, Colson KL, Haung S, Khlor SE, Lee MS, Golik J, Hofstead SJ, Doyle TW, Matson JA (1993) Kedarcidin, a new chromoprotein antitumor antibiotic: structure elucidation of kedarcidin chromophore. J Am Chem Soc 115:8432–8443
41. Yoshida K-I, Minami Y, Azuma R, Saeki M, Otani T (1993) Structure of cycloaromatization of a novel enediyne, C-1027 chromophore. Tetrahedron Lett 34:2637–2640
42. Pletnev VZ, Kuzin AP, Trakhanov SD, Kostetsky PV (1982) Three-dimensional structure of actinoxanthin. IV. 2.5 Å resolution. Biopolymers 21:287–300
43. Van Roey P, Beerman TA (1989) Crystal structure analysis of auromomycin apoprotein (macromomycin) shows importance of protein side chains to chromophore binding selectivity. Proc Natl Acad Sci USA 86:6587–6591
44. Constantine KL, Colson KL, Wittekind M, Friedrichs MS, Zein N, Tuttle J, Langley DR, Leet JE, Schroeder DR, Lam KS, Farmer BT II, Metzler WJ, Bruccoleri RE, Mueller L (1994) Sequential ^{1}H, ^{13}C, and ^{15}N NMR assignments and solution conformation of apokedarcidin. Biochemistry 33:11438–11452

45. Gao X (1992) Three-dimensional solution structure of apo-neocarzinostatin. J Mol Biol 225:125–135
46. Adjadj E, Quiniou E, Mispelter J, Favaudon V, Lhoste J-M (1992) Three-dimensional solution structure of apo-neocarzinostatin determined by NMR spectroscopy. Eur J Biochem 203:505–511
47. Lyndsay Remerowski G, Glaser SJ, Sieker LC, Anantha Samy TS, Drobny GP (1990) Sequential ^{1}H NMR assignments and secondary structure of apo-neocarzinostatin in solution. Biochemistry 29:8401–8409
48. Teplyakov A, Obmolova G, Wilson K, Kuromizu K (1993) Crystal structure of apo-neocarzinostatin at 0.15 nm resolution. Eur J Biochem 213:737–741
49. Sieker LC, Jensen LH, Anantha Samy TS (1976) Neocarzinostatin: an antitumor protein. A preliminary x-ray diffraction study. Biochem Biophys Res Commun 68:358–362
50. Kim K-H, Kwon B-M, Myers AG, Rees DC (1993) Crystal structure of an antitumor antibiotic neocarzinostatin complex. Science 262:1042–1046
51. Luzzati PV (1952) Traitement statistique des erreurs dans la determination des structures crystallines. Acta Crystallogr 5:802–810
52. Deisenhofer J (1981) Crystallographic refinement and atomic models of a human Fc fragment and its complex with fragment B of protein A from *Staphylococcus aureus* at 2.9- and 2.8-Å resolution. Biochemistry 20:2361–2370
53. Ryu S-E, Kwong PD, Truneh A, Porter TG, Arthos J, Rosenberg M, Dai X, Xuong N-H, Axel R, Sweet RW, Hendrickson WA (1990) Crystal-structure of an HIV-binding recombinant fragment of human CD4. Nature 348:419–426
54. Holmgren A, Branden CI (1989) Crystal-structure of chaperone protein PapD reveals an immunoglobulin fold. Nature 342:248–251
55. Leahy DJ, Hendrickson WA, Aukhil I, Erickson HP (1992) Structure of a fibronectin type-III domain from tenascin phased by MAD analysis of the selenomethionyl protein. Science 258:987–991
56. DeVos AM, Ultsch M, Kossakoff AA (1992) Human growth-hormone and extracellular domain of its receptor—crystal-structure of the complex. Science 255:306–312
57. Bork P, Doolittle RF (1992) Proposed acquisition of an animal protein domain by bacteria. Proc Natl Acad Sci USA 89:8990–8994
58. Schonlau F, Kohnlein W, Garnett MC (1994) The acidic groups of the neocarzinostatin protein play an important role in its biological activity. Mol Pharmacol 45:1268–1272
59. Anantha Samy TA (1977) Neocarzinostatin: effect of modification of side chain amino and carboxyl groups on chemical and biological properties. Biochemistry 16:5573–5578
60. Edo K, Saito K, Matsuda Y, Akiyama-Murai Y, Mizugaki M, Koide Y, Ishida N (1991) Neocarzinostatin: selective Trp oxidation and neocarzinostatin chromophore binding to apo-neocarzinostatin. Chem & Pharm Bull (Tokyo) 39:170–176
61. Myers AG, Harrington PM, Kwon B-M (1992) Evidence for aminoglycoside participation in thiol activation of neocarzinostatin chromophore synthesis and reactivity of the epoxy dienediyne core. J Am Chem Soc 114:1086–1087
62. Burley SK, Petsko GA (1985) Aromatic-aromatic interaction: a mechanism of protein structure stabilization. Science 229:23–28
63. Dougherty DA, Stauffer DA (1990) Acetylcholine binding by a synthetic receptor—implications for biological recognition. Science 250:1558–1560

64. Sussman JL, Harel M, Frolow F, Oefner C, Goldman A, Toker L, Silman I (1991) Atomic structure of acetylcholinesterase from *Torpedo califronica*: a prototypic acetylcholine-binding protein. Science 253:872–879
65. Rodham DA, Suzuki S, Suenram RD, Lovas FJ, Dasgupta S, Goddard WA III, Blake GA (1993) Hydrogen bonding in the benzene-ammonia dimer. Nature 362:735–737
66. Levitt M, Perutz MF (1988) Aromatic rings act as hydrogen bond acceptors. J Mol Biol 201:751–754
67. Lee B, Richards FM (1971) The interpretation of protein structures: estimation of static accessibility. J Mol Biol 55:379–400
68. Chothia C, Lesk AM (1986) The relation between the divergence of sequence and structure in proteins. EMBO J 5:823–826
69. Kabsch W, Sander C (1983) Dictionary of protein secondary structure—pattern-recognition of hydrogen-bonded and geometrical features. Biopolymers 22:2577–2637
70. Kraulis PJ (1991) Molscript—a program to produce both detailed and schematic plots of protein structures. J Appl Crystallogr 24:946–950
71. Anantha Samy TS, Hahm K-S, Modesr EHJ, Lampman GW, Keutmann HT, Umezawa H, Herlihy WC, Gibson BW, Carr SA, Biemann K (1983) Primary structure of macromomycin, an antitumor antibiotic protein. J Biol Chem 258:183–191
72. Hofstead SJ, Matson JA, Malacko AR, Marquardt H (1992) Kedarcidin, a new chromoprotein antitumor antibiotic. II. Isolation, purification and physico-chemical properties. J Antibiot (Tokyo) 45:1250–1254

The Biochemical Mechanisms Involved in the Biological Effects Induced by Neocarzinostatin (NCS) and the NCS Chromophore

KENZO OHTSUKI[1] and YASUSHI ONO[2]

Summary. Treatment of bacterial and mammalian cells with neocarzinostatin (NCS) results in selective inhibition of DNA synthesis and induction of DNA degradation. Before the discovery of the NCS chromophore (NCS-chrom), NCS was defined as an antitumor antibiotic composed of an acidic polypeptide (approximately 11000; pI 3.3). In 1990, a NCS-chrom associated with the NCS polypeptide (NCS-apoprotein) was discovered and identified as the molecule responsible for the biological effects of NCS. In 1993, the research group of Goldberg (Harvard University, USA) proposed a biochemical mechanism of NCS-chrom DNA degradation in vitro. In 1995, they also found the base-catalyzed (bc) cleavage of single-stranded DNA (ssDNA) with a bulge structure by NCS-chrom in the absence of thiols. Therefore, it is concluded that the ability of NCS-chrom to degrade DNA is its primary action, which selectively inhibits DNA synthesis in cultured cells. In addition, NCS-chrom induces blocking of the metaphase of the cell cycle and also inhibits protein phosphorylation involved in transcriptional regulation. Disruption of the repair systems by NCS-chrom results in the positive induction of apoptosis, because NCS-chrom (i) is greatly activated by thiols at the intracellular level and (ii) inhibits activation of transcriptional factors through their specific phosphorylation. Taken together, all these biological and biochemical data suggest that NCS-chrom could be an effective chemotherapeutic drug for human cancer if its toxicity can be appropriately controlled.

Key words: Neocarzinostatin—Neocarzinostatin chromophore—Selective inhibition of DNA synthesis—DNA degradation—Site-specific cleavage of ssDNA—Inhibition of protein kinase activity—G_2 block—Apoptosis

[1] Laboratory of Genetical Biochemistry, Kitasato University, School of Allied Health Sciences, 1-15-1 Kitasato, Sagamihara, 228 Japan
[2] Department of Microbiology, Nihon University School of Medicine, 30-1 Kamicho, Ohyaguchi, Itabashi-ku, Tokyo, 173 Japan

First Test of the Biological Activities of Neocarzinostatin In Vitro

In 1965, Ishida and his co-workers found a new antitumor antibiotic in a culture filtrate of *Streptomyces carzinostaticus* F-41, which was isolated from soil in Sendai, Japan. This antibiotic, named neocarzinostatin (NCS), was purified from the culture filtrate to an apparent homogenous acidic polypeptide (pI 3.3) with a molecular weight of approximately 11000 [1,2]. The antimicrobial activity of NCS was tested by the agar disk method. *Sarcina lutea* (now named *Micrococcus lutea*), a gram-positive organism, was found to be one of the most sensitive bacteria to NCS. However, no significant growth inhibition of a number of gram-negative bacteria by NCS was observed at doses greater than 1 μg/ml.

In mammalian cells, Kumagai et al. [3] originally demonstrated the antiproliferative effect of NCS in cultured HeLa S3 cells. Treatment of HeLa S3 cells with NCS at 0.2 μg/ml caused a significant inhibition of cell growth, and 50% inhibition of the cell growth was obtained at 1.0 μg/ml NCS (Fig. 1). In addition, microscopic examination revealed that exposure of HeLa S3 cells to NCS (5 μg/ml) for 48 h resulted in induction of morphological changes, such as formation of multipolar mitosis, chromosome bridges, and scattering of chromosomes. The effect of NCS on the cell cycle of HeLa S3 cells has also been studied quantitatively by measurement of the percentage of metaphase in cells previously labeled with [^{3}H]-thymidine. It was found that treatment of HeLa S3 cells with NCS (0.1–0.5 μg/ml) results in delayed entry into metaphase G_2.

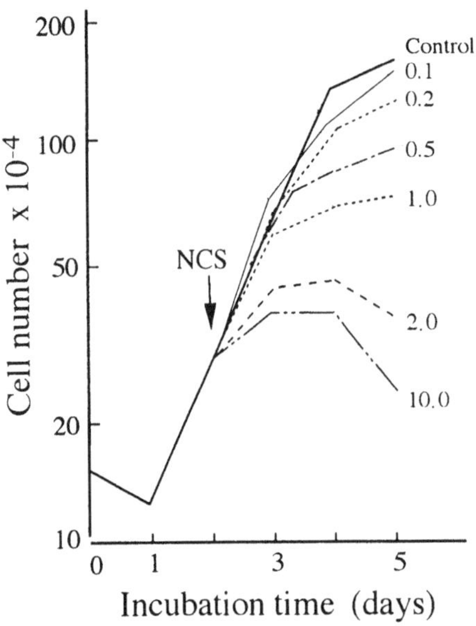

Fig. 1. Effect of neocarzinostatin (*NCS*) on the growth of HeLa S3 cells [3]. The cells were grown on a coverslip in a CO_2 incubator at 37°C. At 48 h after cell seeding, NCS at the indicated doses (μg/ml) was added. (From [3], with permission)

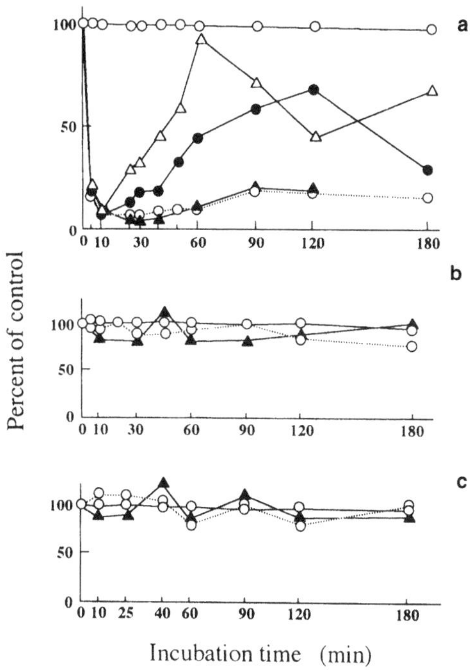

FIG. 2a–c. Effect of NCS on the incorporation of three radioactive precursors ([^{14}C]thymidine, [^{14}C]uridine, and [^{14}C]leucine) into the synthesis of macromolecules [DNA (**a**), RNA (**b**), and protein (**c**)] in *Sarcina lutea* [4]. NCS at the indicated doses was added to the bacterial suspensions at time 0. *Open circles* and *solid line*, control (100%), absence of NCS; *open triangles* and *solid line*, 0.1 μg/ml NCS; *solid circles* and *solid line*, 0.2 μg/ml; *solid triangles* and *solid line*, 0.5 μg/ml; *open circles* and *dotted line*, 2.0 μg/ml; *open triangles* and *dotted line*, 10 μg/ml. (From [4], with permission)

These results suggested that low levels of NCS preferentially block the premitotic phase of the HeLa S3 cell cycle [3].

Selective Inhibition of DNA Synthesis by NCS

In 1996, Ono and his co-workers studied the primary action of NCS by determining its effect on the incorporation of three distinct radioactive precursors ([^{14}C]thymidine, [^{14}C]uridine, and [^{14}C]leucine) during the synthesis of macromolecules (DNA, RNA, and protein), using a NCS-sensitive bacterium, *Sarcina lutea* (Fig. 2) [4]. It was found that NCS selectively inhibits DNA synthesis, but no effect on the synthesis of RNA and protein was detected. A similar inhibitory effect of NCS on DNA synthesis was also observed in HeLa S3 cells [5,6]. From these results, it was concluded that the primary action of NCS involves the selective inhibition of DNA synthesis in vivo. In this bacterial system, it should be noted that the NCS-induced inhibition of DNA synthesis in the cells treated with low doses (0.1–0.2 μg/ml) is recovered after further incubation in the absence of NCS (Fig. 2a). A similar observation was also observed in mammalian cells [5].

DNA Degradation Induced by NCS In Vivo and Vitro

Ono (Nihon University, Tokyo) and his co-workers originally reported that treatment of *Sarcina lutea* with NCS induces DNA degradation [producing trichloroacetic acid-(TCA-) soluble fragments from the TCA-insoluble DNA fragments] [4,7]. This NCS-induced DNA degradation was prevented by the addition of inhibitors of protein synthesis such as puromycin (Fig. 3). A similar DNA degradation by NCS was observed in other gram-positive bacteria (*Bacillus subtilis*) [8]. These observations suggested that the newly synthesized proteins are required for the induction of DNA degradation caused by NCS. Moreover, double-labeling experiments using *B. subtilis* showed that NCS preferentially causes DNA degradation at a region near the growing point of DNA replication (Table 1) [8]. Using an 5%–25% alkaline sucrose density gradient method, it was found that the NCS-induced single-strand DNA (ssDNA) breaks in *B. subtilis* are detectable within 30 min and also that (i) the damaged DNA is repaired by further incubation after the removal of NCS; and (ii) ssDNA is broken down to

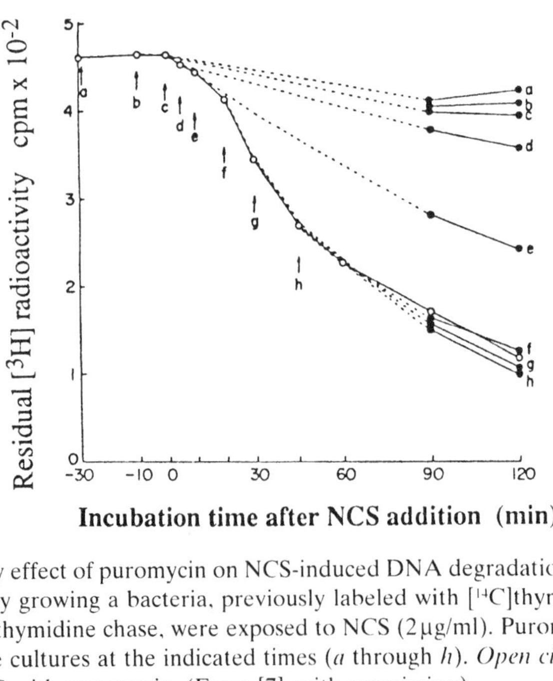

FIG. 3. Inhibitory effect of puromycin on NCS-induced DNA degradation in *Sarcina lutea* [7]. Exponentially growing a bacteria, previously labeled with [^{14}C]thymidine for 2 h followed by a cold thymidine chase, were exposed to NCS (2 μg/ml). Puromycin (200 μg/ml) was added to the cultures at the indicated times (*a* through *h*). *Open circles*, NCS alone; *solid circles*, NCS with puromycin. (From [7], with permission)

TABLE 1. Neocarzinostatin-(NCS-) induced DNA degradation in double-labeled cells (From [8], with permission)

Incubation time (min)	Radioactivity of the TCA-insoluble fraction (%)			
	Control (NCS-untreated)		NCS-treated	
	^{3}H	^{14}C	^{3}H	^{14}C
0	100	100	100	100
30	104	107	61.9	85.7
60	97.1	97.6	20.6	81.5
90	97.0	97.2	12.6	76.3
120	90.0	95.4	10.7	69.7

TCA, Tricholoroacetic acid.
Bacillus subtilus was first labeled with [^{14}C]thymidine for 5 min at 30°C. After chasing with cold thymidine for 15 min, the microorganisms were labeled again with [^{3}H]thymidine for a further 5 min [8]. The double-labeled cells were exposed to NCS (5 μg/ml) for the incubation periods. Radioactivity of [^{14}C] and [^{3}H] remaining in the TCA-insoluble fractions in cells treated or not treated with NCS was determined.

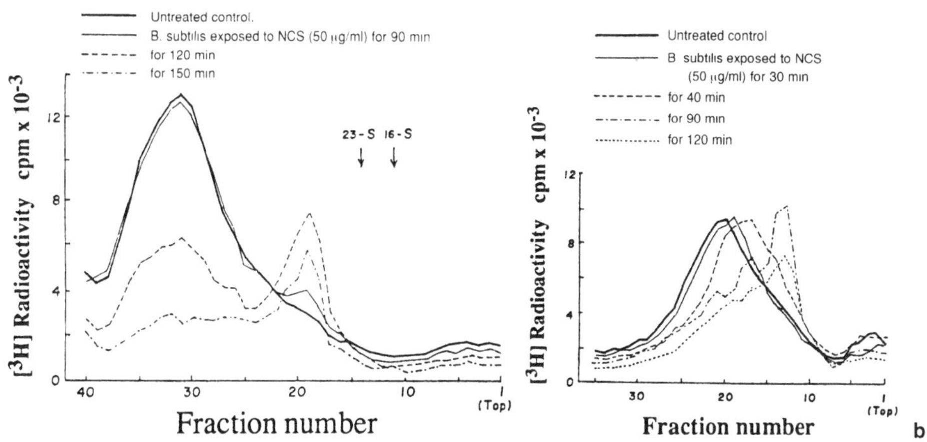

FIG. 4a,b. Neutral (5%–25%) (**a**) and alkaline (**b**) sedimentation ananlysis (105 000 × *g* for 2 h at 18°C of [^{3}H]-labeled DNA extracted from *Bacillus subtilis* after exposure to NCS for the indicated periods. (From [8], with permission)

the size of the replication unit before parental DNA is degraded to the same size (Fig. 4) [8].

Similar degradation of DNA by NCS and repair of the NCS-induced ssDNA breaks were also observed in HeLa S3 [5] and BHK (baby hamster kidney) cells [9]. In 1979, Ishida, R. et al. (Aichi Cancer Research Institute, Nagoya) demonstrated that (i) DNA strand scission takes place at doses of NCS even lower than those which induce G_2 arrest of BHK cells; and (ii) the NCS-induced ssDNA breaks in the cells are repaired by further incubation in the absence of the drug [9]. From these experimental data, it was speculated that DNA degradation is a primary action of NCS at the cellular level.

In 1974, Goldberg's group (Harvard University, USA) [10] and Tatsumi et al. (Kyoto University) [11] independently reported that NCS induces DNA degradation in vitro and that the NCS-induced DNA degradation at cellular level is greatly stimulated in the presence of 2-mercaptoethanol (2-ME). A similar stimulatory effect of 2-ME on NCS-induced DNA degradation was also observed in vitro [12]. By analytical experiments, acid-soluble products from DNA incubated with NCS in the presence of 2-ME in vitro were identified as thymine and oligonucleotide, but no release of adenine or guanine from DNA treated with NCS was detected [13]. All these observations in vitro and in vivo were very important for understanding the primary action of NCS on DNA degradation, and they suggested the possible existence of a nonprotein component (NCS-chrom) responsible for DNA degradation in vitro as a molecule associated with NCS apoprotein.

Before the discovery of NCS-chrom [14], it was difficult to explain why NCS induced DNA degradation in vitro and selectively inhibited DNA synthesis in cultured bacterial and mammalian cells, because NCS was defined as an acidic single polypeptide (pI 3.3) with a molecular weight of approximately 11000: (i) the primary sequence of NCS did not correspond to any DNases; (ii) NCS did not inhibit the activity of DNA polymerases in vitro; and (iii) all biological activities of NCS were rapidly inactivated by UV irradiation or alkaline pH treatment. Determination of the full chemical structure of NCS-chrom isolated from native NCS by Edo et al. (Fukushima Medical College, Fukushima) [15–17] quickly led to a clear explanation of the biochemical mechanism of the NCS-induced DNA degradation in vitro.

NCS-Induced G_2 Block

In 1975, the effects of NCS on cell division and DNA synthesis in synchronized HeLa S3 cells were examined by using various methods to locate the point in the cell cycle at which the NCS-induced block occurred. An NCS-sensitive site in the cell cycle was detected early in S phase (Fig. 5a) and another was observed at the G_2 phase (Fig. 5b), which is almost 1 h ahead of the sensitive site for vinblastin sulfate [18]. The site of the block on the G_2 phase by NCS is different from that of NCS-induced inhibition of DNA synthesis in HeLa S3 cells. These two NCS-sensitive sites are similar to those observed with phleomycin and X-ray irradiation. However, the inhibitory action of NCS is different from that reported for phleomycin, because NCS does not directly inhibit the activity of DNA polymerases by its interaction with primer DNA in vitro. The X-ray-induced G_2 block is similar to that observed with

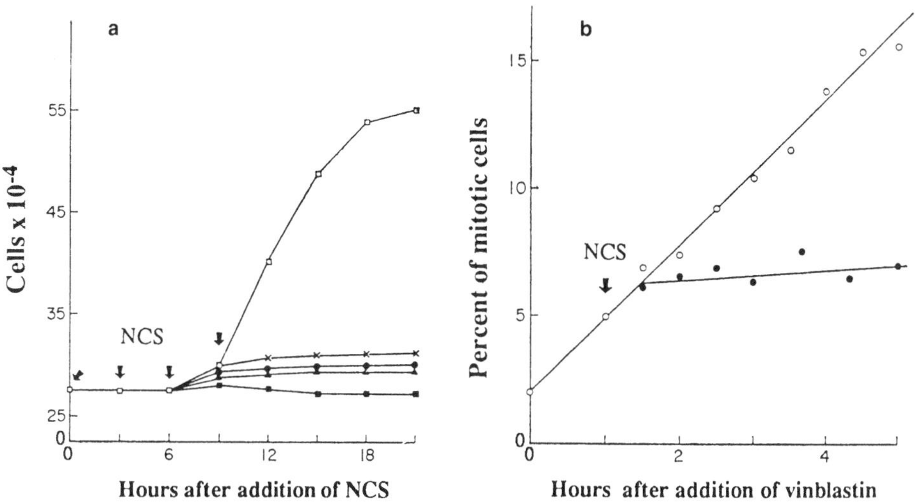

FIG. 5a,b. Effect of NCS on DNA synthesis and accumulation of mitotic cells in synchronized HeLa S3 cells. **a** HeLa S3 cells were synchronized by the double thymidine method and NCS (5 μg/ml) was added to the cell cultures at the times indicated by *arrows*. *Open circles*, time 0; *solid circles*, 2 h; *open squares*, 4 h, X symbols, 6 h. **b** Random cultures of HeLa S3 cells were exposed to NCS-chrom (5 μg/ml) (*arrow*) 1 h after treatment with 0.5 μg/ml vinblastin sulfate. At the times indicated, cultured cells were removed and stained with May–Grunwald–Giemsa solution and measured for mitotic frequency. *Open circles*, control; *solid circles*, NCS-chrom; *triangles*, NCS-chrom with 2-ME (2-mercaptoethanol). (From [6], with permission)

NCS. The G_2 block by NCS is observed at low doses relative to its inhibitory effect on the S phase.

There is a possibility that the NCS-caused inhibition of the activity of microtubulin-associated protein kinase may be involved in the G_2 block in the mammalian cell cycle. This possibility is supported by our observations (Table 2) that (i) NCS as well as NCS-chrom inhibits protein phosphorylation by the protein kinase, but no effect of NCS apoprotein on the kinase activity is detected; and (ii) no significant inhibition of protein phosphorylation by the kinases is observed with other tested G_2 blockers, including mitotic inhibitors such as colcemid and vinblastin sulfate, even at concentrations tenfold higher than that of NCS-chrom [19]. In addition, kinetic and analytical experiments showed that the direct binding of NCS-chrom to the kinase results in the inhibition of its activity in vitro. These results suggested that (i) inhibition of protein kinase activity by NCS-chrom is involved in the arrest of microtubulin assembly, which plays an important role in the biochemical progression of G_2 phase; and (ii) the NCS-chrom-induced G_2 block is another effect of NCS-chrom, which is quite different from NCS-induced DNA degradation and is also different from the mechanism of actions of other G_2 blockers. Moreover, NCS-chrom also inhibits protein phosphorylation by nuclear cyclic adenosine monophosphate-(cAMP-) indepen-

TABLE 2. The effect of G_2 blockers on the activity of microtubule-associated protein kinase in vitro (From [20], with permission)

G_2 Blockers	Concentration (μg/ml)	Inhibition (%)
None (absence of drugs)	—	0.0
G_2-block antibiotics		
Native NCS	1	21.3
	10	60.8
NCS-chrom	1	78.2
	10	98.2
Bleomycin	1	5.4
	10	12.5
Mitomycin C	1	2.3
	10	18.8
Adriamycin	1	8.3
	10	19.6
Cytochalasin B	1	6.5
	10	25.0
Actinomycin D	1	9.6
	10	15.0
Puromycin	1	4.8
	10	15.6
Mitotic inhibitors		
Vinblastin sulfate	1	0.1
	10	1.8
Colcemid	1	3.1
	10	17.5

Microtubulin-associated proteins were highly purified from rabbit brain. The kinase activity was assayed for 10 min at 37°C, using histone H2A as a phosphate acceptor [26]; 100% activity of the kinase corresponds to the incorporation of 96.0 pmol of [γ-^{32}P]ATP into histone H2A.

dent protein kinases involved in transcriptional regulation in vitro [20]. Because a similar inhibitory effect of naphthalenecarboxylic acid (NA) of NCS-chrom on the activities of protein kinases (A-kinase, C-kinase, and growth factor receptor tyrosine kinases) in vitro is observed, NA may be a responsible site for the inhibitory effect of NCS-chrom against the activities of these kinases in vitro.

Biochemical Characterization of NCS-Induced DNA Degradation In Vitro

Before the discovery of NCS-chrom on the NCS molecule, it was believed that NCS is simply an acidic polypeptide that induces DNA degradation in vivo and in vitro, as described above. To characterize the mode of the NCS-induced DNA breaks in vitro, DNA fragments generated from DNA incubated with NCS in the

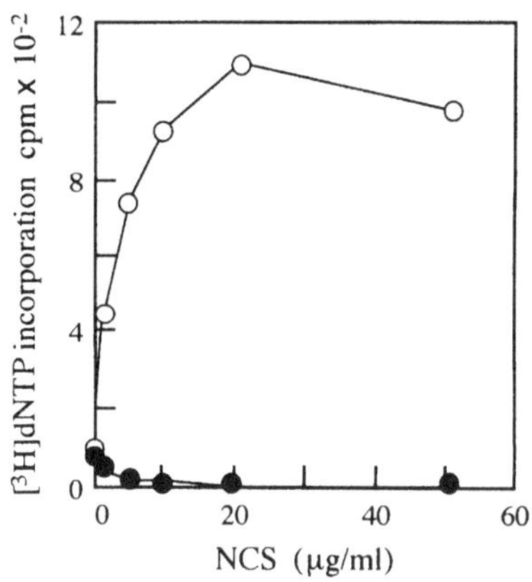

FIG. 6. The primer activity ([^{3}H]dTNP incorporation) of DNA nicked by NCS after treatment with bacterial alkaline phosphatase. Bacterial T4 phage DNA was incubated with NCS (10μg/ml) in the presence of 2 m*M* dithiothreitol (DTT); the nicked DNA in the solution was then precipitated with ethanol. The primer activity of the NCS-damaged DNA for DNA polymerase I was determined after treatment with bacterial alkaline phosphatase. *Solid circles*, Untreated with alkaline phosphatase; *open circles*, treated with alkaline phosphatase. *dNTP*, Deoxyribonucleoside 5′-triphosphate

presence of thiols were biochemically examined in vitro. It was found that NCS-induced DNA degradation in vitro is greatly stimulated under aerobic conditions or in the presence of thiols, such as 2-ME, dithiothreitol (DTT), and glutathione (GSH) [10,11]. These observations are consistent with DNA degradation induced by macromomycin (a macromolecule antibiotic similar to NCS). However, the mode of action of NCS is different from that reported for bleomycin and mitomycin C, which induce DNA degradation by their intercalation with DNA.

To characterize biochemically the cleaved sites on DNA fragments generated from DNA treated with NCS in the presence of thiols in vitro, the primer activity of the damaged DNA fragments for DNA polymerase I was determined before and after treatment with bacterial alkaline phosphatase. The primer activity of DNA fragments degraded by NCS was found to be extremely high (Fig. 6) [21]. However, DNA fragments from DNA treated with bleomycin had very low primer activity (Fig. 6) [21]. These results suggest that the cleaved sites of the DNA fragments generated after NCS treatment have 3′-phosphate with either 5′-hydroxyl or 5′-phosphate, but not 3′-hydroxyl with 5′-phosphate. The 3′-phosphate of the DNA fragments generated by NCS is different from that observed with bleomycin, but is similar to the 3′- and 5′-ends of DNA cleaved by X-ray irradiation. As NCS causes the release of thymidine from DNA, it is reasonable to find both phosphatidyl sites (3′-phosphate and 5′-phosphate) of DNA fragments damaged by NCS in vitro.

In 1978, Hatayama et al. (Kyoto College of Pharmacology) originally reported the nucleotide specificity in DNA damaged by NCS, using $\phi\chi174$ phage DNA restriction fragments [22]. By the DNA-sequencing technique of Maxam and Gilbert, it was found that deoxythymidylic acid residues are attacked much more frequently by NCS than deoxyadenylic acid residues and ssDNA ($\phi\chi174$ phage DNA fragment) is a very poor substrate for NCS. Moreover, Hatayama and Goldberg extensively studied the cleaved site for NCS on native DNA, and they reported that NCS in the presence of 2-ME in vitro causes sugar damage and forms the 5′-phosphate end of the DNA gap [23].

Mechanism of DNA Degradation Induced by NCS-Chrom In Vitro

Activation of NCS-Chrom by Thiols

In 1979, Kappen and Goldberg reported that various organic solvents (hexa-methylphosphoramide, isopropanol, butanol, and methanol) and urea significantly stimulate NCS-induced DNA degradation in vitro [24]. Other properties of NCS include a shoulder in its UV spectrum at 310–360 nm and inactivation of its biological activities by UV irradiation or by alkaline pH treatment [14]. These findings suggested the existence of a nonprotein component (NCS-chrom) associated with the NCS protein (NCS apoprotein) that is responsible for the biological effects of NCS. In 1980, Koide et al. (Kitasato University) succeed in extracting NCS-chrom from native NCS with 1 *N* HCl-MeOH (1 : 5) in the dark: it was then partially purified from the methanol extract by means of gel filtration on Sephadex G-25 followed by CM-cellulose column chromatography [14]. Independently, Goldberg's group, using a procedure similar to that reported by Koide et al. [14], also reported the isolation of a biologically active chromophore from native NCS [25].

After the discovery of NCS-chrom on the NCS molecule, a clear explanation of the selective inhibition of DNA synthesis by NCS was possible because (i) NCS-chrom induces DNA degradation in cultured HeLa S3 cells in a dose-dependent manner (Fig. 7a); (ii) the drug is greatly activated in the presence of 2-ME, and effectively cleaves T4 phage DNA in vitro (Fig. 7b); and (iii) no DNA degradation in HeLa S3 cells by NCS apoprotein, even at 500 μg/ml, is detected (Fig. 7a) [26]. In addition, Ohtsuki et al. confirmed that NCS-chrom inhibits protein phosphorylation by protein kinases in vitro [20] and the NCS-chrom-induced inhibition of protein kinase activity is involved in the G_2 block [19]. On the basis of these data, it was concluded that (i) NCS-chrom is the molecule responsible for all the biological activities observed with native NCS; (ii) NCS apoprotein is a carrier to protect NCS-chrom from inactivation of its biological properties by UV irradiation or alkaline pH treatment; and (iii) NCS-chrom exists in nature as a 1 : 1 complex (holo-NCS) with NCS apoprotein.

FIG. 7. **a** DNA degradation induced by NCS-chrom in growing HeLa S3 cells. HeLa S3 cells were previously labeled with [^{3}H]thymidine and then exposed to NCS-chrom (*open circles*) or NCS-apoprotein (*solid circles*) at the indicated doses. The [^{3}H] radioactivity in the trichloroacetic acid-(TCA-) soluble fractions was determined after incubation for 12 h at 37°C. **b** Alkaline (5%–25%) sedimentation analysis (105 000 × *g* for 4.5 h at 4°C) of [^{3}H]-labeled T4 phage DNA after incubation for 20 min at 37°C with NCS-chrom (5 μg/ml) in the presence (*solid circles*) or absence (*open triangles*) of 2 m*M* 2-ME. *Open circles*, control (absence of NCS-chrom). (From [19], with permission)

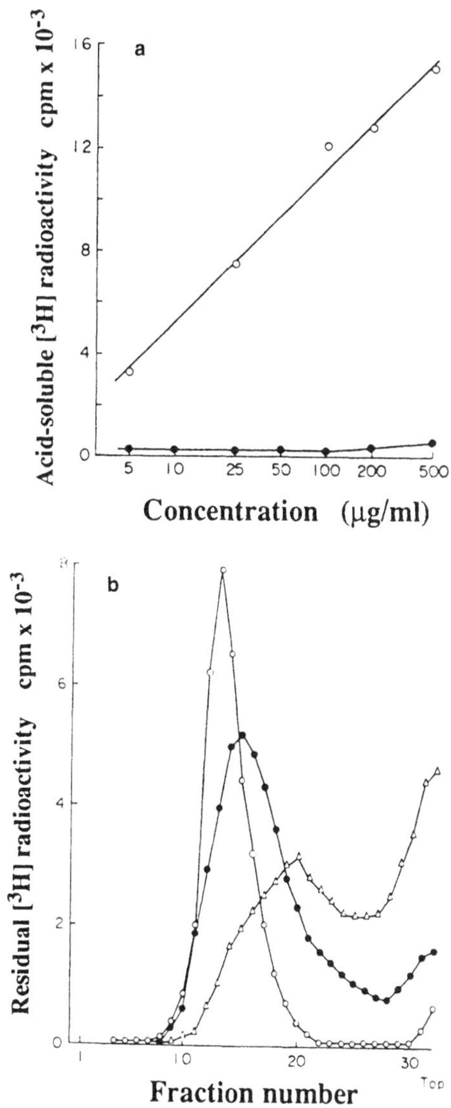

After the discovery of NCS-chrom on the native NCS molecule, research on the biological properties of NCS shifted to focus on the chemical structure of NCS-chrom and the biochemical mechanism involved in DNA degradation induced by NCS-chrom in vitro [21]. In 1992, Dedon and Goldberg established a quantitative assay to determine the stimulatory effect of various thiols on

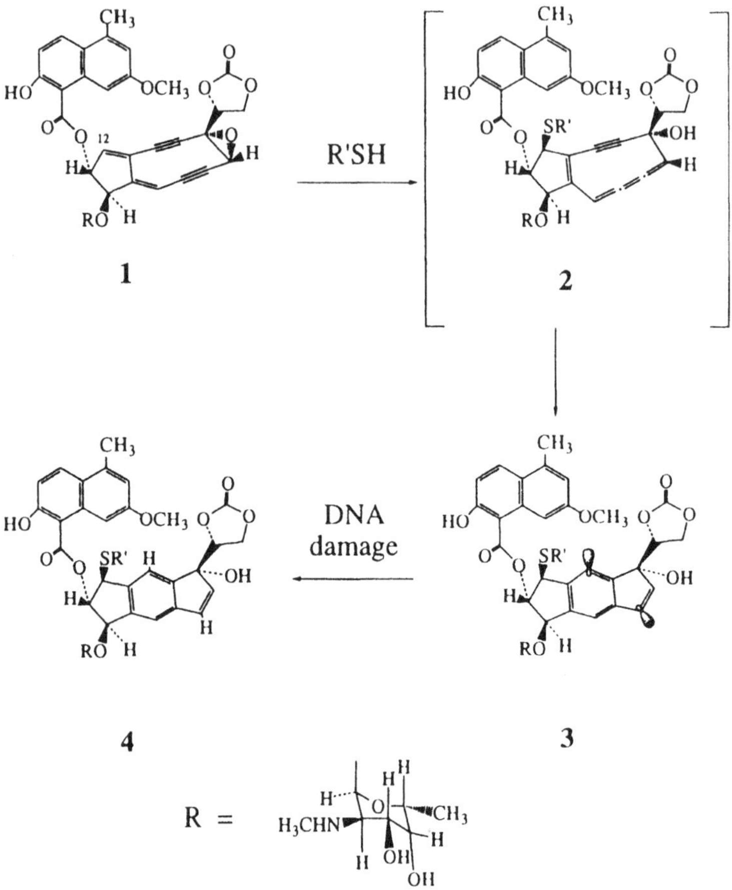

R' = $CH_2CH(CONHCH_2CO_2H)NHCOCH_2CH_2CH(NH_2)CO_2H$

FIG. 8. Structural representation of the activation (formation of a biradical species) of NCS-chrom by glutathione (GSH). (From [40], with permission)

NCS-chrom-mediated ssDNA and dsDNA (double-strand DNA) damages, using plasmid pBR322 as a substrate for the drug in vitro [27]. In the subsequent 15 years since the discovery of NCS-chrom as the molecule responsible for the biological activity of NCS, Goldberg's group in particular has made an extensive study on the biochemical mechanisms involving in DNA degradation induced by NCS-chrom in vitro, contributing greatly to the current understanding of the biochemistry of the activation mechanisms of NCS–NCS-chrom in the presence or absence of thiols [28].

In 1989, Mayers and Proteau (California Institute of Technology, USA) originally reported an important finding to explain the biochemical mechanism of the effective cleavages of ssDNA and dsDNA by NCS-chrom in the presence of GSH in vitro [29]. As shown in Fig. 8, the activation of NCS-chrom by GSH occurs through a nucleophilic attack at C-12, which results in the formation of the cumulene **2**. The labile cumulene **2** undergoes a spontaneous Bergman-like cyclization to the biradical **3**, which after abstraction of hydrogen atoms from deoxyribose of DNA becomes the stable aromatic postactivated form 4 (Fig. 8). The formation of a NCS-chrom radical is in a manner similar to that reported for other enediyne antibiotics, such as C-1027 [30], esperamicin A_1 [31], calicheamicin γ_1 [32], dynemicin A [33], and kedarcidin [34]. In 1991, both Frank's (Harvard University) [35] and Saito's (Kyoto University) [36] groups independently reported a biochemical mechanism involved in the DNA degradation induced by the activated NCS-chrom: sequence-specific hydrogen atom abstraction by the biradical from the C-5′, C-1′, or C-4′ positions of deoxyribose in the minor groove of dsDNA ultimately leads to strand cleavage (Fig. 9). After activation of NCS-chrom by thiols and molecular oxygen to form peroxyradicals, the carbon-centered radicals of NCS-chrom undergo degradation to unique sugar fragments, depending on the location of the abstracted hydrogen. At the C-4′ position, cleavage of the peroxyradical partitions forms a 4′-hydroxylated abasic site [36] or produces a strand break with a 3′-phosphoglycolate residue and a

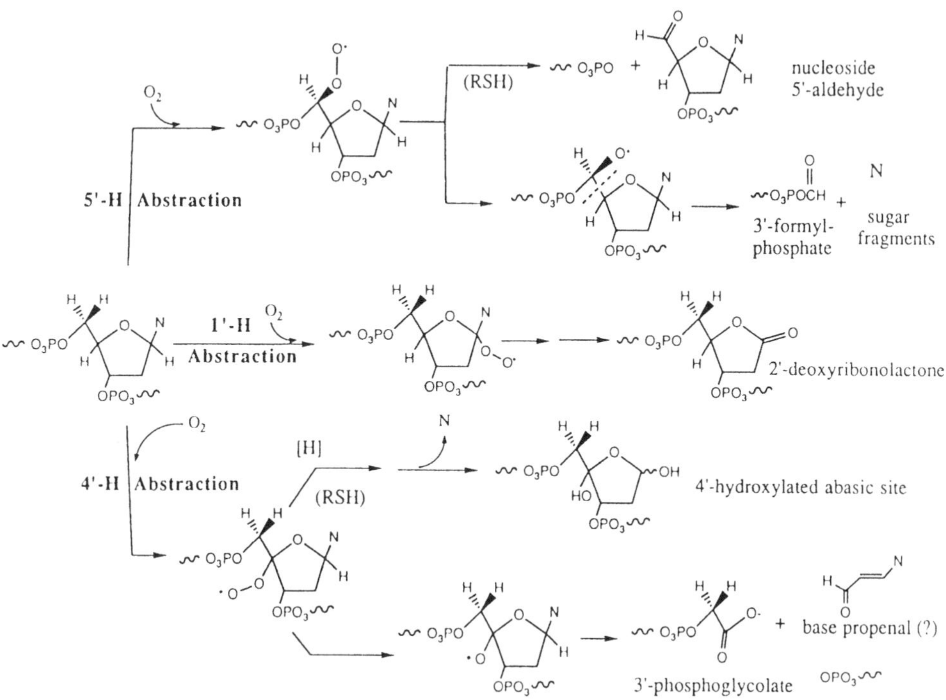

FIG. 9. Proposed mechanisms of DNA cleavage induced by NCS-chrom under aerobic conditions in the presence of RSH. (From [40], with permission)

5′-phosphate [37]. The strand break resulting from DNA degradation of the C-5′ peroxyradical involves either a nucleoside 5′-aldehyde at the 5′-end and a 3′-phosphate [38] or a labile 3′-formylphosphate-ended DNA fragment and a 5′-phosphate [39]. In addition, abstraction of the C-1′ hydrogen atom, the third hydrogen atom accessible from the minor groove, and formation of the peroxyradical results in the production of an alkali-label 2′-deoxyribonolactone abasic site [39]. The DNA cleavage by NCS-chrom is at least 1000 times greater in the presence of thiol than in its absence. In 1995, Stassinopoulos and Goldberg summarized all products generated from deoxyribose of DNA treated with NCS-chrom in vitro, and they also showed that the drug abstracts mainly the hydrogen atom from the C-5′ position on deoxyribose of dsDNA in the presence of thiols [40].

Base Specificity of DNA Cleavage by NCS-Chrom

The specific sites at which DNA is cleaved by NCS-chrom have been studied in detail by Goldberg's group, using various oligonucleotides as substrates for NCS-chrom in the presence of (RSH) in vitro [41]. When trinucleotides AGX · YCT (X = C or T and Y = G or A) were used as substrates for NCS-chrom in the presence of thiols, the drug induced sugar damage at the C-5′ or C-4′ position of the T residue and also induced damage at the C-1′ position of the X residue. Preferential T release from nucleotides containing AGT and TCA by NCS-chrom in the presence of thiols was observed, as shown long ago in studies by a group at the Aichi Cancer Center (Nagoya) [13].

Preferential release of T from these trinucleotides induced by NCS-chrom is an interesting observation, because the DNA sequence TGACTCA is a specific binding site for the DNA-binding transcriptional factor Ap1, which forms a complex with c-Jun and c-Fos [42,43]. It has been reported that the transcriptional activity of Ap1 is regulated by its specific phosphorylation by casein kinase II (CK-II) [44]. CK-II, a cAMP- and Ca^{2+}/phospholipid-independent nuclear protein kinase, is characterized as one of the key protein kinases responsible for the transcriptional regulation through specific phosphorylation of DNA-binding transcriptional factors [Sp1, serum response factor (SRF) and upstream binding factor (UBF)] by the kinase [45,46]. Our recent in vitro experiments revealed that NCS-chrom inhibits phosphorylation of these transcriptional factors by CK-II in a dose-dependent manner, as shown in the inhibition of protein phosphorylation by microtubulin-associated kinase (see Table 2). These indirect data suggest that NCS-chrom directly inhibits the expression of specific genes that are regulated by Ap1 by the NCS-chrom-induced impairment of the specific binding site for the transcriptional factor, and that the drug also prevents phosphorylation of some transcriptional factors, including Ap1, by CK-II at the cellular level.

Goldberg's group revealed the frequency of bases attacked by four enediyne antibiotics [NCS-chrom (T > A >> C > G), C-1027 (A ≧ T >>C > G), calicheamicin

γ_1 (C >> T > A ≧ G), and esperamicin A_1 (T > C > A > G)] and bleomycin (C > T > A > G) [47]. The base specificity of DNA attacked by NCS-chrom is similar to that of C-1027: the major damage caused by these two chromophores appears to be strand scission, caused by hydrogen abstraction from the C-4′ position of the deoxyribose of T and A. Analysis by agarose gel electrophoresis using pBR322 DNA as a substrate showed that the ratio of double-strand to single-strand breaks induced by C1027 is about 1:4, which is slightly higher than that obtained with NCS-chrom in the presence of GSH as a thiol activator (1:5) [47].

In 1995, Gao and his co-workers (University of Houston, USA) used nuclear magnetic resonance (NMR) spectroscopic analysis to demonstrate direct evidence for the formation of a complex between postactivated NCS-chrom [NCS-glutathione (GSH postactivated NCS-chrom)] and DNA fragment (d[GGAGCGC]·[GCGCTCC]) [48]. NCS-glutathione interacts with the GCTC tetranucleotides on one strand and with the AGC trinucleotides on the other strand through the unique interaction at the 5′-CT/5′-AG step in the minor groove of the DNA fragment [49]. These analytical observations suggest that the ss- and ds clavages by the GSH-activated NCS-chrom specifically occur at the AGC and its related sites.

Site-Specific Cleavage of DNA with Bulge Structure by NCS-Chrom

In 1993, Kappen and Goldberg first demonstrated that, in the absence of thiols, NCS-chrom induces a site-specific cleavage at the 3′-side of ssDNA containing a bulge structure [50]. Because ssDNA is a very poor substrate for NCS in the absence of thiols [22], ssDNA containing a bulge structure is prepared as the minimum structural requirements of DNA substrates for NCS-chrom. A series of 31-based DNA substrates, which exactly correspond to the 3′-terminals of yeast tRNA ($tRNA^{Phe}$) and their deletions, including double-point mutations, are prepared. These DNA substrates for NCS-chrom can be folded into a hairpin structure with an apical loop and a 2-base-pair stem hinged to a longer base-paired region by a bulge containing T_{22} on the 3′-side [50]. Experimental results show that the NCS-chrom-induced DNA cleavage in the absence of thiols occurs exclusively by oxidation at C-5′ of deoxyribose at T_{22}. However, no DNA cleavage is detected even at high concentrations (50 μ*M*) in the absence of thiols when the DNA substrate is attacked with calicheamicin γ_1 or esperamine A_1. In addition, there is no cleavage by NCS-chrom when a $T_{23}C$ mutation is used as a substrate. By contrast, about 60% of substrate containing a $T_{22}C$ mutation (site within the bulge) is cleaved by NCS-chrom. These data clearly suggest that NCS-chrom induces a site-specific cleavage at the 3′-side of a bulge in ssDNA in the absence of thiols.

Specific Cleavage of RNA · DNA Hybrids and Transcriptional Response Element DNA by NCS-Chrom

In 1995, Kappen and Goldberg revealed that NCS-chrom, in the absence of thiols, cleaves at the 3′-side of the bulge in ssDNA that is capable of folding into a hairpin structure [51]. A similar cleavage occurs when transactivation response element (TAR) DNA is used as a substrate. It is therefore interesting to examine the effect of NCS-chrom attack on RNA · DNA hybrids and on RNA with a bulge structure. For these experiments, a series of shorter oligoribonucleotides, containing the target sequence r(AGAAUUC)·(GAATTCT), was used as a substrate. As expected, it was found that major damage occurs in the form of an abasic site at the U residues and induces U-base release via C-1′ hydrogen abstraction of the oligonucleotides. In contrast, NCS-chrom cleaves at the T residues on the complementary DNA strand involved with C-5′ hydrogen abstraction. Formation of this abasic site on the RNA strand was confirmed by sequencing gel analysis and measurement of uracil base release [52]. The site containing the substitution does not function as a substrate for cleavage or abasic site formation by NCS-chrom. A similar RNA lesion generated on the RNA by C-1′ hydrogen abstraction with bleomycin has also been reported, but bleomycin does not cleave the RNA strand of a RNA · DNA hybrid [53].

The Harvard group further investigated the thiol-independent site-specifc cleavage of ssDNA by NCS-chrom at a signal site on the 3′-side of a bulge in the DNA [51]. For this experiment, 29-mer and 27-mer oligoribonucleotides, corresponding to the Tat (transactivator) binding protein of the 59-residue TAR of human immunodecifiency virus 1 (HIV-1), were prepared as potential substrates for NCS-chrom. To determine the sequence specificity of the site on TAR DNA, the DNA substrate was substituted with a G at the position of T_{25} and at T_{23} with an A. The results showed that (i) a 29-mer TAR DNA acts as a substrate for NCS-chrom, which cleaves selectively at T residues at T_{24} and T_{25} in the bulges, which are much more susceptible than at T_{23}, and (ii) lesser strand cleavage also occurs at the bases (between A_{22} and G_{26}) flanking the bulge (Fig. 10).

In 1994, Hensens et al. (Merck Research Laboratory, USA) originally reported a novel mode of the thiol-independent cleavage of ssDNA with bulge structures by NCS-chrom and a spontanous conversion of the drug into a stable compound at alkaline pH conditions [54]. In 1995, Stassinopoulos and Goldberg studied the mechanism in detail of the base-catalyzed (bc) bulge-specific reaction by NCS-chrom in the absence of thiols [40] (Fig. 11). During the activation, instead of the attack by thiol group at C-12, an intramolecular Michael addition of the phenol enolate at C1″ to C-12 occurs and forms the cumulene **2**, which undergoes a Bergman-like rearrangement to the biradical **3**. This intermediate is responsible for abstraction of the hydrogen atom from the C-5′ position of the deoxyribose residue (see Fig. 9), and this biradical **3** attacks at the C-8″ of the naphthoate group. The biradical **3** transforms into the aromatic compound **5**, through the proposed intermediary of the radical **3a** and compound **3b**. Finally, it is proposed that there are two mechanistic pathways (thiol-dependent and thiol-independent) for formation of a biradical species of NCS-chrom (Fig. 12).

FIG. 10a–g. DNA substrates for NCS-chrom. (**a**) Transactivation response element (TAR) DNA, (**b**) DNA analog of TAR RNA, and (**c**) bulged DNA duplex made up of two linear oligomers (a 13-mer and a 10-mer), and effect of sequence changes on the NCS-chrom-induced cleavage of DNA with bulge structures (**c** through **g**) in the absence of thiols. *Arrows* indicate the sites of cleavage for NCS-chrom. (From [51], with permission)

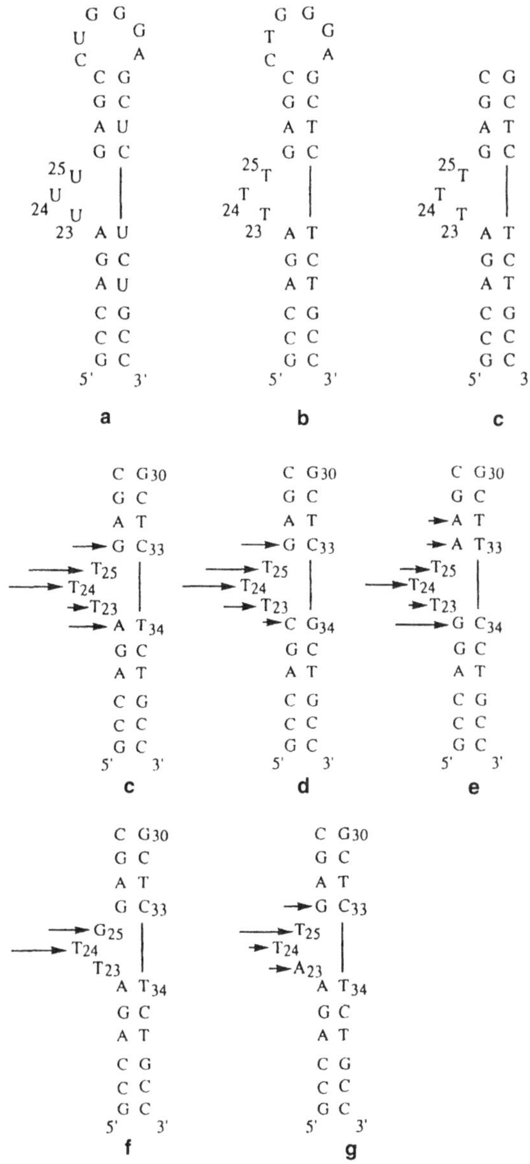

In the absence of thiols using long native ssDNA substrates, such as oligonucleotides, $\phi\chi174$ phage (approximately 5000 bases) and M13mp18 phage (approximately 75000 bases) DNAs, and long ssDNA fragments (150–450 bases), Stassinopoulos and Goldberg revealed that (i) NCS-chrom cleaves $\phi\chi174$ and M13mp18 phage DNAs in pH- and dose-dependent manner; (ii) NCS-chrom also cleaves ssDNA fragments with bulge structures at T-rich non-ds structural sequences with minimal energy secondary structures; and (iii) the minimal

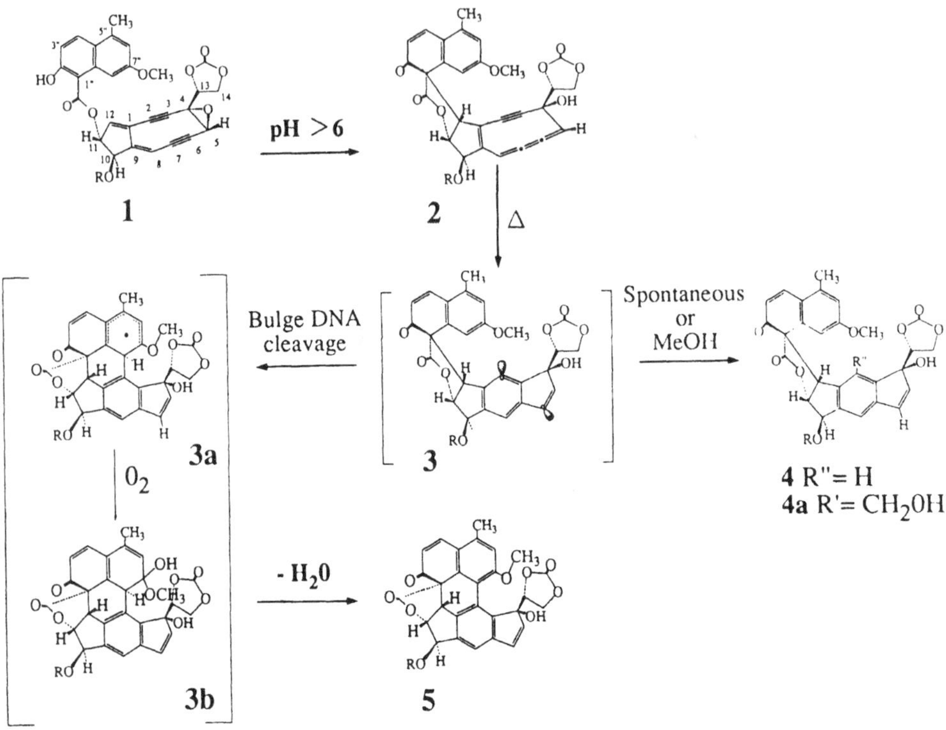

FIG. 11. Proposed thiol-independent mechanism (*1–8*) for the intramolecular activation (formation of a biradical species) of NCS-chrom at basic pH in the absence or presence of bulged DNA. (From [40], with permission)

sequence requirement for the cleaved site by NCS-chrom is a 19-mer oligonucleotide (sequence 3′TACTGAGTCTCCT TTGTA5′; attacked residue in bold) [40]. In addition, a combination of thermodynamic and kinetic study confirmed that NCS-chrom has a low binding of the cleaving species to RNAs.

From the biochemical characteristics of the NCS-chrom-induced DNA damage, it is speculated that the most sensitive target DNA for NCS-chrom may be an initiation site, particularly an autonomously replicating sequence [ARS; 5′-(T/A)TTA(A/G)TTT(T/A)-3′], in the replication origin of bacterial and mammalian cells [55,56], because the ARS contains an AT-rich region with an apical loop structure. This speculation is consistent with our earlier observation that DNA of the replication origin is preferentially degradated in synchronized HeLa S3 cells [18]. Although there are no reports demonstrating the substrate activity of the ARS region for NCS-chrom at present, there is a possibility that the NCS-chrom-induced preferential cleavage of the replication origins of some DNA viruses (SV-40, adenoviruses, herpes simplex, and vaccinia viruses) results in the inhibition of these viruses.

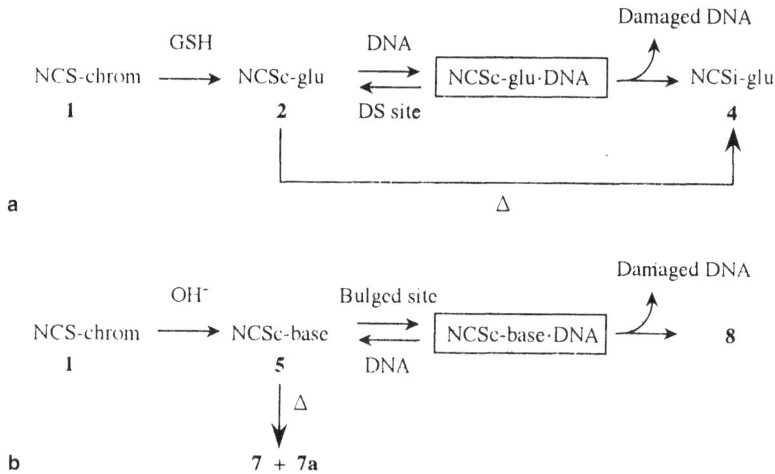

FIG. 12a,b. Reaction schemes for (**a**) the activation of NCS-chrom by GSH resulting in DNA cleavage and (**b**) the base-catalyzed (bc) activation of NCS-chrom resulting in bulge-specific DNA cleavage. *NCSc-glu*, GSH-generated cumulene (**2** in Fig. 8); *NCSc-base*, base-generated cumulene (**5** in Fig. 11); *NCSi-glu*, glutathione postactivated NCS-chrom; *DS site*, dsDNA cleavage site; Δ, thermodynamically spontaneous reaction. (From [40], with permission)

Other Biological Effects of NCS-Chrom: The Redox Reaction and Apoptosis

Stimulatory Effect of NCS-Chrom on the Redox Reaction

During electron transport to molecular oxygen via the mitochondrial chain, as well as in various hydroxylation and oxygenation reactions in aerobic cells, toxic partial reduction products of oxygen, such as superoxide [$O_2 \cdot^-$] may be formed as transient intermediates on the active sites of certain enzymes. The most important are the superoxide anion (O_2^-) and hydroxyl radical (·OH) radicals, which are extremely reactive and capable of irreversible damages to various biomolecules, including DNA, RNA, and proteins, at the intracellular level. Generally, cells contain an enzyme, superoxide dismutase, that converts superoxide into hydrogen peroxide, which is processed to H_2O and molecular oxygen by catalase. Under certain stress conditions, such as UV irradiation, presence of carcinogens, and virus infection, the formation of superoxide anion (O_2^-) and hydroxyl radicals is greatly elevated and results in the induction of cellular damage and DNA degradation, which are serious threats to the continuing viability of any kind of cells [57,58] including virus-infected cells [59,60].

Usually, ssDNA damage is rapidly repaired by a SOS system, which requires several proteins to be newly synthesized after external signals, as shown by treating cells with antitumor antibiotics that directly cause DNA degradation

[61]. In the case of NCS-chrom, it will be immediately converted into an active (biradical) form, which effectively damages not only dsDNA but also cleaves ssDNA with bulge structures after reaction with thiols such as GSH and cysteine at the intracellular level. At this point, the NCS-chrom molecule acts as an effective DNA scissor, which is different from the mechanical actions of other DNA-degrading antibiotics such as bleomycin and mitomycin C. Moreover, the research group of Maeda (Kumamoto University School of Medicine, Kumamoto) recently reported that (i) NCS-chrom produces O_2^- in the presence or absence of RSH in vitro and (ii) the ability of NCS-chrom to degrade DNA is greatly enhanced in the NADPH/cytochrome P-450 reductase assay system [62]. These results suggest that the ubiquitous cellular enzyme, NADPH/cytochrome P-450 reductase, is involved in the activation of NCS-chrom. This finding is very important to explain the far more potent cytotoxic activity of NCS-chrom against various cancer cells as compared with that observed in the cell-free system. This account is described in Chapter of this monograph by Akaike et al.

NCS-chrom acts as a very toxic substance not only in cancer cells but also in the cells that manufacture NCS-chrom and NCS-apoprotein. In the NCS-chrom-producing microorganism (*Streptomyces carzinostaticus* var. F-41), it may be neccessary to rapidly transport NCS-chrom outside the producing cells, presumably after neutralization of its toxicity with the specific polypeptide, NCS apoprotein. Production of native NCS probably results from a neutralizing reaction to reduce the toxicity of NCS-chrom in the producing cells themselves. For a better understanding of NCS, it will therefore be necessary to study (i) the metabolic pathway of production of NCS-chrom in the NCS-chrom-producing microorganism, (ii) the biological significance of its production in the producing cells, and (iii) the physiological correlation between the two production systems for NCS-chrom and the gene expression of NCS apoprotein. These investigations will be widely covered among enediyne antibiotics (such as macromomycin, actinoxanthin, kedarcidin, and C-1027), because they are produced by different microorganisms as a complex with each specific protein (chromoprotein) that is similar to NCS apoprotein [63]. A comparison of the amino acid sequences shows that NCS apoprotein shares 38.7% identity with kedarcidin chromoprotein in a 111-residue overlap (four gaps) (see chapter by K. Kuromizu and H. Maeda). Recently, Zein et al. (Bristol-Myers Squibb, USA) reported an interesting observation that these enediyne-containing chromoproteins, including NCS apoprotein, exhibit proteolytic activity when basic polypeptides, such as histone H1, are used as substrates [64]. This report suggests that these natural products exhibit the dual functions of specific DNA cleavage and selective proteolytic activity in cancer cells.

NCS-Induced Apoptosis

It is now widely accepted that apoptosis is a genetically programmed cell death defined by characteristic morphological and biochemical changes. A multitude of factors modulate apoptosis induction, including growth factors, intracellular mediators of signal transduction, and nuclear proteins regulating gene expression,

DNA replication, and the cell cycle. These diverse factors and pathways likely convert on common late steps, resulting in the characteristic apoptosis phenotype. Apoptosis is a pathway that may be potentially disrupted in tumor cells, conferring a survival advantage [65].

The external signals that lead to apoptosis are probably as varied as those that lead to differentiation and proliferation and can include the withdrawal of extracellular signals as well as their appearance. Indeed, dependence on essential survival factors for suppression of apoptosis seem to be widespread. Several membrane-associated proteins (such as Fas, TNFR, TCR, and DP-1), cytoplasmic proteins (such as gramzyme, fragmetin, ubiqutin, and proteases), nuclear proteins (such as NF-κB, c-Fos, c-Jun, p53, E1A, RB, and DNase γ), and suppressing proteins (such as Bcl-2, Bcl-x, Bax, and E1B) are identified as regulatory proteins involved in the induction of apoptosis in cells treated with DNA-degrading antibiotics [65] and ceramide [66]. Recently, it has been reported that the intracellular O_2^- plays a key role in regulating cell sensitivity to a potentially lethal signal and provide tumor cells with a natural, inducible mechanism of resistance to Fas-mediated apoptosis [67].

When cancer cells are treated with native NCS or polystyrene-co-maleic acid-conjugated NCS (SMANCS) [68], the NCS-chrom molecule will immediately be dissociated from NCS apoprotein at the intracellular level. Maeda's group has reported that SMANCS is accumulated more selectively in tumor tissue than in normal tissue, is more stable than NCS in blood, and exhibits an enhanced permeability and retention (EPR) effect but not in NCS [69]. The reaction of NCS-chrom with thiols in the treated cells results in its activation (Fig. 13). The activated form (biradical) of NCS-chrom directly cleaves DNA

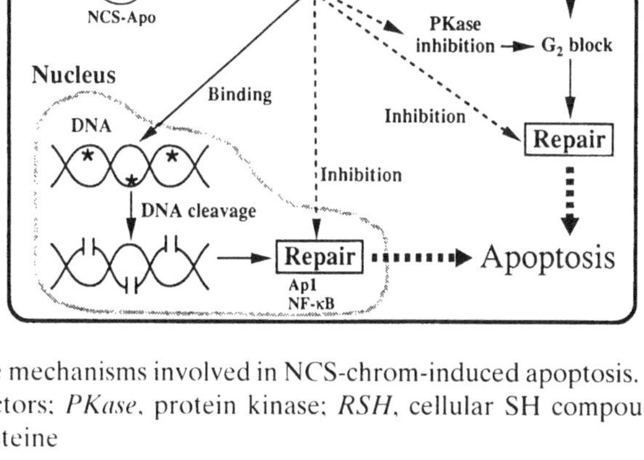

FIG. 13. Possible mechanisms involved in NCS-chrom-induced apoptosis. *Apl* and *NF-κB*, transcription factors; *PKase*, protein kinase; *RSH*, cellular SH compounds such as glutathione and cysteine

and also induces G_2 block, as described above. At an early stage of the NCS-chrom-induced ssDNA damage, the repair systems will be signaled by formation of the damaged ssDNA itself: the repair system involved in the NCS-chrom-induced ssDNA damage has been shown to require the newly synthesized proteins (see Fig. 3).

Furthermore, four indirect observations suggest the possibility that the activated NCS-chrom directly induces apoptosis by impairment of DNA repair systems: (i) NCS-chrom produces O_2^- itself after its reaction with thiols at the intracellular level [61]; (ii) the activated NCS-chrom directly inhibits activation of transcriptional factors through their phosphorylation by second-messenger-independent nuclear protein kinases, such as CK-II, involved in the regulation of gene expression for the cellular repair systems [20]; (iii) the drug cleaves preferentially the ARS site of the replication origin and the specific binding site for transcriptional factor Ap1 on DNA [41]; and (iv) antitumor drugs, such as bleomycin and mitomycin C with the abilities to degrade DNA similar to NCS, induce apoptosis [65].

Our preliminary experiments using two-dimensional gel electrophoresis (2DE) of a cell extract from NCS-treated HeLa S3 cells detected several new proteins, including DNase, heat shock proteins (HSPs), and DNA-binding proteins. The experimental data observed in the NCS-induced DNA degradation suggest that some gene expressions are required for the repair systems and induction of apoptosis will be signaled. These results are consistent with our earlier observations that puromycin or cyclohexiamide (inhibitors of protein syntheis) prevents the NCS-induced DNA degradation (see Figs. 3 and 4) [7]. However, the physiological role of these newly synthesized proteins in the repair systems and the regulatory mechanisms involved in their gene expression in NCS-treated cells remain to be elucidated.

Concluding Remarks

The enediyne compound NCS-chrom is a potent antitumor drug. This conclusion is supported by following unique characteristics of NCS-chrom. First, it is immediately activated by intracellular reaction with or without thiols. Second, the thiol-activated NCS-chrom prefentially damages dsDNAs, particularly the DNA replication origin and the specific binding site for transcriptional factors, and NCS-chrom also cleaves thiol independently ssDNA and RNA with bulge structures. Third, the drug inhibits protein phosphorylation by protein kinases involved in G_2-phase processing and in transcriptional regulation. Fourth, treatment of cancer cells with NCS-chrom or SMANCS results in the induction of apoptosis.

However, the signaling mechanisms involved in the initiation of the DNA repair system and in the induction of the NCS-induced apoptosis remain to be elucidated. Because several proteins are newly induced when the cultured cells are treated with NCS (Ohtsuki, unpublished result), further analytical studies on their cellular functions and on the transcriptional regulation mediated by protein kinases may provide direct evidence to explain the biochemical mechanisms involved in the DNA repair system and the induction of apoptosis by NCS.

Acknowledgments. We would like to thank Professor H. Maeda (Kumamoto University, Kumamoto) who provided the opportunity to summarize this review. We thank also Emeritus Professors N. Ishida and K. Kumagai for their interest and support concerning the study on the mechanism of action of NCS and NCS-chrom.

References

1. Ishida N, Miyazaki K, Kumagai K, Rikimaru M (1965) Neocarzinostatin, an antitumor antibiotic of high molecular weight. J Antibiot (Tokyo) Ser A 18:68–76
2. Maeda H, Kumagai K, Ishida N (1966) Characterization of neocarzinostatin. J Antibiot Tokyo 19:253–259
3. Kumagai K, Ono Y, Nishikawa T, Ishida N (1966) Cytological studies on the effect of neocarzinostation on HeLa cells. J Antibiot (Tokyo) 19:69–74
4. Ono Y, Watanabe Y, Ishida N (1966) Mode of action of neocarzinostatin: inhibition of DNA synthesis and degradation of DNA in *Sarcina lutea*. Biochim Biophys Acta 119:46–58
5. Ohtsuki K, Ishida N (1975) Neocarzinostatin-induced breakdown of dexyribonucleic acid in HeLa S3 cells. J Antibiot (Tokyo) 28:143–148
6. Ishida N, Ebina T (1977) Neocarzinostatin (NCS). Jap. J Antibiot (Tokyo) 30:S-93–103
7. Ono Y, Ito Y, Maeda H, Ishida N (1967) Mode of action of neocarzinostatin: requirement of protein synthesis for neocarzinostatin-mediated DNA degradation in *Sarcina lutea*. Biochim Biophys Acta 155:616–618
8. Ohtsuki K, Ishida N (1975) Mechanism of DNA degradation induced by neocarzinostatin in *Bacillus subtilis*. J Antibiot (Tokyo) 28:229–236
9. Ishida R, Takahashi T (1978) Role of 2-mercaptoethanol in vitro DNA degradation by neocarzinostatin. Cancer Res 38:2617–2620
10. Beerman TA, Goldberg IH (1974) DNA strand scission by the antitumor protein neocarzinostatin. Biochem Biophys Res Commun 59:1254–1261
11. Tatsumi K, Nakamura T, Wakisaka G (1974) Damage of mammalian cell DNA in vivo and in vitro induced by neocarzinostatin. Gann 65:459–461
12. Ishida R, Nishimoto T, Takahashi T (1979) DNA strand scission by neocarzinostatin and its relation to the inhibition of cell cycle traverse and DNA synthesis. Cell Struct Funct 4:235–250
13. Ishida R, Nishimoto T, Takahashi T (1976) In vitro release of thymine from DNA by neocarzinostatin. Biochem Biophys Res Commun 68:256–261
14. Koide Y, Ishii F, Hasuda K, Koyama Y, Edo K, Katamine S, Kitame F, Ishida N (1980) Isolation of non-protein component and a protein component from neocarzinostatin (NCS) and their biological activities. J Antibiot (Tokyo) 33:342–345

15. Edo K, Katamine S, Kitame F, Ishida N, Koide Y, Kusano G, Nozoe S (1980) Naphthalene carboxylic acid from neocarzinostatin (NCS). J Antibiot (Tokyo) 33:347–351
16. Edo K, Mizugaki N, Koide Y, Seto H, Furuhata K, Otake N, Ishida N (1985) The chemical structure of antitumor polypeptide antibiotic neocarzinostatin chromophore. Tetrahedron Lett 26:331–334
17. Edo K, Akiyama Y, Saito K, Mizugaki M, Koide Y, Ishida N (1986) Absolute configuration of the amino sugar moiety of the neocarzinostatin chromophore. J Antibiot (Tokyo) 34:1615–1619
18. Ebina T, Ohtsuki K, Seto M, Ishida N (1975) Specific G_2 block in HeLa S3 cells by neocarzinostatin. Eur J Cancer 11:155–158
19. Ohtsuki K, Koike T, Sato T, Koyama K, Ishida N, Satake N (1980) Neocarzinostatin (NCS) and microtubule-associated kinase. J Antibiot (Tokyo) 33:1590–1593
20. Ohtsuki K, Sato T, Koike T, Koyama K, Ishida N (1980) The inhibitory mechanism of in vitro protein phosphorylation by a nonprotein chromophore removed from neocarzinostatin. Biochim Biophys Acta 673:147–156
21. Ohtsuki K, Ishida N (1981) Mechanisms of actions of neocarzinostatin (NCS) and NCS-associated nonprotein chromophore. Protein Nucleic Acid Enzyme 26:937–949
22. Hatayama T, Goldberg IH, Takeshita M, Grollman AP (1978) Nucleotide specificity in DNA scission by neocarzinostatin. Proc Natl Acad Sci USA 8:3603–3607
23. Hatayama T, Goldberg IH (1980) Deoxyribonucleic acid sugar damage in the action of neocarzinostatin. Biochemistry 19:5890–5898
24. Kappen LS, Goldberg IH (1979) Mechanism of the effect of organic solvents and other protein denaturants on neocarzinostatin activity. Biochemistry 18:5647–5653
25. Kappen LS, Napier MA, Goldberg IH (1980) Roles of chromophore and apoprotein in neocarzinostatin action. Proc Natl Acad Sci USA 77:1970–1974
26. Ohtsuki K, Ishida N (1980) The biological effect of a nonprotein component removed from neocarzinostatin. J Antibiot (Tokyo) 33:744–750
27. Dedon PC, Goldberg IH (1992) Influence of thiol structure on neocarzinostatin activation and expression of DNA damage. Biochemistry 31:1909–1917
28. Goldberg IH, Kappen LZ (1995) Neocarzinostatin: chemical and biological basis of oxidative DNA damage. In: Borders DB, Doyle TW (eds) Enediyne antibiotics as antitumor agants. Dekker, New York, pp 327–362
29. Mayers AG, Proteau PL (1989) Evidence for spontaneous low-temperature biradical formation from a highly reactive neocarzinostatin chromophore-thiol conjugate. J Am Chem Soc 111:1146–1147
30. Yoshida K, Minami Y, Azuma R, Saeki M, Otani T (1993) Structure and cycloaromatization of a novel enediyne, C-1027 chromophore. Tetrahedron Lett 34:2637–2640
31. Golik J, Dubay G, Groenewold G, Kawauchi H, Konishi M, Krishnan B, Ohkuma H, Saitoh K, Doyle TW (1987) Espermicins, a novel class of potent antitumor antibiotics: structures of espermicins A1, A2 and A1b. J Am Chem Soc 109:3462–3464
32. Lee MD, Dunne TS, Chang CC, Elestad GA, Siegel MM, Morton GO, McGahren WJ, Borders DB (1987) Calichemicins, a novel family of antitumor antibiotic: chemistry and structure of calichemicin γ1. J Am Chem Soc 109:3466–3468
33. Sugiura T, Arakawa T, Uesugi M, Shiraki T, Ohkuma H, Konishi M (1991) Reduction and nucleophilic activation products of dynemicin A with methyl thioglycolate: a rational mechanism for DNA cleavage of the thiol-activated dynemicin A. Biochemistry 30:2989–2992

34. Leet JE, Schroeder DR, Hofstead SJ, Golik J, Colson KL, Huang S, Klohr SE, Doyle TW, Matson JA (1992) Kedarcidin, a new chromophore antitumor antibiotic: structure elucidation of kedarcidin chromophore. J Am Chem Soc 114:7946–7947
35. Frank BL, Worth L Jr, Christner DF, Kozarich JW, Stubbe J, Kappen LS, Goldberg IH (1991) Isotope effects on the sequence-specific cleavage of DNA by neocarzinostatin: kinetic partitioning between 4′- and 5′-hydrogen abstraction at unique thymidine sites. J Am Chem Soc 113:2271–2275
36. Saito I, Kawabata H, Fujiwara T, Sugiyama H, Matsuura T (1989) A novel ribose C4′ hydroxylation pathway in neocarzinostatin-mediated degradation of oligonucleotides. J Am Chem Soc 111:8302–8303
37. Kappen LS, Goldberg IH (1983) Deoxyribonucleic acid damage by neocarzinostatin chromophore: strand breaks generated by selective oxidation of C-5′ of deoxyribose. Biochemistry 22:4872–4878
38. Chin DH, Kappen LS, Goldberg IH (1987) 3′-Formyl phosphate-ended DNA: high-energy intermediate in antibiotic-induced DNA sugar damage. Proc Natl Acad Sci USA 84:7070–7074
39. Kappen LS, Goldberg IH (1989) Identification of 2-deoxyribonolactone at the site of neocarzinostatin-induced cytosine release in the sequence d(AGC). Biochemistry 28:1027–1032
40. Stassinopoulos A, Goldberg IH (1995) Probing the structure of long single-stranded DNA fragments with neocarzinostatin chromophore. Extension of the base-catalyzed bulge-specific reaction. Biochemistry 34:15359–15374
41. Dedon PC, Jiang ZW, Goldberg IH (1992) Neocarzinostatin-mediated DNA damage in a model AGT·ACT site: mechanistic studies of thiol-sensitive partitioning of C4′ DNA damage products. Biochemistry 31:1917–1927
42. Lee W, Mitchell P, Tjian R (1987) Purified transcriptional factor AP-1 interacts with TPA-inducible enhancer elements. Cell 49:741–752
43. Turner R, Tjian R (1989) Leucine repeats and an adjacent DNA binding domain mediate the formation of functional cFos-cJun heterodimers. Science 31:1689–1694
44. Lin A, Frost J, Deng T, Smeal T, Al-Alawi N, Kikkawa U, Hunter T, Brenner D, Karin M (1992) Casein kinase II is a negative regulator of c-Jun DNA binding and Ap1 activity. Cell 70:777–789
45. Pinna LA (1990) Casein kinase 2: an "eminence grise" in cellular regulation? Biochim Biophys Acta 1054:267–284
46. Ohtsuki K, Nishikawa Y, Saito H, Kato T (1996) DNA-binding sperm proteins with oligo-arginine clusters function as potent activators for egg CK-II. FEBS Lett 378:115-120
47. Xu Y-J, Zhen Y-S, Goldberg IH (1994) C1027 chromophore, a potent new enediyne antitumor antibiotic, induces sequences-specific double-strand DNA cleavage. Biochemistry 33:5947–5954
48. Gao X, Stassionpoulos A, Gu J, Goldberg IH (1995) NMR studies of the post-activated neocarzinostatin chromophore-DNA complex: conformational changes induced in drug and DNA. Bioorg Med Chem Lett 3:795–809
49. Gao X, Stassionopoulos A, Rice JS, Goldberg IH (1995) Structural basis for the sequence-specific DNA strand cleavage by the enediyne neocarzinostatin chromophore: structure of the post-activated chromophore-DNA complex. Biochemistry 34:40–49

50. Kappen LS, Goldberg IH (1993) DNA conformation-induced activation of an enediyne for site-specific cleavage. Science 261:1319–1321
51. Kappen LS, Goldberg IH (1995) Bulge-specific cleavage in transactivation response region RNA and its DNA analogue by neocarzinostatin chromophore. Biochemistry 34:5997–6002
52. Zeng X, Xi Z, Lizzy S, Kappen LS, Tan W, Goldberg IH (1995) Double-stranded damage of RNA·DNA hybrids by neocarzinostatin chromophore: selective C-1′ chemistry on the RNA strand. Biochemistry 34:12435–12444
53. Duff RJ, De Vroom E, Geluk A, Hecht SM (1993) Evidence for C′-1 hydrogen abstraction from modified oligonucleotides by Fe-bleomycin. J Am Chem Soc 115:3350–3351
54. Hensens OD, Chin D-H, Stassinopoulos A, Zink DL, Kappen LS, Goldberg IH (1994) Spontaneous generation of a biradical species of neocarzinostatin chromophore: role in DNA bulge-specific cleavage. Proc Natl Acad Sci USA 91:4534–4538
55. Zyskind JW, Smith DW (1992) DNA replication, the bacterial cell cycle, and cell growth. Cell 69:5–8
56. Depamphilis ML (1993) Origins of DNA replication in metazoan chromosomes. J Biol Chem 268:1–4
57. Taniguchi N, Fujii Y, Suzuki K (1994) The biochemical mechanisms involved in the redox reaction induced by various stresses. J Exp Med 12:23–29
58. Wiseman H, Halliwell B (1996) Damage to DNA by reactive oxgen and nitrogen species: role in inflammatory disease and progression to cancer. Bichem J 313:17–29
59. Oda T, Akaike T, Hamamoto T, Suzuki F, Hirano T, Maeda H (1989) Oxegen radicals in influenza-induced pathogenesis and treatment with pyran polymer-conjugated SOD. Science 244:974–976
60. Maeda H, Akaike T (1989) Minireview: Oxygen free radicals as pathogenic molecules in viral diseases. Proc Soc Exp Biol Med 198:721–727
61. Williams GT, Smith CA (1993) Molecular regulation of apoptosis: genetic controls on cell death. Cell 74:777–779
62. Sato K, Akaike T, Suga M, Ando M, Maeda H (1994) Generation of free radicals from neocarzinostatin mediated by NADPH/cytochrome P-450 reductase via activation of enediyne chromophore. Biochem Biophys Res Commun 205:1716–1723
63. Doyle TW, Borders DB (1995) Enediyne antitumor antibiotics. In: Borders DB, Doyle TW (eds) Enediyne antibiotics as antitumor agents. Dekker, New York, pp 1–15
64. Zein N, Reiss P, Bernatowicz M, Bolgar M (1995) The proteolytic specificity of the natural enedine-containing chromoproteins unique to each chromoprotein. Chem Biol 2:451–455
65. Fisher DE (1994) Apoptosis in cancer therapy: crossing the threshold. Cell 78:539–542
66. Jarris WD, Grant S, Kolesnick RN (1996) Ceramide and the induction of apoptosis. Clin Cancer Res 2:1–6
67. Clement M-V, Stamenkovic I (1996) Superoxide anion is a natural inhibitor of Fas-mediated cell death. EMBO J 15:216–225
68. Kimoto A, Konno T, Kawaguchi T, Miyauchi Y, Maeda H (1992) Antitumor effects of SMANCS on rat mammary tumor induced by 7,12-dimethylbenz[α]anthracene. Cancer Res 52:1013–1017
69. Maeda H (1991) SMANCS, and polymer-conjugated macromolecular drugs: advantages in cancer chemotheraphy. Adv Drug Delivery Syst 6:181–202

Activation of Neocarzinostatin by NADPH/Cytochrome P-450 Reductase: Proposal of an Ultimate Mode of Action in Cells Involving Oxygen Free Radicals

TAKAAKI AKAIKE, KEIZO SATO, and HIROSHI MAEDA

Summary. Neocarzinostatin (NCS), a potent proteinaceous anticancer agent, is a single-chain polypeptide (MW, 11 700) with two disulfide bonds that contains a nonprotein chromophore (NPC) with a unique enediyne structure as described elsewhere in this volume. In this chapter, a new activation mechanism of NCS was explored by analyzing oxygen radical generation from NCS in a reduced nicotinamide adenine dinucleotide (NADPH)/cytochrome (cyt) P-450 reductase system. Electron spin resonance (ESR) spin-trapping was performed to identify oxygen radicals such as superoxide ($O_2\cdot^-$) and hydroxyl radical (OH·) produced in the reaction mixture with NCS + NADPH/cyt P-450 reductase in the presence of NADPH. 5,5-Dimethyl-1-pyrroline *N*-oxide (DMPO) was used as a spin-trap for ESR study. Based on the Lineweaver–Burk plot of the DMPO-OOH, a spin adduct of $O_2\cdot^-$ generated in the reaction system against the concentration of NCS, mitomycin C, and adriamycin, K_m values of cyt P-450 reductase for these antitumor agents were determined to be $2.89 \times 10^{-5} M$, $4.46 \times 10^{-6} M$, and $1.82 \times 10^{-6} M$, respectively. The K_m value for NCS was comparable to those of mitomycin C and adriamycin. The most important observation in this study is that DNA cleavage of plasmid pBR322 occurred most effectively in the system with picomolar ($10^{-12} M$) concentrations of NCS in the presence of NADPH/cyt P-450 reductase system as revealed by agarose gel electrophoresis of the plasmid DNA, whereas at least 10^{-5}~$10^{-6} M$ of NCS is needed without NADPH/P450 reductase system. These results indicate the importance of the NADPH/cyt P-450 reductase system in activation of NCS, and the present mechanism may explain the potent cytotoxic action of NCS in cells because NADPH/cyt P-450 is a ubiquitous cellular component.

Key words: Non-protein chromophore—Enediyne structure—Superoxide ($O_2\cdot^-$)—Hydroxyl radical (OH·)—Oxygen radicals—DNA cleavage—NADPH/cyt P-450 reductase

---

Departments of Microbiology and Medicine I, Kumamoto University School of Medicine, 2-2-1 Honjo, Kumamoto, 860 Japan

Introduction

The mechanism of action of neocarzinostatin (NCS) were described in detail in an earlier chapter of this volume (chapter by Ohtsuki and Ono). NCS, which has a nonprotein chromophore (NPC) with an enediyne structure [1–6], has been reported to be one of the most potent antitumor agents ever known [7]. It is believed that enediyne antitumor antibiotics (e.g., NCS chromophore, dynemicin A, calicheamicin γ_1, and esperamicin A_1) exhibit their cytotoxic action through a chemical modification of nuclear DNA in which intrastrand cross-links to DNA (DNA adduct) are formed by the active organic diradical(s) in the enediyne moiety generated via the Masamune–Bergman reaction [8–11]. Edo et al. [12] observed free radical generation (NCS or NPC radical) from NPC in the presence of 2-mercaptoethanol using an electron spin resonance (ESR) spin-trap method with *t*-butyl phenylnitrone as a spin trap [12]. Similarly, Chin and Goldberg [13] reported that $O_2\cdot^-$ is spontaneously generated through autooxidation of NPC, as evidenced by reduction of nitroblue tetrazolium chloride monohydrate (NBT) and a ferricytochrome *c* reduction method.

Oxygen free radicals interact with nucleic acids, lipids, proteins, and other vital molecules [14–17]. It was recently reported that cytochrome (cyt) P-450 reductase (also called cytochrome *c* reductase; EC 1.6.2.4) together with reduced nicotinamide dinucleotide (NADPH) can generate oxygen radicals, particularly superoxide radical ($O_2\cdot^-$), from adriamycin and mitomycin C [18–21].

In this chapter, the mechanism of NCS activation is described, focusing on generation of oxygen radicals from NCS in the system of NADPH/cyt P-450 reductase [22]. Generation of oxygen radicals from NCS was identified by using the ESR spin-trapping method as described recently [23–25]. In addition, potency of activation of NCS was tested by measuring DNA cleavage using plasmid pBR322 with or without NADPH/cyt P-450 reductase. Our data provide insight into the actual action and activation mechanism of NCS in the cell in which NADPH/cyt P-450 reductase is involved. Because the present mode of action of NCS takes place at far lower (10 000 fold or more) concentrations than a nonenzymatic cell-free system, we propose this mechanism as an ultimate mode of NCS action that depends on NADPH/cyt P-450 reductase.

Materials and Methods

Chemicals

We used 5,5-dimethyl-1-pyrroline *N*-oxide (DMPO) throughout the experiments as a spin-trap. DMPO of the highest purity, diethylenetriaminepentaacetic acid (DTPA), and tetracyanoquinodimethane lithium salt (TCNQ-Li) were products of Dojindo Laboratories (Kumamoto, Japan); nicotinamide adenine nucleotide diphosphate, reduced form (NADPH, coenzyme-II) was obtained from Oriental Yeast (Tokyo, Japan); dimethyl sulfoxide was purchased from

Nacalai Tesque (Kyoto, Japan); and NCS was a gift from Kayaku Antibiotics Research Laboratories (Tokyo, Japan). Mitomycin C and adriamycin were supplied by Kyowa Hakko (Tokyo, Japan); glutathione (GSH, reduced form) was from Kohjin Tokyo, Japan); and 2,2,6,6-tetramethyl-4-hydroxylpiperidine (Tempol) was obtained from Sigma Chemical (St. Louis, MO, USA).

Enzyme and Proteins

Cyt P-450 reductase was purified according to the method of Ardies et al. [26] from rat liver after induction by injecting phenobarbital sodium. This protein preparation was more than 95% pure as judged by sodium dodecyl sulfate-polyacrylamide gel electrophoresis (SDS-PAGE). Catalase was from Boehringer Mannheim GmbH (Mannheim, Germany), and human recombinant Cu,Zn-superoxide dismutase (SOD) was s gift from Nippon Kayaku (Tokyo, Japan).

Quantification of Various Radicals Using ESR Spectroscopy

ESR spin-trapping with spin-trap DMPO was performed to identify the radical species generated in the reaction systems of NADPH/cyt P-450 reductase with NCS by using a JEOL ESR spectrometer (JES-RE1X, Akishima, Tokyo, Japan) with a quartz flat cell of 160 μl effective volume [22–25]. $O_2 \cdot^-$ is entrapped by DMPO, and DMPO-OOH(-$O_2 \cdot^-$) as being formed was quantified by normalization of the DMPO-OOH signal height against the standard manganese oxide (MnO) signal. ESR spectra were recorded at room temperature under the following conditions: modulation frequency, 100 kHz; modulation amplitude, 0.079 mT; scanning field, 336.1 ± 5 mT; receiver gain, 100 or 500; time constant, 0.3 s; sweep time, 2 min; microwave power, 20 mW; and microwave frequency, 9.421 GHz.

The radical-generating reaction was initiated by addition of NADPH at 0.3 m*M* to the reaction mixture containing 1.4 μ*M* cyt P-450 reductase, 9.3 μ*M* NCS, 0.5 m*M* DTPA, and 90 m*M* DMPO in the presence or absence of either 500 U/ml Cu,Zn-SOD or 500 U/ml catalase in 10 m*M* sodium phosphate buffer (pH 7.5). After various incubation periods, ESR measurements were carried out as just described. Absolute concentrations of the spin adduct of the $O_2 \cdot^-$ radical were determined by double integration of the ESR spectrum, in which 1.0 μ*M* of Tempol solution was used for a primary standard of ESR absorption [22–25].

In addition, the NCS-derived radical was directly detected by ESR measurement at 110 K. Specifically, NCS at 0.8 m*M* was reacted with 3.0 μ*M* cyt P-450 reductase, 10 μ*M* NADPH, and 0.5 m*M* DTPA in 50 m*M* sodium phosphate buffer (pH 7.5) for 3 min at room temperature. All reaction mixtures were deoxygenated by N_2 gas bubbling before initiating the reaction. The reaction mixtures (0.5 ml) were then transferred to 3.5-mm (inner diameter) quartz ESR tubes under ambient conditions; these were then subjected to ESR spectroscopy at 110 K by using an ESR spectrometer (ESR 380E, Bruker Instruments, (Reinstetten, Germany)) under the following conditions: modulation frequency,

100 kHz; modulation amplitude, 0.3 mT; scanning field, 342 ± 15 mT; receiver gain, 20 000; time constant, 0.7 s; microwave power, 10 mW; microwave frequency, 9.600 GHz.

Preparation of Plasmid pBR322 DNA

The pBR322 DNA in *E. coli* HB101 cultured in the presence of aminobenzylpenicillin was extracted by denaturing with alkaline and SDS treatment followed by phenol extraction and RNase digestion. Finally, plasmid DNA was purified by CsCl density gradient centrifugation in the presence of ethidium bromide (0.35 mg/ml) at 45 000 × g in a vertical rotor (RP65VT, Hitachi Koki, Tokyo, Japan) for 18 h at 15°C. After ethidium bromide was removed from the DNA solution by extraction with isoamyl alcohol, DNA was precipitated with 70% ethanol.

DNA Cleavage and Agarose Gel Electrophoresis

Unless otherwise specified, the standard reaction mixture containing 0.8 mg/ml pBR322 DNA, 0.3 m*M* NADPH, 0.5 m*M* DTPA, and various concentrations of NCS with or without 1.4 μ*M* cyt P-450 reductase or GSH in 25 m*M* sodium phosphate buffer (pH 7.5), was incubated at room temperature for 15 min. Then, 10 μl of the reaction mixture was electrophoresed on 1% agarose gel.

Results

$O_2^{\cdot -}$ Generation in the System of NCS and NADPH/cyt P-450 Reductase

NCS was reacted with NADPH and cyt P-450 reductase, and the reaction mixture after adding DMPO was subjected to ESR spin-trapping. The ESR spectrum obtained for NCS reacted with NADPH/cyt P-450 reductase (Fig. 1A) indicates generation of $O_2^{\cdot -}$ and OH· based on the formation of DMPO-OOH (a_N = 1.43 mT, a_H^{β} = 1.15 mT, a_H^{γ} = 0.13 mT) and DMPO-OH (a_N = 1.49 mT, a_H = 1.49 mT) signals; their generation was completely inhibited by Cu,Zn-SOD (Fig. 1B), but was not inhibited by catalase (Fig. 1C). On the contrary, only traces of ESR signals of DMPO-OOH and DMPO-OH were produced in the reaction of NADPH and cyt P-450 reductase without NCS (Fig. 1D). Thus, NADPH/cyt P-450 reductase alone seems to be ineffective in $O_2^{\cdot -}$ and OH· generation even in the presence of 0.3 m*M* NADPH. No spin adduct signal was observed in the reaction mixture with NCS but without NADPH/P-450 reductase (Fig. 1E), suggesting that the amount of $O_2^{\cdot -}$ generation by autooxidation of NCS per se does not occur in the reaction without the P-450 reductase system. In addition, $O_2^{\cdot -}$ and OH· generation increased in a dose-dependent manner with NCS in this system with NADPH/P-450 reductase (Fig. 2).

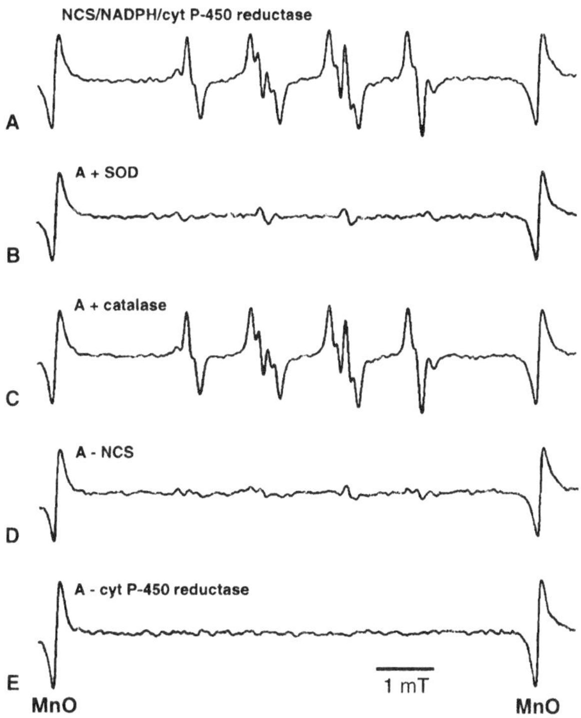

FIG. 1A–E. Electron spin resonance (ESR) spectra of 5,5-dimethyl-1-pyrroline-*N*-oxide (DMPO) spin adducts of superoxide ($O_2{\cdot}^-$) generated in the reaction system of neocarzinostatin (NCS) with reduced nicotinamide adenine nucleotide NADPH/cytochrome (cyt) P-450 reductase. **A** The spectrum obtained in the reaction mixture of 0.3 m*M* NADPH, 1.4 μ*M* cyt P-450 reductase, 9.3 μ*M* NCS, 0.5 m*M* DTPA (diethylene triaminepentaacetic acid), and 90 m*M* DMPO in 10 m*M* sodium phosphate buffer (pH 7.5). **B** The spectrum obtained in the same reaction mixture as **A** with 500 U/ml Cu,Zn-SOD (superoxide dismutase). **C** The spectrum in the same reaction mixture as **A** with 500 U/ml catalase. **D** and **E** Spectra observed with the same reaction mixture as **A** without NCS and cyt P-450 reductase, respectively. All spectra were recorded at 1 min after initiating the reaction

The K_m value for $O_2{\cdot}^-$ generation of NCS in the NADPH/cyt P-450 reductase system was determined on the basis of a Lineweaver–Burk plot of DMPO-OOH generation against the concentration of NCS. Similarly, each K_m value for $O_2{\cdot}^-$ generation was obtained with mitomycin C and adriamycin as well. The K_m values of cytochrome P-450 reductase for mitomycin C, adriamycin, and NCS were 4.46×10^{-6} M, 1.82×10^{-6} M, and 2.89×10^{-5} M, respectively. These results indicate that the K_m value for NCS was comparable to those for other quinone-containing antitumor agents such as mitomycin C and adriamycin. Thus, NCS is effective as a substrate for NADPH cyt P-450 reductase to generate $O_2{\cdot}^-$.

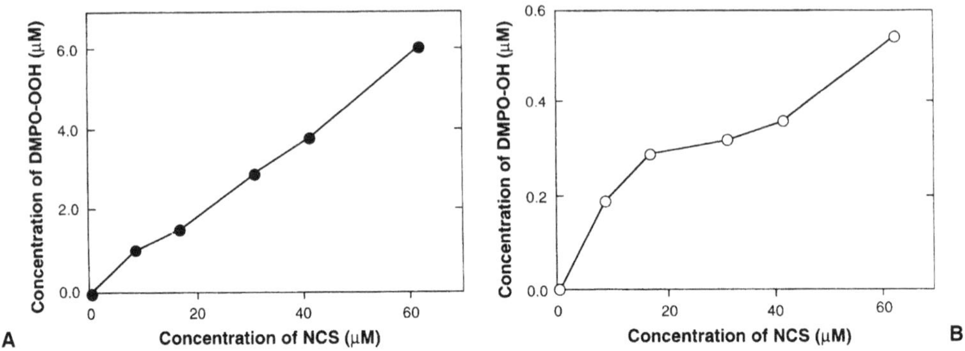

FIG. 2A,B. Generation of $O_2^{\cdot -}$ spin adduct (*DMPO-OOH*) (**A**) and OH· spin adduct (*DMPO-OH*) (**B**) in NADPH/cyt P-450 reductase and NCS. ESR measurement was performed in the same manner as described in Fig. 1 except that the concentration of NCS was varied in the reaction mixture of NADPH/cyt P-450 reductase in 10 m*M* sodium phosphate buffer (pH 7.5). In the ordinate of both (**A**) and (**B**), the amount of each spin adduct generated was quantified by ESR spectroscopy. Each spectrum was recorded at 1 min after initiating the reaction, and the absolute concentration of spin adduct was determined as described in the text

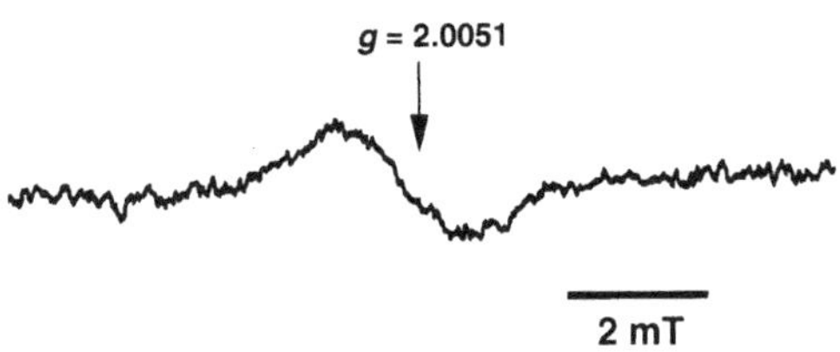

FIG. 3. Generation of NCS radical in the system of NADPH/cyt P-450 reductase and NCS. Reaction mixtures contained 10 μ*M* NADPH, 3.0 μ*M* cyt P-450 reductase, 0.5 m*M* DTPA, and 0.8 m*M* NCS in 50 m*M* sodium phosphate buffer (pH 7.5). The spectra were recorded at 3.5 min after incubation. An ESR signal with $g = 2.0051$ was produced in the reaction system of NCS with NADPH/Cyt P-450 reductase

Direct Measurement of NCS Radical Generated in the System of NCS and NADPH/Cyt P-450 Reductase

Generation of the NCS radical was further investigated in the system of NADPH/P-450 reductase in anerobic conditions by using the direct ESR method. As shown in Fig. 3, an organic radical derived from NCS was indeed generated in the reaction mixture containing 800 μ*M* NCS, 3.0 μ*M* P-450 reductase, 10 μ*M* NADPH, and 0.5 m*M* DTPA in 50 m*M* sodium phosphate buffer (pH 7.5). A similar reaction was carried out with NCS except that 5.0 m*M* GSH was used instead of NADPH/cyt P-450 reductase; however, generation of this NCS-derived radical was not observed in this reaction mixture without addition of

P-450 reductase. The *g* value of the NCS radical was determined to be 2.0051 by using TCNQ-Li salt as a *g* marker.

DNA Cleavage Caused by the NCS/NADPH/Cyt P-450 Reductase System

Using pBR322 plasmid DNA (closed circular/supercoiled form), DNA cleavage in P-450 reductase plus NCS was examined based on the change in the gel electrophoretic pattern of the plasmid DNA. It is notable that activity of NCS was augmented enormously in the presence of NADPH/cyt P-450 reductase, exhibiting 100–100 000 times more potent DNA-cleaving activity than without NADPH/cyt P-450 reductase, i.e., NCS with GSH (Fig. 4). DNA cleavage was evidenced by the mobility shift of supercoiled (SC) DNA to open-circular (OC) DNA and further to extensively damaged DNA that showed a broad and inhomogeneous electrophoretic pattern in the upper area of the gel. More specifically, NCS at a concentration as low as 10^{-12} *M* (about pg/ml) showed DNA cleavage in the presence of NADPH/cyt P-450 reductase, while more than 10^{-7} *M* of NCS was needed to obtain an appreciable formation of OC DNA.

Discussion

We have demonstrated here that the enzyme cyt P-450 reductase, which is a ubiquitous intracellular component of the microsomal cyt P-450 system, does participate in the production of $O_2^{\cdot -}$ and OH· from NCS.

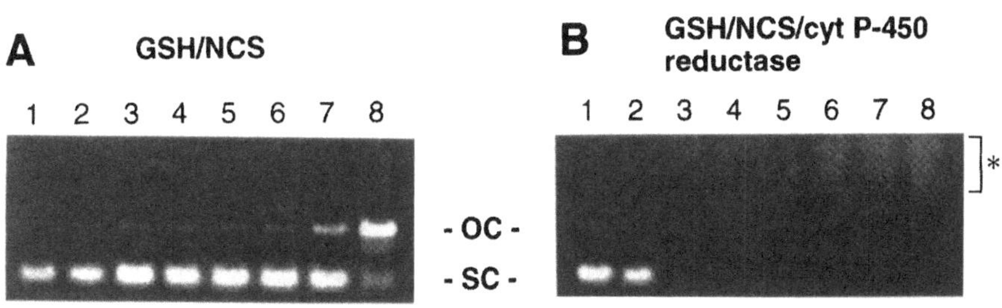

FIG. 4A,B. DNA-cleaving potential of NCS in NADPH/cyt P-450 reductase system. **A** DNA cleavage induced by GSH/NCS. *Lane 1*, pBR 322 DNA alone; *lanes 2, 3, 4, 5, 6, 7*, and *8*, pBR 322 DNA treated with 10^{-12}, 10^{-11}, 10^{-10}, 10^{-9}, 10^{-8}, 10^{-7}, and 10^{-6} *M* NCS, respectively, in the presence of NADPH and GSH but without cyt P-450 reductase. **B** DNA cleavage caused by GSH/NCS/cyt P-450 reductase. *Lane 1*, pBR 322 DNA alone; *lanes 2, 3, 4, 5, 6, 7*, and *8*, pBR 322 DNA treated with 10^{-12}, 10^{-11}, 10^{-10}, 10^{-9}, 10^{-8}, 10^{-7}, and 10^{-6} *M* NCS, respectively, in the presence of NADPH, GSH, and cyt P-450 reductase. Note that the DNA treated with NCS/NADPH/cyt P-450 reductase showed a broad and inhomogeneous band, as marked with the *asterisk* in **B**. *SC*, Supercoiled DNA; *OC*, open-circular DNA

There is accumulating evidence that oxygen free radicals interact with nucleic acids, proteins, lipids, and other vital molecules. This realization suggests that NCS as well as copoly (styrene-maleic acid)-conjugated NCS (SMANCS) or other enediyne-containing compounds may induce not only DNA damage but also protein and cell membrane injury via generation of oxygen radicals. Zein et al. [27,27a] showed a unique and previously undescribed action of enediyne-antitumor agents on proteins, i.e., proteolytic cleavage of histone proteins by enediyne compounds, which may be difficult to interpret only by the mechanism attributed to the interaction between NCS radical and DNA. Oxygen free radicals thus generated may affect proteins similarly.

It is generally appreciated that the cytotoxic activity of highly purified NCS against various tumor cells in culture and susceptible bacteria (such as *Micrococcus luteus* and drug-resistant *Shigella flexneri*) ranges from 0.01 to 0.1 μg/ml (~10 p*M*). Further, it was also reported earlier and later confirmed that in vivo activity of NCS against the ascitic form of murine tumors (e.g., S-180, L-1210, P-388, leukemia, etc.) is observed at doses as low as 0.1 mg/kg or higher when the drug is given intraperitoneally, which is again almost equivalent to in vitro cytotoxic activity (see chapter by H. Maeda, this volume).

The molecular mechanism of NCS action in a cell-free system usually was worked out by using a concentration of 10–100 μg/ml (1–10 μ*M*) of NCS with DNA, and in such a system DNA cleavage by NCS at 0.1 μg/ml is not apparent. Therefore, the present notion, that cyt P-450 and NADPH are involved the action of NCS at extremely low concentrations, fills the gap of this discrepancy between a cellular system and a cell-free system without cyt P-450 reductase.

The microsomal membrane in eukaryotic cells contains a substantial amount of cyt P-450 reductase. A wide variety of compounds such as anticancer drugs and other chemicals, as well as mutagens, can be activated by this enzyme [18–23]. When NCS is added to cells, it will be internalized into the target [28] and may react with cyt P-450 reductase; consequently, oxygen radicals are generated as identified in this experiment with the NCS/NADPH/cyt P-450 reductase system. As discussed in the earlier chapter, internalization is essential for the cytotoxicity of NCS. In this connection, NCS or SMANCS is resistant to *p*-glycoprotein-dependent multidrug resistance in which macromolecular drugs seem not to be the subject of *p*-glycoprotein-dependent efflux, although the chromophore of NCS appeared to be pumped out of the cell by this mechanism [29].

Chin and Goldberg [13] previously reported spontaneous generation of $O_2^{\cdot -}$ through an autooxidation process based on reduction of NBT and ferricytochrome *c*. They also tested the effect of Cu,Zn-SOD on DNA cleavage induced by NCS at a very high concentration range (50–200 μg/ml) and showed that DNA cleavage was significantly inhibited by Cu,Zn-SOD. We have shown

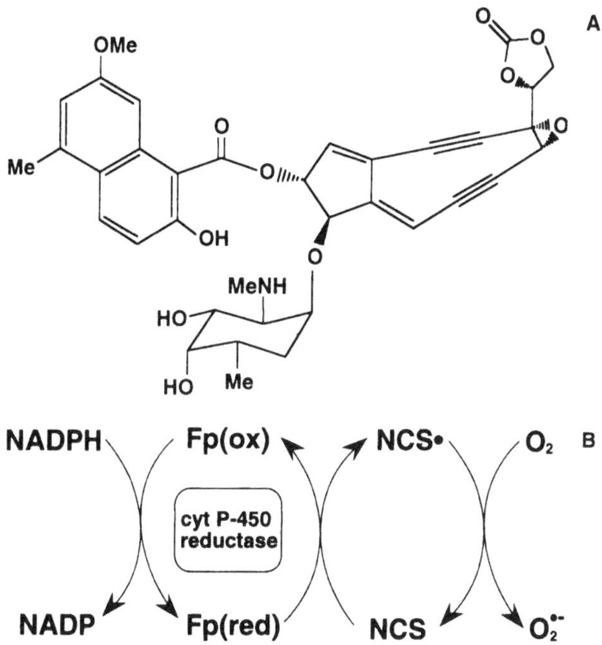

FIG. 5A,B. Proposed reaction mechanism of NCS involving the NADPH/cyt P-450 reductase system. **A** Chemical structure of NCS-nonprotein chromophore. **B** Electron transfer reaction in NCS activation by NADPH/cyt P-450 reductase. Fp (OX) and Fp (red), oxidized and reduced forms of cyt P-450 reductase, respectively; NCS·, NCS chromophore radical

here, however, that the amount of $O_2{\cdot}^-$ generated in such a system was much smaller than that of the NCS/NADPH/cyt P-450 reductase system. More importantly, the amount of NCS required to cleave DNA in the system without cyt P-450 reductase was 1000–100 000 times higher than that in the NADPH/cyt P-450 reductase system, as already discussed. We have tested another enediyne antitumor agent CP-1027 [30] for this mechanism and found very similar results (unpublished observations).

We have also examined generation of oxygen radicals by the NADPH/cyt P-450 reductase system in some anticancer agents such as mitomycin C, adriamycin, and heterocyclic amines [22,23] and found that all of these generated an appreciable amount of oxygen radicals. These results indicate that such anticancer agents and heterocyclic amines may exert cytotoxicity, DNA cleavage, and mutagenesis via activation with the cyt P-450 reductase system.

Miyamoto et al. [31] measured the content of antioxidants in both an NCS-resistant cell line (CHRC5) and an NCS-nonresistant cell line (AUXB1), and found that only Mn-SOD in the resistant cells was about twofold higher than that in nonresistant cells [31]. Specifically, levels of other enzymes or antioxidants such as glutathione (SH form) content, glutathione peroxidase, catalase, and glutathione-*S*-transferase are not changed remarkably. This observation is also consistent with the idea that superoxide generation is important for the cytotoxicity of NCS or SMANCS in the cell.

Conclusion

We have described here a novel mechanism for the action of NCS, involving generating a superoxide, which is mediated by NADPH/cyt P-450 reductase, and it is 1000–100 000 times more potent than NCS alone or NCS with thiols. A scheme for this hypothetical activation mechanism of NCS is shown in Fig. 5. This concept may explain the extremely potent cytocidal activity of NCS and SMANCS in cells, with a clear contrast to the mode of action of a cell-free system devoid of cyt P-450 reductase. To date, considerable attention has been paid to the unique characteristics of enediyne anticancer antibiotics (e.g., NCS chromophore, dynemicin A, calicheamicin γ_1, esperamicin A_1, C-1027) [11,32,33]. A similar activation mechanism appears to be operative in the cytotoxic mechanism of all these enediyne anticancer agents.

Acknowledgment. We thank Rie Yoshimoto for preparing the manuscript. Thanks are also due to Dr. H. Kosaka for his assistance and valuable discussion.

References

1. Ishida N, Miyazaki K, Kumagai K, Rikimaru M (1965) Neocarzinostatin, an antitumor antibiotic of high molecular weight. J Antibiot Ser A (Tokyo) 18:68–76
2. Meienhoffer J, Maeda H, Glaser CB, Czomos J, Kuromizu K (1972) Primary structure of neocarzinostatin, an antitumor protein. Science 178:875–876
3. Kuromizu K, Tsunasawa S, Maeda H, Abe O, Sakiyama F (1986) Reexamination of the primary structure of antitumor protein, neocarzinostatin. Arch Biochem Biophys 246:199–205
4. Kuromizu K, Abe O, Maeda H (1991) Location of the disulfide bonds in the antitumor protein, neocarzinostatin. Arch Biochem Biophys 286:569–573
5. Koide Y, Ishii F, Hasuda K, Koyama Y, Edo K, Katamine S, Kitame F, Ishida N (1980) Isolation of a non-protein component and a protein component from neocarzinostatin (NCS) and their biological activities. J Antibiot (Tokyo) 33: 342–346
6. Napier MA, Kappen LS, Goldberg IH (1980) Effect of non-protein chromophore removal on neocarzinostatin action. Biochemistry 19:1767–1773
7. Maeda H (1981) Neocarzinostatin in cancer chemotherapy. Anticancer Res 1:175–186
8. Kappen LS, Goldberg IH (1993) DNA conformation-induced activation of an enediyne for site specific cleavage. Science 261:1319–1321
9. Myers AG, Proteau PJ, Handel TM (1988) Stereochemical assignment of neocarzinostatin chromophore. J Am Chem Soc 110:7212–7214

10. Hirama M, Gomibuchi T, Fujiwara K, Sugiura Y, Uesugi M (1991) Synthesis and DNA-cleaving abilities of functional neocarzinostatin chromophore analogues. J Am Chem Soc 113:9851–9853
11. Goldberg IH, Kappen LS (1995) Neocarzinostatin: chemical and biological basis of oxidative DNA damage. In: Borders DB, Doyle TW (eds) Enediyne antibiotics as antitumor agents. Dekker, New York, pp 327–362
12. Edo K, Iseki S, Ishida N, Horie T, Kusano G, Nozoe S (1980) An electron spin resonance study of spin adduct of the non-protein component (NPC) of neocarzinostatin. J Antibiot (Tokyo) 33:1586–1589
13. Chin D-H, Goldberg IH (1986) Generation of superoxide free radical by neocarzinostatin and its possible role in DNA damage. Biochemistry 25:1009–1015
14. Fridovich I (1978) The biology of oxygen radicals. Science 201:875–880
15. Kasai H, Okada Y, Nishimura S, Rao MS, Reddy JK (1989) Formation of 8-hydroxydeoxyguanosine in liver DNA of rats following long-term exposure to a peroxisome proliferater. Cancer Res 49:2603–2605
16. Oda T, Akaike T, Hamamoto T, Suzuki F, Hirano T, Maeda H (1989) Oxygen radicals in influenza-induced pathogenesis and treatment with pyran-polymer conjugated SOD. Science 224:974–976
17. Forrest VJ, Kang Y-H, McClain DE, Robinson DH, Ramakrishnan N (1994) Oxidative stress-induced apoptosis prevented by Trolox. Free Radical Biol Med 16:675–684
18. Berlin V, Haseltine WA (1981) Reduction of adriamycin to a semiquinone-free radical by NADPH cytochrome P-450 reductase produces DNA cleavage in a reaction mediated by molecular oxygen. J Biol Chem 256:4747–4756
19. Kalyanaraman B, Perez-Reyes E, Mason RP (1980) Spin-trapping and direct electron spin resonance investigation of the redox metabolism of quinone anticancer drugs. Biochim Biophys Acta 630:119–130
20. Bligh HFJ, Bartoszek A, Robson CN, Hickson ID, Kasper CB, Beggs JD, Wolf CR (1990) Activation of mitomycin C by NADPH: cytochrome P-450 reductase. Cancer Res 50:7789–7792
21. Hoban PR, Walton MI, Robson CN, Godden J, Stratford IJ, Workman P, Harris AL, Hickson ID (1990) Decreased NADPH: cytochrome P-450 reductase activity and impaired drug activation in a mammalian cell line resistant to mitomycin C under aerobic but not hypoxic conditions. Cancer Res 50:4692–4697
22. Sato K, Akaike T, Suga M, Ando M, Maeda H (1994) Generation of free radicals from neocarzinostatin mediated by NADPH/cytochrome P-450 reductase via activation of enediyne chromophore. Biochem Biophys Res Commun 205:1716–1723
23. Sato K, Akaike T, Kojima Y, Ando M, Nagao M, Maeda H (1992) Evidence of direct generation of oxygen free radicals from heterocyclic amines by NADPH/cytochrome P-450 reductase. Jpn J Cancer Res 83:1204–1209
24. Akaike T, Sato K, Ijiri S, Miyamoto Y, Kohno M, Ando M, Maeda H (1992) Bactericidal activity of alkylperoxyl radicals generated by heme-iron-catalyzed decomposition of organic peroxides. Arch Biochem Biophys 294:55–63
25. Sato K, Akaike T, Kohno M, Ando M, Maeda H (1992) Hydroxyl radical production by H_2O_2 plus Cu,Zn-superoxide dismutase reflects the activity of free copper released from the oxidatively damaged enzyme. J Biol Chem 267:25371–25377
26. Ardies CM, Lasker JM, Bloswick BP, Lieber CS (1987) Purification of NADPH: cytochrome c (cytochrome P-450) reductase from hamster liver microsomes by detergent extraction and affinity chromatography. Anal Biochem 162:39–46

27. Zein N, Casazza AM, Doyle TW, Leet JE, Schroeder DR, Solomon W, Hadler SG (1993) Selective proteolytic activity of the antitumor agent kedaridin. Proc Natl Acad Sci USA 90:8009–8012
27a. Zein N, Reiss P, Bernatowicz M, Bolgar M (1995) The proteolytic specificity of the natural enediyne-containing chromoproteins is unique to each chromoprotein. Chem Biol 2:451–455
28. Oda T, Maeda H (1987) Binding to and internalization by cultured cells of neocarzinostatin and enhancement of its actions by conjugation with lipophilic styrene-maleic acid copolymer. Cancer Res 47:3206–3211
29. Miyamoto Y, Maeda H (1991) Enhancement by verapamil of neocarzinostatin action on multiple-resistant Chinese hamster ovary cells: possible release of nonprotein chromophore in cells. Jpn J Cancer Res 82:351–356
30. Cobuzzi RJ Jr, Kotsopoulos SK, Otani T, Beerman TA (1995) Effects of the enediyne C-1027 on intracellular DNA targets. Biochemistry 34:583–592
31. Miyamoto Y, Oda T, Maeda H (1990) Comparison of the cytotoxic effects of the high- and low-molecular-weight anticancer agents on multidrug-resistant Chinese hamster ovary cells in vitro. Cancer Res 50:1571–1575
32. Nicolou KC, Dai W-M, Tsay S-C, Estevez VA, Wrasidlo W (1992) Designed enediynes: a new class of DNA-cleaving molecules with potent and selective antitumor activity. Science 256:1172–1178
33. Mann J (1993) Nature's anticancer agents. Nature 361:21–22

Immunomodulating Antitumor Mechanisms of SMANCS

FUJIO SUZUKI and MAKIKO KOBAYASHI

Summary. In the first part of this chapter, the antitumor effects of poly(styrene-maleic acid)-conjugated neocarzinostatin (SMANCS) are compared with those of neocarzinostatin against various transplanted solid tumors in mouse models. As an antitumor mechanism of SMANCS in mice bearing solid tumors, the antitumor immunity stimulated by SMANCS was focused on these descriptions rather than the direct cytotoxic effect of SMANCS against tumor cells shown by neocarzinostatin. SMANCS stimulated the production of interferon-γ (IFN-γ), and activated macrophages and cytotoxic T lymphocytes were generated by stimulation from SMANCS-induced IFN-γ. IFN-γ played a key role in the elimination of solid tumors when mice were treated with SMANCS. The next part of this review discusses the antitumor efficacy of orally administered, oily formulated SMANCS (oily SMANCS). Because SMANCS is lipophilic and has a high affinity to fatty acids such as medium-chain triglycerides (MCT), the oily formulated SMANCS is easy to prepare. Active proteins commonly cannot be administered orally because they will be inactivated in the acidic environment of the digestive tract or digestive enzymes. However, orally administered oily SMANCS is absorbed by the intestine. Orally administered oily SMANCS was shown to have antitumor effects equal to aqueous SMANCS administered i.v. Furthermore, orally administered oily SMANCS inhibited pulmonary metastasis of B16 melanoma as a result of unique mechanisms that control the activity of T-helper type 2 cells (Th2 cells) in the host. The efficacy of SMANCS described here suggests that the oral administration of oily SMANCS to cancer patients may be beneficial.

Key words: Antitumor activity—Antimetastatic activity—IFN induction—Oral administration—SMANCS—Immunopotentiation

---

Department of Internal Medicine, University of Texas Medical Branch, 301 University Boulevard, Galveston, TX 77555-0835, USA

Introduction

Poly(styrene-maleic acid)-conjugated neocarzinostatin (SMANCS) [1–3] is a lipophilic derivative of the high molecular weight proteinaceous anticancer drug neocarzinostatin [4]. The antitumor activity of this conjugate against various experimental tumors in vivo and in vitro has been well described [5–9]. Compared with neocarzinostatin, the conjugate has been reported to have several considerable pharmacological advantages, including (1) decreased toxic side effects, (2) extension of the biological half-life, (3) enhanced tropism for tumor tissues, and (4) increased hydrophobicity [1–3,5–7]. Like the parental neocarzinostatin [10], SMANCS has been shown to retain the properties for arresting DNA synthesis of mammalian cells by direct DNA stand scission [2,6]. SMANCS was also shown to stimulate the host's nonspecific resistance against tumors [11–13]. In addition, the oily formulated SMANCS (oily SMANCS) [14] exhibits antitumor activity that is similar to aqueous SMANCS dissolved in phosphate-buffered saline[15]. Because oily SMANCS is absorbed by the intestine in its active form (in spite of proteinaceous agents) [14], it is possible to use it orally [15–17]. This review discusses the antitumor activities and antitumor mechanisms of SMANCS that are produced through stimulation of the host's antitumor immunity. In addition, the possibility of SMANCS as an oral anticancer drug is discussed.

Immunomodulating Antitumor Effect

The antitumor activity of SMANCS is exhibited both directly [2,6], through the inhibition of DNA synthesis by neocarzinostatin, and indirectly [11–13], through the stimulation of nonspecific resistance to tumors by the conjugated macromolecule. In this section, the effect of SMANCS on the host's antitumor immunity and the mechanisms involved in the regression of tumors in mice are discussed.

It is well known that polyanions, such as polyphosphates, polysulfates, and polycarboxylates, stimulate the production of interferon (IFN) [18–21]. In its molecule, SMANCS contains polymaleic-acid residues (polycarboxylate) [1,2], suggesting that SMANCS may have the potential to induce IFN production. Therefore, we examined the IFN-inducing activity of SMANCS in mice. Sera from mice at various times after treatment with or without SMANCS were assayed for IFN activity in L-929 cells infected with vesicular stomatitis virus. The IFN-inducing activity of SMANCS in mice is shown in Fig. 1. The IFN activity appeared in the sera of mice 12 h after administration of SMANCS (8 mg/kg, i.v.) and reached maximum levels (240 U/ml) after 20 h [11]. However, IFN-inducing activity was not detected in sera of mice treated with neocarzinostatin (1 mg/kg) or poly(styrene-maleic-acid) of low molecular size, i.e., about 1.5 kDa (SMA, 25 mg/kg) alone [11].

This suggests that when both neocarzinostatin and SMA were conjugated together, the conjugates have the ability to stimulate IFN production. The IFN

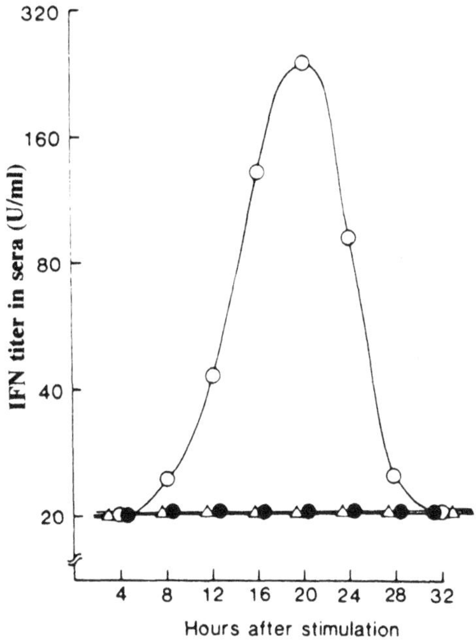

FIG. 1. The production of interferon (*IFN*) in serum of mice treated with polystyrene-maleic acid (SMA), neocarzinostatin, or poly(styrene-maleic acid)-conjugated neocarzinostatin (SMANCS). Mice were stimulated i.v. with SMA (25 mg/kg, *solid circles*), neocarzinostatin (1 mg/kg, *triangles*), or SMANCS (8 mg/kg, *open circles*); sera were obtained from these mice at various times after stimulation. Serum specimens were assayed for IFN activity in L-929 cells infected with vesicular stomatitis virus. (From [11], with permission)

TABLE 1. Activity of various interferons (IFNs) after treatment with anti-IFN antisera

	IFN titers (U/ml) after treatment with[b]		
IFN[a]	Rabbit serum	Anti-IFN-α/β	Anti-IFN-γ
Murine IFN α/β	200 (0)[c]	<14 (>93)	198 (1)
Murine IFN-γ	200 (0)	196 (2)	<14 (>93)
Serum IFN	108 (2)	65 (41)	47 (57)

[a] Serum IFN was prepared from mice 20h after poly(styrene-maleic acid)-conjugated neocarzinostatin (SMANCS) administration (1.6mg/kg, i.v.). Murine IFN-α/β and IFN-γ were purchased commercially.

[b] Murine IFN-α/β (0.5 ml) (400 U/ml), IFN-γ (400 U/ml), and serum IFN (220 U/ml) were mixed with an equal volume of 1:16 diluted normal rabbit serum, 400 IFN-neutralizing U/ml of anti-IFN-α/β and 400 IFN-neutralizing U/ml of anti-IFN-γ antisera, respectively, and assayed for their antiviral activities in L-929 cells infected with vesicular stomatitis virus.

[c] Loss of IFN activity (%) in parentheses.

TABLE 2. Induction of IFN by SMANCS in mice pretreated with various monoclonal antibodies (mAb)

Pretreatment[a]	IFN titer (U/ml)[b]	Suppression (%)
Anti-Thy 1.2 mAb	74	69[c]
Anti-Lyt 1.2 mAb	248	0
Anti-Lyt 2.2 mAb	65	65[c]
Saline	240	—

[a] Mice were pretreated s.c. with anti-Thy 1.2 mAb (2.4×10^4 U/kg), anti-Lyt 1.2 mAb (1.7×10^4 U/kg), or anti-Lyt 2.2 mAb (1.0×10^4 U/kg) 24 h before stimulation with SMANCS (8 mg/kg, i.v.).
[b] Serum specimens prepared from mice 20 h after SMANCS administration were assayed for IFN activity, as described in Fig. 1.
[c] Student's *t*-test, $P < .005$.

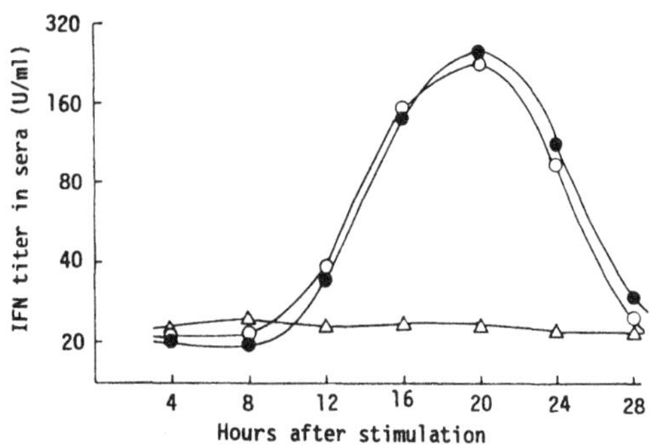

FIG. 2. IFN production by SMANCS in mice treated with anti-asialo GM1 antiserum. Mice were treated with anti-asialo GM1 antiserum (5 mg/kg, *triangles*), SMANCS (8 mg/kg, *open circles*), or the same amount of SMANCS and the antiserum (*solid circles*); sera obtained from these mice at various times after stimulation were assayed for IFN activity as shown in Fig. 1. (From [11], with permission)

induced in the sera of mice treated with SMANCS was shown to be 57% IFN-γ and 41% IFN-α/β when the serum specimens were treated with the respective antisera (Table 1) [12]. IFN production stimulated by SMANCS was demonstrated in mice treated with anti-Lyt 1.2 monoclonal antibody (mAb) or anti-asialo GM1 antiserum, while it disappeared in mice treated with carrageenan, anti-Thy 1.2 mAb, or anti-Lyt 2.2 mAb (Table 2; Fig. 2) [11]. This indicates that macrophages and $CD8^+$ T cells play an important role in the production of IFN-γ in mice stimulated with SMANCS.

In the next set of experiments, we attempted to determine whether there was a correlation between the appearance of IFN in the serum and the nonspecific resistance to tumors produced by SMANCS. The growth of solid tumors (Meth A fibrosarcoma and RL♂1 leukemia) in mice treated with a single

TABLE 3. Growth of solid tumors in mice pretreated with SMANCS

Tumor[a]	Agent	Dose	No. of mice	Tumor weight (mg)[b]	Inhibition (%)
Meth A	SMANCS	1.6 mg/kg	20	255 ± 30	82[c]
	Neocarzinostatin	1.0 mg/kg	20	1220 ± 89	13
	Saline	0.2 ml/mouse	20	1410 ± 110	0
RL♂ 1	SMANCS	1.6 mg/kg	20	549 ± 45	56[c]
	Neocarzinostatin	1.0 mg/kg	20	1134 ± 76	9
	Saline	0.2 ml/mouse	20	1247 ± 93	0

[a] At 24 h after i.v. administration of the agents, 1×10^6 cells/mouse of tumor cells were implanted s.c. into the shaved left side of the abdomen of BALB/c mice.
[b] Mice were killed 3 weeks after tumor implantation, and tumor nodules were excised and weighed.
[c] Student's *t*-test, $P < .001$.

TABLE 4. Antitumor activity of sera from SMANCS-treated mice on RL♂ 1 solid tumors in mice

Treatment[a]	No. of mice	Tumor size[b] ($\sqrt{a \times b}$ mm)	Inhibition (%)
Sera from SMANCS-treated mice[c]	10	12.7 ± 2	43[d]
Sera from normal mice[c]	10	21.9 ± 2	2
Saline	20	22.4 ± 4	0

[a] Tumor-bearing mice inoculated s.c. with 1×10^6 RL♂1 tumor cells/mouse were treated i.p. with 0.5 ml/mouse of a 1:10 dilution of sera or saline (0.5 ml/mouse). Treatments were administered to mice once daily for 10 consecutive days beginning 24 h after tumor inoculation.
[b] Tumor sizes, measured 14 days after tumor inoculation, were expressed as the mean of diameters that were determined by the largest diameter (a) and the smallest diameter (b) of the solid tumor, i.e., $\sqrt{a \times b}$ millimeters.
[c] Serum specimens were obtained from mice 20 h after administrations of SMANCS (1.6 mg/kg) or saline (0.2 ml/mouse).
[d] Student's *t*-test, $P < .01$.

i.v. dose of SMANCS 24 h before tumor implantation is shown in Table 3. Results are shown as the weight of solid tumors 3 weeks after s.c. tumor implantation. Pretreatment of mice with SMANCS resulted in significant inhibition of the growth of solid tumors [12]; one pretreatment with neocarzinostatin did not inhibit the growth of tumors. Because SMANCS was shown to be undetectable within 90 min after i.v. administration in mice [14], this suggests that the antitumor efficacy might be expressed through the host's nonspecific tumor immunity, which is stimulated by the compound. Therefore, the role of serum IFN in the suppression of tumor growth in mice pretreated with SMANCS was investigated, because IFN-γ is one of the representative effector cytokines for host antitumor immunity.

As shown in Table 4, when mice bearing RL ♂1 solid tumors were treated with IFN-containing sera from mice 20 h after administration of SMANCS, tumor growth was significantly reduced [12]. When serum IFN and anti-IFN-α/β anti-

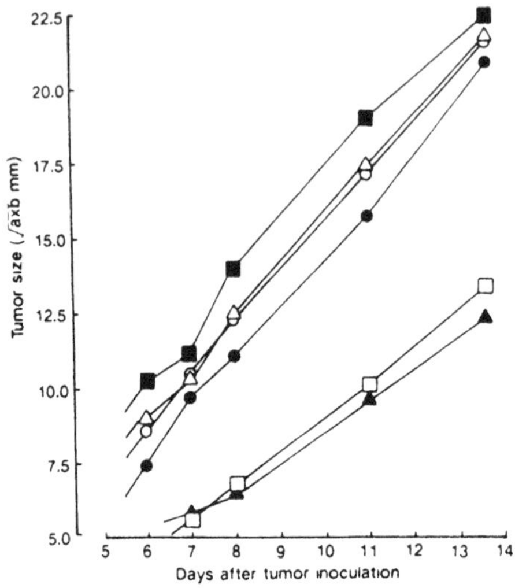

FIG. 3. Antitumor activity of SMANCS against RL ♂1 solid tumors in mice treated with anti-IFN antibodies. Mice treated s.c. with a 2500 IFN-neutralizing U/kg of anti-IFN-γ monoclonal antibody (mAb) (*solid squares*), anti-IFN-α/β antisera (*open triangles*), anti-IFN-γ mAb and SMANCS (*solid circles*), anti-IFN-α/β antisera and SMANCS (*solid triangles*), or SMANCS alone (1.6 mg/kg, i.v., *open squares*) were inoculated with RL ♂1 tumor cells (1×10^6 cells/mouse) into the shaved left side of the abdomen 24 h after various treatments. As a tumor control, mice were injected i.v. with 0.2 ml saline/mouse (*open circles*). (From [12], with permission)

sera were administered together to mice bearing RL ♂1 solid tumors, the antitumor activity of serum IFN was shown. However, this antitumor activity disappeared when mice bearing solid tumors were treated with a combination of serum IFN and anti-IFN-γ mAb (Fig. 3) [12]. These results indicated that IFN-γ provided nonspecific resistance to tumors in mice after stimulation by the conjugate.

What mechanisms are involved in the inhibition of tumor growth by SMANCS-induced IFN-γ? It is well known that IFN stimulates the host's nonspecific resistance against tumors through an increase of natural killer (NK) cell activity, activation of macrophages against tumor cells, increased expression of major histocompatibility complex class II antigens, and increased expression of major histocompatibility complex class I antigen [22–24]. The expression of histocompatibility complex class I antigen makes target cells more susceptible to lysis by cytotxic T lymphocytes [22–24]. To determine which effector cells were required for the development of nonspecific antitumor resistance of mice stimulated with SMANCS, tumor-bearing mice were treated with both SMANCS and blockers of macrophages (carrageenan) or NK cells (anti-asialo GM1 antiserum), and tumor growth was observed in these mice (Figs. 4 and 5). The depletion of macrophages from tumor-bearing mice by treatment with carrageenan resulted in complete elimination of the nonspecific resistance to tumors of mice stimu-

4

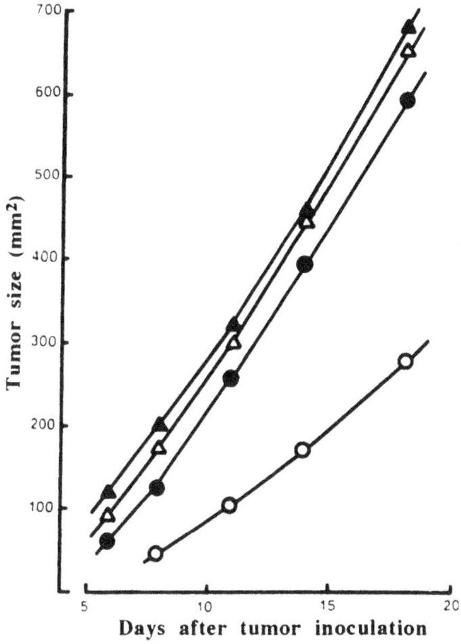

5

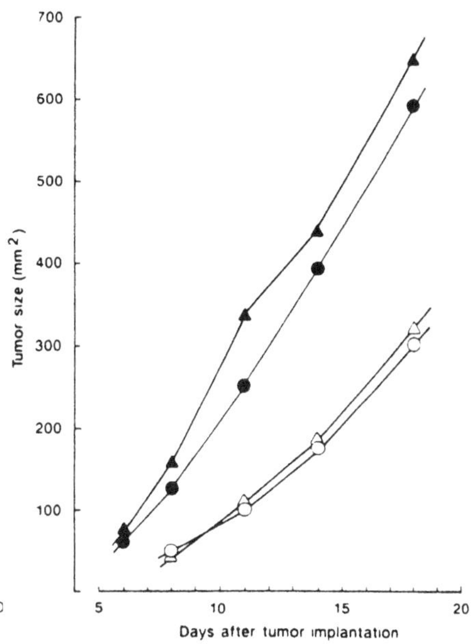

FIG. 4. Effect of carrageenan on the nonspecific resistance to tumors of mice stimulated with SMANCS. Mice bearing RL♂1 solid tumors were treated with saline (i.v., 0.2 ml/mouse, *solid circles*), carrageenan (50 mg/kg, i.p., *solid triangles*), SMANCS (1.6 mg/kg, i.v., *open circles*), or SMANCS and carrageenan (*open triangles*). Carrageenan was administered to mice 2 days before and 1 day after tumor inoculation and once weekly for 2 weeks; SMANCS and saline were administered to mice 24 h before tumor inoculation (From [13], with permission)

FIG. 5. Effect of anti-asialo GM1 antiserum treatment on the nonspecific resistance to tumors of mice stimulated with SMANCS. Mice bearing RL♂1 solid tumors were treated with saline (i.v., 0.2 mg/mouse, *solid circles*), anti-asialo GM1 antiserum (5 mg/kg, i.v., *solid triangles*), SMANCS (1.6 mg/kg, i.v., *open circles*), or SMANCS and the antiserum (*open triangles*). Antiserum treatment began 1 day before tumor inoculation and was repeated 4 times at 5-day intervals; SMANCS and saline were administered to mice 24 h before tumor inoculation. (From [13], with permission)

lated by SMANCS [13]. When tumor-bearing mice were depleted of NK cells by treatment with anti-asialo-GM1 antiserum, the antitumor activity of SMANCS was unchanged, suggesting that NK cells did not play a role in the development of nonspecific resistance to tumors of mice stimulated with SMANCS. In addition, the antitumor activity of SMANCS was also demonstrated in NK cell-deficient mice (beige mice) just as it was demonstrated in normal mice (nonbeige littermates) when NK cell-resistant EL-4 thymoma cells were used as a tumor target (Table 5) [13]. These results suggest that SMANCS requires activated macrophages, but not NK cells to express nonspecific resistance to tumors of mice bearing solid tumors.

In conclusion, macrophages and $CD8^+$ T cells participated in the production of IFN-γ in mice after treatment with SMANCS. IFN-γ then activated macrophages and the growth of transplanted solid tumors was inhibited by the action of these activated macrophages. The findings are interesting in that IFN-γ induced by SMANCS activates nonspecific tumor resistance, and activated macrophages act as the final effector cells.

Antitumor Activity of Oily SMANCS

SMANCS suspended or dissolved in a lipid contrast medium (Lipiodol) has been widely administered to patients with liver and other solid cancers by catheterizing the respective feeding arteries under X-ray monitoring [6,25,26]. Using this method, SMANCS can be delivered and deposited in the tumor tissue over a long period of time without major systemic side effects [6,26]. Obvious clinical effectiveness in patients with unresectable liver cancer has been demonstrated [27]. It has been available commercially in Japan since 1994.

Oily formulated SMANCS (oily SMANCS) is also prepared using medium-chain triglycerides (MCT) [14]. Although protein drugs are often ineffective when they are administered orally because of their susceptibility to enzymes of the digestive tract, the possibility of intestinal absorption of oily SMANCS was examined after oral administration in mice. Blood concentrations of SMANCS after oral administration are shown in Fig. 6. After oral administration of 100 mg/kg of SMANCS dissolved in phosphate-buffered saline (PBS) (aqueous SMANCS), the blood level of the compound reached a maximum of 0.1 μg/ml 1 h after administration, then gradually decreased and disappeared 5 h after administration [14]. In contrast, the same amount of oily SMANCS reached a maximum of 1 μg/ml 5 h after oral administration, and the activity in plasma disappeared

TABLE 5. Antitumor activities of SMANCS beige mice and nonbeige littermates implanted with EL-4 thymoma cells

		Tumor size ($\sqrt{a \times b}$ mm)[b]		
Mice[a]	Agent	Day 6	Day 10	Day 14
Nonbeige littermate	SMANCS	6.5 ± 3	11.6 ± 4[c]	13.1 ± 4[c]
Nonbeige littermate	Saline	8.4 ± 4	16.7 ± 5	21.9 ± 5
Beige	SMANCS	6.5 ± 3	12.3 ± 4[c]	15.8 ± 5[c]
Beige	Saline	10.1 ± 4	21.3 ± 5	28.3 ± 7

[a] At 24 h after i.v. administration of SMANCS (1.6 mg/kg), 1×10^6 cells/mouse of EL-4 thymoma cells were implanted s.c. into the shaved left side of the abdomen of beige mice and nonbeige littermates (10 mice each). As a control, saline (0.2 ml/mouse, i.v.) was administered to mice using the same schedule.

[b] Tumor size was determined as described in Table 4.

[c] Student's *t*-test, $P < .001$.

6

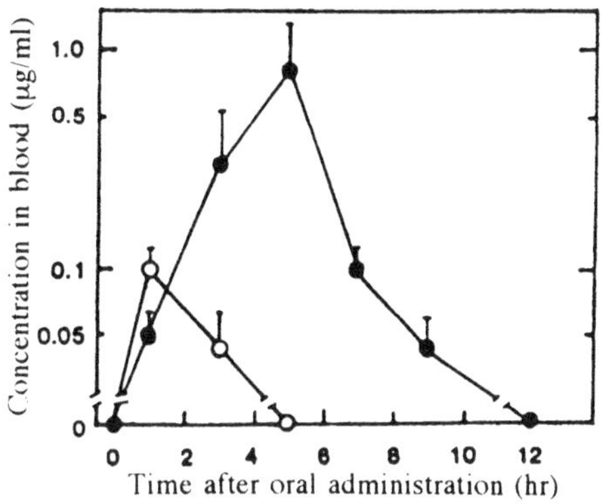

7

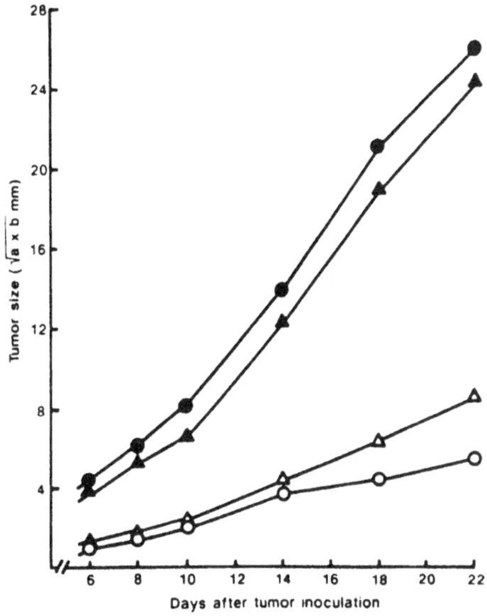

FIG. 6. Plasma concentrations of SMANCS in mice treated orally with the aqueous or oily form. A 100-mg/kg dose of aqueous (*open circles*) or oily (*solid circles*) SMANCS was administered to mice orally. At various times after administration, serum specimens were prepared from these mice. The amount of the agents contained in these serum specimens was measured by the in vitro growth inhibition test on TK/B cells, which are highly sensitive to the cytotoxicity of neocarzinostatin and SMANCS. Data are means ± SE (n = 5). (From [14], with permission)

FIG. 7. Effect of oral administration of oily SMANCS on the growth of Meth A solid tumors in mice. Mice bearing Meth A solid tumors were treated with MCT (0.2 ml/mouse, p.o., *solid circles*), oily SMANCS (10 mg/kg, p.o., *open triangles*), aqueous SMANCS (0.1 mg/kg, i.v., *open circles*), or aqueous SMANCS (0.1 mg/kg, p.o., *solid triangles*) once daily for 7 consecutive days. Tumor sizes were determined as described in Table 4. (From [15], with permission)

12 h after administration [14]. Neocarzinostatin, whether in PBS or in MCT, did not demonstrate detectable levels in the plasma after oral administration (data not shown).

Thus, oily SMANCS is absorbed much more efficiently than aqueous SMANCS when administered orally. In addition, orally administered oily SMANCS has been shown to accumulate in tumor tissues rather than in normal tissues [14]. Therefore, the antitumor activity of oily SMANCS after oral administration to mice bearing solid tumors was examined. The effect of oily SMANCS administered orally on the growth of Meth A solid tumors is shown in Fig. 7. When mice bearing solid tumors were treated with a 10 mg/kg oral dose of oily SMANCS, significant reductions in tumor growth were demonstrated as compared with that of controls treated with MCT [15]. The antitumor potential of orally administered oily SMANCS was comparable to i.v. administration of aque-

8

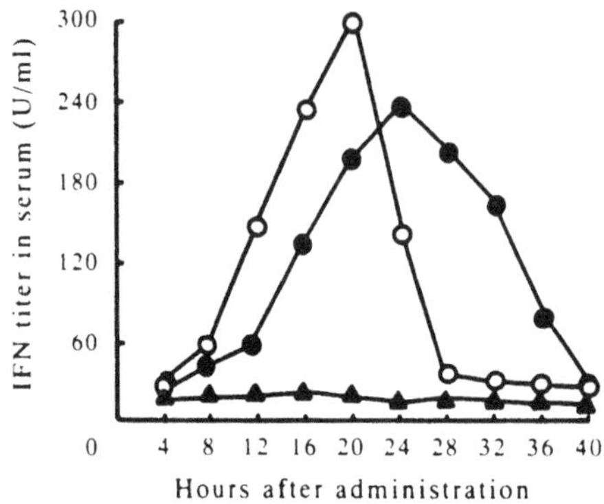

9

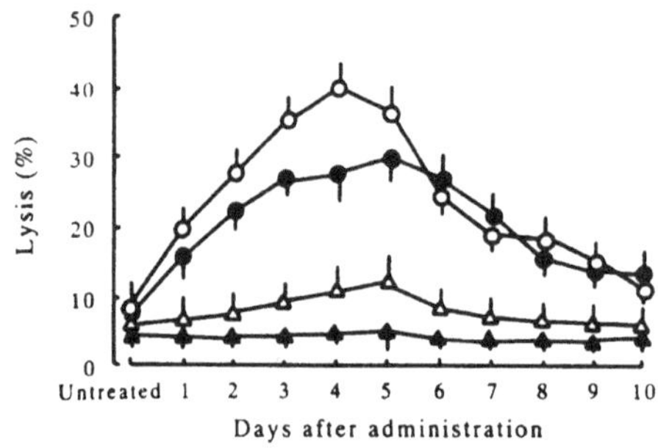

FIG. 8. IFN production in mice treated orally with oily SMANCS. Mice were stimulated with either aqueous (1.6 mg/kg, i.v., *open circles*) or oily (10 mg/kg, oral, *solid circles*) SMANCS. At various times after stimulation, serum specimens were obtained and assayed for their antiviral activities in L-Galveston cells infected with vesicular stomatitis virus. As a control, mice were treated with medium-chain triglyceride (MCT) solution (*solid triangles*). (From [16], with permission)

FIG. 9. Cytostasis of peritoneal macrophages derived from oily SMANCS-treated mice. Peritoneal exudate cells were prepared from mice treated with aqueous (i.v., 1.6 mg/kg, *open circles*) or oily (p.o., 10 mg/kg, *solid circles*) SMANCS. Cytostatic activity of macrophages against RL♂1 target cells was assayed by a [^{3}H]thymidine incorporation inhibition test at an effector-to-target cell ratio of 10 : 1. As controls, mice were treated with a 10 mg/kg oral dose of aqueous SMANCS (*open triangles*) or MCT solution (*solid triangles*). (From [16], with permission)

ous SMANCS [15]. Similar antitumor activity of oily SMANCS was shown in mice bearing solid RL♂1 tumors [15].

Does orally administered oily SMANCS express the same antitumor process as i.v. administered aqueous SMANCS? To answer this question, the immunomodulating activity of orally administered oily SMANCS was investigated in the following experiments. Because aqueous SMANCS inhibits growth of solid tumors through the induction of IFN-γ [11–13], the IFN-inducing activity of oily SMANCS after oral administration was examined first. When a 10 mg/kg dose of oily SMANCS was administered to mice orally, IFN activity was demonstrated in sera [16] (Fig. 8). The peak levels of activity were found in sera of mice 24 h after oral treatment and were similar to the IFN activity induced in mice 20 h after i.v. treatment with aqueous SMANCS [16]. Serum IFN induced by oily SMANCS contained about 55% IFN-α/β and 48% IFN-γ activity (Table 6) [16]. These results indicate that the IFN-inducing activity of orally administered oily SMANCS was comparable to the IFN-inducing activity of aqueous SMANCS administered i.v.

In the previous section, we described the role of macrophages on the antitumor activity of SMANCS administered in the parenteral cavity. Cytotoxic effects of macrophages activated by SMANCS-induced IFN-γ were found to inhibit the

TABLE 6. Antiviral activity of various IFNs after treatment with anti-IFN antisera

IFN[a]	IFN titer[b] (U/ml) after treatment with		
	Anti-IFN-α/β	Anti-IFN-γ	Both antibodies
IFN-α/β, 100 U/ml	<8 (>92)[c]	99 (1)	<8 (>92)
IFN-γ, 100 U/ml	98 (2)	<8 (>92)	<8 (>92)
Serum IFN, 120 U/ml	54 (55)	63 (48)	<8 (>93)

[a] Serum IFN was prepared from mice 24 h after oily SMANCS administration (10 mg/kg, oral). Murine IFN-α/β and IFN-γ antisera were purchased commercially.
[b] Murine IFN-α/β (0.5 ml, 200 U/ml), murine IFN-γ (0.5 ml, 200 U/ml), and serum IFN (0.5 ml, 240 U/ml) were mixed with equal volumes of anti-IFN-α/β, anti-IFN-γ, and their mixture (200 IFN-neutralizing U/ml), respectively. Their antiviral activities were then assayed in L-Galveston cells infected with vesicular stomatitis virus.
[c] Loss of IFN activity (%).

growth of solid tumors. Therefore, the appearance of cytotoxic macrophages in mice treated orally with oily SMANCS was examined next. As shown in Fig. 9, peritoneal macrophages from mice treated with aqueous SMANCS (i.v.) or oily SMANCS (p.o.) showed cytotoxicities for RL♂1 target cells when they were cocultured. However, peritoneal macrophages from mice treated orally with aqueous SMANCS or MCT solution did not inhibit the growth of target cells significantly [16]. These results suggest that activated macrophages were also generated in mice treated orally with oily SMANCS.

In conclusion, orally administered oily SMANCS was shown to be equally effective as aqueous SMANCS administered i.v. on the inhibition of solid tumors in mice. Similar to the antitumor mechanism of aqueous SMANCS, oily SMANCS stimulates the generation of activated macrophages through the induction of IFN-γ. An interesting aspect of this experiment is the effectiveness of orally administered SMANCS in spite of its proteinaceous nature. The oily formulation of SMANCS for the oral route would provide greater usefulness.

New Immunomodulating Effects of Oily SMANCS

Because of various developments in medical techniques, equipment, and chemotherapeutic agents, mortality rates for patients with primary tumors have been decreasing [28]. However, cancer patients still have high mortality rates if they develop metastasis [28], suggesting that control of metastasis is one of the most important factors for decreasing mortality rates. For this reason, immunotherapies to induce cytotoxic T lymphocytes against tumors and monoclonal antibodies for adhesion molecules on the surface of metastatic tumor cells have been studied [29–32]. However, the successful treatment of cancer metastasis is the greatest limitation because the metastatic process of tumor cells is complex [33].

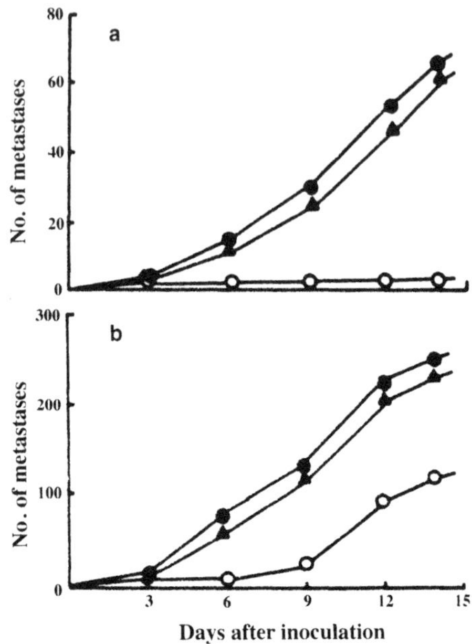

FIG. 10a,b. Pulmonary metastases of B16F10 melanoma cells in mice treated with oily SMANCS. Mice were inoculated i.v. with (**a**) 1×10^5 or (**b**) 2×10^5 cells/mouse of B16F10 melanoma cells. These mice were treated orally with MCT (0.2 ml/mouse, *solid circles*), oily SMANCS (10 mg/kg, *open circles*), or aqueous SMANCS (10 mg/kg, *solid triangles*) once daily for 7 consecutive days. At various days after tumor inoculation, metastases in lungs of these mice were determined, as described previously

It has been documented [34] that development of cancer metastasis is linked with the impairment of the host's antitumor resistance. Because SMANCS has a direct cytotoxic effect against tumor cells [2,6] and stimulates the host's antitumor resistance [11–13], it is considered that treatment with SMANCS of individuals bearing metastatic tumors may inhibit growth of the primary tumor as well as the metastatic tumor foci or nodules. Therefore, the antimetastatic effect of orally administered oily SMANCS was investigated using a pulmonary metastatic model of murine B16 melanoma [33].

Figure 10 shows the number of metastatic colonies in lungs of mice treated orally with oily SMANCS or MCT vehicles at various days after i.v. inoculation with metastatic B16F10 melanoma cells [17]. When oily SMANCS (10 mg/kg) was administered orally to mice inoculated with B16F10 cells, 100% of pulmonary metastasis was inhibited as compared to that of controls (treated with MCT alone) [17]. However, orally administered aqueous SMANCS did not inhibit pulmonary metastasis of melanoma (Fig. 10). As shown in Table 7, pulmonary metastasis in mice bearing B16F10 melanoma cells was significantly reduced in all groups treated orally with oily SMANCS at various schedules. The maximum

TABLE 7. Effect of various schedules of oily SMANCS administration on pulmonary metastasis in mice bearing B16F10 melanoma cells

Oral treatments[a]	No. of metastasis (mean ± SE)[b]	Inhibition (%)
Medium-chain tryglycerides (MCT) (0.2 ml/mouse)	68 ± 7	—
Oily SMANCS (10 mg/kg)		
1, 2, 3, 4, 5, 6, 7 days after tumor inoculation	2 ± 1	97
1, 3, 5, 7, 9 days after tumor inoculation	15 ± 2	77
3, 4, 5, 6, 7 days after tumor inoculation	19 ± 5	72
7 days after tumor inoculation	16 ± 3	76
Aqueous SMANCS (10 mg/kg)	72 ± 9	0

[a] Oily SMANCS was administered to mice bearing B16F10 melanoma cells (1×10^5 cells/mouse, i.v.) at the various schedules listed in the table. Aqueous SMANCS was administered to mice 7 consecutive days beginning 1 day after tumor inoculation.

[b] At 14 days after tumor inoculation, the number of pulmonary metastatic colonies was determined under a dissecting microscope.

antimetastatic effect of oily SMANCS was demonstrated when 10 mg/kg was given to mice once a day for 7 consecutive days beginning 1 day after tumor inoculation [17]. A single oral administration of oily SMANCS to mice 7 days after tumor inoculation also inhibited pulmonary metastasis in mice inoculated with B16F10 melanoma cells (76% inhibition) [17].

In the next step we attempted to clarify the effective mechanism of oily SMANCS administered orally on the inhibition of pulmonary metastasis in mice inoculated with B16F10 melanoma cells. From previous data (see section on *Antitumor Activity*), the inhibitory mechanism of oily SMANCS against RL♂1 or Meth A solid tumors appears to result from stimulation of the host's antitumor immunity. Therefore, it is possible that the antimetastatic activity of oily SMANCS administered orally in mice bearing B16 melanoma cells is also expressed through the host's antitumor functions that are stimulated with the agent. To test this possibility, we attempted to examine the mechanisms involved in the pulmonary metastasis of B16 melanoma.

It has been documented [35–37] that suppression of the host's antitumor immune responses (type 1 T-cell responses) is accelerated in cancer patients. The type 1 T-cell response is an up-regulatory cellular immune response associated with increased levels of IL-2 and IFN-γ (type 1 cytokines) and the generation of $CD8^+$ cytotoxic T lymphocytes (typical type 1 effector T cells). The release of soluble tumor antigens [38], the formation of circulating immune complexes [39], and the appearance of immunosuppressive factors [40–44] and of cytokines produced by tumor cells [45–48] have been described as the causes of tumor-related impaired type 1 T-cell responses. In contrast, the type 2 T-cell response is a down-regulatory cellular immune response associated with increased levels of IL-4 and IL-10 (type 2 cytokines). Tumor-associated suppressor cells (type 2 T cells) have also been shown to be an important factor in the suppression of type 1 T-cell responses in cancer patients [35,36]. Therefore, the correlation between

11

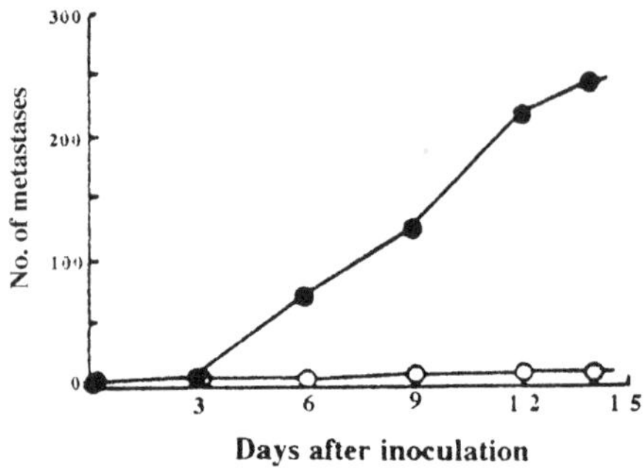

12

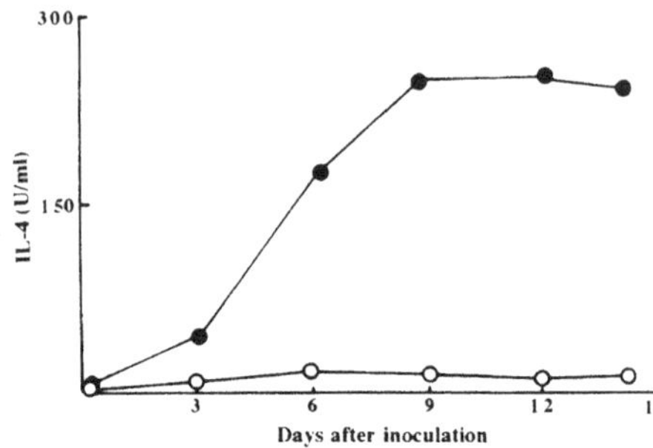

FIG. 11. Pulmonary metastases in mice at various days after the inoculation of B16F1 or B16F10 melanoma cells. Mice were inoculated i.v. with 2×10^5 cells/mouse of B16F1 (*open circles*) or B16F10 (*solid circles*) melanoma cells. Pulmonary metastases in these mice were determined at various days after inoculation of melanoma cells as described in Fig. 10

FIG. 12. IL-4 activity in culture fluids of spleen cells from mice at various days after inoculation of B16F1 or B16F10 melanoma cells. Splenic cells from mice, at various days after inoculation with B16F1 (*open circles*) or B16F10 (*closed circles*) melanoma cells, were stimulated with anti-CD3 mAb (2.5 μ/ml) for 48 h. Culture fluids obtained from these cultures were assayed for IL-4 by bioassay with CTLL-2 cells as described previously

metastasis and the appearance of type 2 T cells in mice bearing B16 melanoma cells was first examined, using poorly metastatic B16F1 cells and highly metastatic B16F10 cells.

Mice inoculated i.v. with cultured B16F1 cells or B16F10 cells were designated as BF1 mice and BF10 mice, respectively. In contrast to pulmonary metastasis in BF10 mice, metastatic colonies were not observed in the lungs of BF1 mice 14 days after inoculation with melanoma cells (Fig. 11). To detect type 2 T-cell activity, spleen cells from BF1 mice or BF10 mice at various times after tumor

TABLE 8. Pulmonary metastasis in BF1 mice inoculated with T-helper type 2 (Th2) cells

Treatment[a]	Number of metastases/lung (mean ± SE)[b]	Index (control/tested)
Saline (control)	12 ± 1	1.0
Th2 cells from BF10 mice		
2×10^5 cells/mouse	11 ± 4	0.9
1×10^6 cells/mouse	100 ± 23[c]	8.3
5×10^6 cells/mouse	>250[c]	>20.8
Splenic T cells from normal mice		
5×10^6 cells/mouse	11 ± 2	0.9

[a] Various numbers of Th2 cells, prepared from spleens of BF10 mice (14 days after tumor inoculation), were adoptively transferred into BF1 mice 1 day after tumor inoculation.
[b] At 14 days after tumor inoculation, the number of pulmonary colonies in these mice (10 mice each) was determined under a dissecting microscope.
[c] Student's *t*-test, $P < .001$.

inoculation were cultured for 48 h, and culture fluids harvested were assayed for their IL-4 activity by bioassay with CTLL-2 cells [49] (Fig. 12). The IL-4 activity was demonstrated in culture fluids of spleen cells from BF10 mice starting 3 days after tumor inoculation. However, IL-4 activities were not observed in culture fluids of spleen cells from BF1 mice until 14 days after tumor inoculation. Cells responsible for the type 2 T-cell activity were characterized as $CD4^+$ $CD28^+$ $TCR\alpha/\beta^+$ IL-4- and IL-10-producing T cells (data not shown). This suggests that the development of pulmonary metastasis of B16 melanoma might be associated with the appearance of $CD4^+$ type 2 T cells.

To determine the role of T-helper type 2 (Th2) cells on the development of pulmonary metastasis, Th2 cells obtained from BF10 mice were adoptively transferred to BF1 mice. Although significant numbers of metastatic colonies were not demonstrated in the lungs of BF1 mice, the pulmonary metastasis of B16F1 cells increased to the same levels shown in BF10 mice (see Fig. 11) when BF1 mice were inoculated with Th2 cells derived from BF10 mice (Table 8) [17]. These results indicate that tumor-associated Th2 cells may play an important role in the pulmonary metastasis of B16 melanoma, suggesting that if Th2 cells could be controlled, pulmonary metastasis in BF10 mice might be inhibited. Because the antimetastatic effect of orally administered oily SMANCS is considered to appear through the host's immune function, we have hypothesized that oral administration of oily SMANCS may inhibit the Th2 cell activity. To examine this hypothesis, we tried to detect antagonistic cells for Th2 cells induced by the SMANCS stimulation. Various cells prepared from the spleen of mice 7 days after oral treatment with oily SMANCS were adoptively transferred to mice 7 days after inoculation with B16F10 cells, and the number of metastatic colonies in the lungs were counted 7 days after the transfer (14 days after tumor inoculation).

TABLE 9. Pulmonary metastasis in BF10 mice inoculated with various splenic cells from SMANCS-treated mice

Treatment[a]	Numbers of metastases/lung (mean ± SE)[b]	Inhibition (%)
Saline (control)	136 ± 11	—
Spleen cells	68 ± 8	50
Plastic-adherent cells	132 ± 23	3
B cells	139 ± 16	0
$CD4^+$ T cells	60 ± 8[c]	56
$CD8^+$ T cells	125 ± 4	1

[a] Spleen cells and their fractions (plastic-adherent cells, B cells, $CD4^+$ T cells and $CD8^+$ T cells), which were obtained from mice 1 day after the final oral treatment with oily SMANCS (10 mg/kg, every other day, 3 times), were adoptively transferred to BF10 mice 7 days after inoculation with B16F10 melanoma cells (2×10^5 cells/mouse, i.v.).

[b] At 14 days after tumor inoculation, the number of pulmonary metastatic colonies in recipient mice was determined under a dissecting microscope.

[c] Student's t-test, $P < .01$.

TABLE 10. Inhibition of type 2 cytokine production by Th2 cells from mice bearing B16F10 melanoma cells

Th2 cells cocultured with splenic CD4+ T cells from[a]	IL-4 (U/ml)[b]	IL-10 (U/ml)[b]
Normal mice	163	58
Mice treated with oily SMANCS	<5	<2

[a] Th2 cells were prepared from spleens of BF10 mice. CD4+ T cells, from normal mice treated with SMANCS (10mg/kg), were assayed for their abilities to inhibit IL-4 or IL-10 production from Th2 cells in coculture system at a Th2 cell-to-CD4+ T-cell ratio of 1:1.
[b] At 48h after cocultivation in the presence of anti-CD3 mAb (2.5μg/ml), type 2 cytokine activity in culture fluids was measured by bioassay with CTLL-2 cells (IL-4) or enzyme-linked immunosorbent assay (ELISA) (IL-10) [49].

TABLE 11. Characterization of anti-Th2 T cells induced by SMANCS

	Cytokine production (U/ml)[b]	
Coculture of Th2 cells[a]	IL-4	IL-10
Medium	250	40
CD4+ T cells from normal mice	248	37
CD4+ T cells from SMANCS-treated mice (effector cells)	<5	<2
CD28+ cell-depleted effector cells	242	38
TCRα/β+ cell-depleted effector cells	245	38
TCRγ/δ+ cell-depleted effector cells	12	4
Vicia villosa lectin-adherent cells in CD4+ T cells	<5	<2

[a] Th2 cells were prepared from spleens of BF10 mice. Using affinity columns, CD4+ T cells were prepared from spleens of normal mice or mice 1 day after the final oral treatment with oily SMANCS (10mg/kg, every other day, 3 times). As *Vicia villosa* lectin-adherent cells, CD4+ T cells that showed an affinity to the lectin were harvested, as described previously [49]; 1×10^6 cells/ml of CD4+ T cells treated with mAbs and complement [49] and other cell preparations were assayed for their abilities to inhibit IL-4 or IL-10 production from Th2 cells in a coculture system (as shown in Table 10).
[b] At 48h after cocultivation in the presence of anti-CD3 mAb (2.5μg/ml), type 2 cytokine activity in culture fluids was measured by bioassay with CTLL-2 cells (IL-4) or ELISA (IL-10) [49].

Results obtained (Table 9) showed that pulmonary metastasis of BF10 mice inoculated with spleen cells from SMANCS-treated mice was inhibited as demonstrated in BF10 mice treated with SMANCS, suggesting that the antimetastatic activity of SMANCS was produced through the induction of antagonistic cells against Th2 cells [50]. Phenotypical analysis of spleen cells from mice treated with oily SMANCS showed that the cells responsible for the inhibition of pulmonary metastasis were $CD4^+$ T cells (Table 9). However, neither macrophages, B cells, nor $CD8^+$ T cells showed any antimetastatic activity when they were adoptively transferred (Table 9). These results indicate that $CD4^+$ T cells generated by stimulation with oily SMANCS played a role in the inhibition of pulmonary metastasis.

We next examined the function of splenic $CD4^+$ T cells from mice treated with oily SMANCS. Splenic Th2 cells from BF10 mice 14 days after tumor inoculation produced type 2 cytokines in their culture fluids, but type 2 cytokines were not produced by $CD4^+$ T cells from mice treated with oily SMANCS (Table 10) [50]. Thus, when Th2 cells were cocultured with $CD4^+$ T cells from normal mice treated with oily SMANCS, type 2 cytokine production from Th2 cells was completely inhibited (Table 10). To determine the phenotypic properties of SMANCS-induced anti-Th2 T cells, they were treated with various mAbs followed by complement or *Vicia villosa* lectin-coated petri dishes, and the remaining cells harvested were assayed for their anti-Th2 T-cell activity (Table 11) [50]. When effector cells were treated with anti-CD28 mAb or anti-TCRα/β mAb plus complement, their anti-Th2 T-cell activities were eliminated. When they were treated with anti-TCRγ/δ mAb followed by complement, there was no change in the anti-Th2 T-cell activity of these cells (Table 11).

In addition, the anti-Th2 T-cell activity was only detected in $CD4^+$ T cells that have an affinity to *Vicia villosa* lectin (Table 11). These results suggested that the cell responsible for the anti-Th2 T-cell activity induced by SMANCS was a $CD4^+$ $CD28^+$ $TCR\alpha/\beta^+$ *Vicia villosa* lectin-adherent T cell. $CD4^+$ anti-Th2 T cells induced by oily SMANCS bore a striking resemblance to contrasuppressor T cells described by Gershon and co-workers in 1981 [51]. These cells are able to counteract the suppressor T-cell activity in vitro and were generated in response to the appearance of suppressor T cells [52,53]. Further studies are necessary to clarify the properties of these two cell populations.

In conclusion, the inhibitory effect of oily SMANCS against pulmonary metastasis of B16 melanoma is different from the antitumor mechanisms described in previous sections of this review. It is a unique effect which inhibits the host's negative immune systems, and indicates that SMANCS inhibits pulmonary metastasis of B16 melanoma through correction of the imbalance of Th1 and Th2 responses.

References

1. Maeda H, Takeshita J, Kanamaru R (1979) A lipophilic derivative of neocarzinostatin. A polymer conjugation of antitumor protein antibiotic. Int J Pept Protein Res 14:81–87
2. Maeda H, Ueda M, Morinaga T, Matsumoto T (1985) Conjugation of poly(styrene-co-maleic acid) derivatives to the antitumor protein neocarzinostatin: pronounced improvements in pharmacological properties. J Med Chem 29:455–461
3. Maeda H (1991) SMANCS and polymer-conjugated macromolecular drugs: advantages in cancer chemotherapy. Adv Drug Delivery Rev 6:181–202

4. Maeda H (1981) Neocarzinostatin in cancer chemotherapy. Anticancer Res 1:175–186
5. Takeshita J, Maeda H, Kanamaru R (1982) In vitro mode of action, pharmacokinetics, and organ specificity of poly(maleic acid-styrene)-conjugated neocarzinostatin, SMANCS. Gann 73:278–284
6. Maeda H, Matsumoto T, Konnno T, Iwai K, Ueda M (1984) Tailor making of protein drugs by polymer conjugation for tumor targeting: a brief review on SMANCS. J Protein Chem 3:181–193
7. Iwai K, Maeda H, Konno T, Matsumura Y, Yamashita R, Yamasaki K, Hirayama S, Miyauchi Y (1987) Tumor targeting by arterial administration of lipids: rabbit model with VX2 carcinoma in the liver. Anticancer Res 7:321–328
8. Suzuki F, Okuno Y, Maeda Y, Maeda H (1987) Tumor growth inhibitory effects of SMANCS, a poly(styrene-maleic acid) conjugated derivative of neocarzinostatin, on various tissue culture cells. Jpn J Cancer Chemother 14:3305–3312
9. Kimoto A, Konno T, Kawaguchi T, Miyauchi Y, Maeda H (1992) Antitumor effect of SMANCS on rat mammary tumor induced by 7,12-dimethylbenz[α]anthracene. Cancer Res 52:1013–1017
10. Napier MA, Holmquist B, Strydom DJ, Goldberg IH (1979) Neocarzinostatin: spectral characterization and separation of non-protein chromophore. Biochem Biophys Res Commum 89:635–642
11. Suzuki F, Munakata T, Maeda H (1988) Interferon induction by SMANCS: a polymer-conjugated derivative of neocarzinostatin. Anticancer Res 8:97–104
12. Suzuki F, Pollard RB, Maeda H (1989) Stimulation of non-specific resistance to tumors in the mouse using a poly(maleic-acid-styrene)-conjugated neocarzinostatin. Cancer Immunol Immunother 30:97–104
13. Suzuki F, Pollard RB, Uchimura S, Munakata T, Maeda H (1990) Role of natural killer cells and macrophages in the nonspecific resistance to tumors in mice stimulated with SMANCS, a polymer-conjugated derivative of neocarzinostatin. Cancer Res 50:3897–3904
14. Oka K, Miyamoto Y, Matsubara Y, Tanaka S, Oda T, Suzuki F, Maeda H (1990) Enhanced intestinal absorption of a hydrophobic polymer-conjugated protein drug, SMANCS, in an oily formulation. Pharm Res (NY) 7:852–855
15. Schmitt DA, Kisanuki K, Kimura S, Oka K, Pollard RB, Maeda H, Suzuki F (1992) Antitumor activity of orally administered SMANCS, a polymer-conjugated protein drug, in mice bearing various murine tumors. Anticancer Res 12:2219–2224
16. Suzuki F, Matsumoto M, Schmitt DA, Pollard RB, Maeda H (1993) Immunomodulating activities of orally administered SMANCS, a polymer-conjugated derivative of the proteinaceous antibiotic neocarzinostatin, in an oily formulation. Int J Immunopharmacol 15:175–183
17. Suzuki F, Kobayashi M, Maeda M, Pollard RB (1994) Anti-metastatic activity of SMANCS, a polymer-conjugated proteinaceous antibiotic neocarzinostatin, in mice bearing B16F10 melanoma. Proc Am Assoc Cancer Res 35:399
18. DeClercq E, Merigan TC (1970) Induction of interferon by nonviral agents. Arch Intern Med 126:94–108
19. DeSomer P, DeClercq E, Billiau A, Schonne E, Claesen M (1968) Antiviral activity of polyacrylic and polymethacrylic acid. II. Model of action in vivo. J Virol 2:886–893
20. Merigan TC (1967) Induction of circulating interferon by synthetic anionic polymers of known composition. Nature 214:416–417
21. Ottenbrite RM (1982) The antitumor and antiviral effect of polycarboxylic acid polymers. In: Biological activities of polymers. American Chemical Society, Washington, DC, p 205

22. Hook JJ, Detrick B (1987) The interferon system and disease. In: Pfeffer LM (ed) Mechanisms of interferon actions, vol II. CRC Boca Raton, pp 113–128
23. Faltynek CR, Oppenheim JJ (1988) Interferons in host defense. J Natl Cancer Inst 80:151–153
24. Mannel DN, Falk W (1983) Interferon-γ is required in activation of macrophages for tumor cytotoxicity. Cell Immunol 29:396–402
25. Konno T, Maeda H, Iwai S, Maki S, Tashiro S, Uchida M, Miyauchi Y (1984) Selective targeting of anticancer drug and simultaneous image enhancement in solid tumors by arterially administered lipid contrast medium. Cancer 54:2367–2374
26. Maeda H (1986) Development of cancer selective anticancer agent based on the unique properties of tumor neovasculatures and its clinical outlook (in Japanese). Taisha (Metabolism) 23:249–258
27. Konno T, Maeda H (1987) Targeting chemotherapy of hepatocellular carcinoma. Arterial administration of SMANCS/lipiodol. In: Okuda K, Ishak G (eds) Neoplasms of the liver. Springer-Verlag, Berlin Heidelberg New York, pp 343–352
28. Guerry D, Schuchter LM (1992) Disseminated melanoma. Is there a new standard therapy? New Engl J Med 20:560–561
29. Yang SC, Fry KD, Grimm EA, Roth JA (1991) Phenotype and cytolytic activity of mouse tumor-bearer splenocytes and tumor-infiltrating lymphocytes from K-1735 melanoma metastases following anti-CD3, interleukin-2, and tumor necrosis-α combination immunotherapy. J Immunother 10:326–335
30. Crowley NJ, Vervaert CE, Seigler HF (1992) Human xenograft-nude mouse model of adoptive immunotherapy with human melanoma-specific cytotoxic T-cells. Cancer Res 52:394–399
31. Saiki I, Yoneda J, Kobayashi H, Igarashi Y, Komazawa H, Ishizaki Y, Kato I, Azuma I (1993) Antimetastatic effect by anti-adhesion therapy with cell-adhesive peptide of fibronectin in combination with anticancer drugs. Jpn J Cancer Res 84:326–335
32. Guo Y, Ma J, Wang J, Che X, Narula J, Bigby M, Wu M, Sy M-S (1994) Inhibition of human melanoma growth and metastasis in vivo by anti-CD44 monoclonal antibody. Cancer Res 54:1561–1565
33. Poste G, Fidler IJ (1980) The pathogenesis of cancer metastasis. Nature 283:139–145
34. Yokoyama Y, Yoshie O, Aso H, Ebina T, Ishida N (1986) Inhibition of intravascular mouse melanoma dissemination by recombinant human interferon αA/D. Jpn J Cancer Res 77:80–84
35. Werkmeister J, McCarthy W, Hersey P (1981) Suppressor cell activity in melanoma patients. I. Relation to tumor growth and immunoglobulin levels in vitro. Int J Cancer 28:1–9
36. Mukherji B, Wilhelm SA, Guha A, Ergin MT (1986) Regulation of cellular responses against autologous human melanoma. I. Evidence for cell-mediated suppression of in vitro cytotoxic immune response. J Immunol 136:1888–1892
37. North RJ, Awwad M, Dunn PL (1989) The immune response to tumors. Transplant Proc 21:575–577
38. Vaage J (1974) Circulating tumor antigens versus immune serum factors in depressed concomitant immunity. Cancer Res 34:2979–2983
39. Sjogren HO, Hellstrom I, Bansal SC, Hellstrom KE (1971) Suppressive evidence that the "blocking antibodies" of tumor-bearing individuals may be antigen-antibody complexes. Proc Natl Acad Sci USA 68:1372–1375
40. Roth JA (1983) Tumor-induced immunosuppression. Surg Gynecol & Obstet 156:233–240

41. Ishida N (1985) Isolation of factors responsible for the immunosuppression found in tumor-bearing animals (in Japanese). Yakugaku Zasshi (J Pharm Soc Jpn) 105:91–109
42. Putnam JB, Roth JA (1985) Identification and characterization of a tumor-derived immunosuppressive glycoprotein from murine melanoma K-1735. Cancer Immunol Immunother 19:90–100
43. Guillou JP, Ramsden CW, Somers SS, Sedman PC (1989) Suppression of the generation of lymphokine-activated killer (LAK) cells by serum-free supernatants of in vitro maintained tumor cell lines. Br J Cancer 59:515–521
44. Ebert EC, Roberts AI, Devereux D, Nagase H (1990) Selective immunosuppressive action of a factor produced by colon cancer. Cancer Res 50:6158–6161
45. Romeo DS, Mizel SB (1988) Partial purification of an immunosuppressive protein from human tumor cell line and analysis of its relationship to transforming growth factor. Cell Immunol 122:483–492
46. Fu YX, Watson GA, Kasahara M, Lopez DM (1991) The role of tumor-derived cytokines on the immune system of mice bearing a mammary adenocarcinoma. J Immunol 146:783–789
47. Oghiso Y, Yamada Y, Ando K, Ishihara H, Shibata Y (1993) Differential induction of prostaglandin E_2-dependent and -independent immune suppressor cells by tumor-derived GM-CSF and M-CSF. J Leukocyte Biol 53:86–92
48. Gorelik L, Prokhorova A, Mokyr MB (1994) Low-dose melphalan-induced shift in the production of a Th2-type cytokine to a Th1-type cytokine in mice bearing MOPC-315 tumor. Cancer Immunol Immunother 39:117–126
49. Kobayashi M, Herndon DN, Pollard RB, Suzuki F (1995) $CD4^+$ contrasuppressor T cells improve the resistance of thermally injured mice infected with HSV. J Leukocyte Biol 57:159–167
50. Suzuki F, Kobayashi M, Pollard RB, Maeda H (1995) SMANCS, a polymer-conjugated neocarzinostatin, regulates tumor-induced Th2 cells causing inhibition of pulmonary metastases of B16 melanoma in mice. Proc Am Assoc Cancer Res 36:461
51. Gershon RK, Eardley DD, Durum S, Green DR, Shen FW, Yamauchi K, Cantor H, Murphy DB (1981) Contrasuppression. A novel immunoregulatory activity. J Exp Med 153:1533–1546
52. Suzuki I, Kiyono H, Kitamura K, Green DR, McGhee JR (1986) Abrogation of oral tolerance by contrasuppressor T cells suggests the presence of regulatory T cell networks in the mucosal immune system. Nature 320:451–454
53. Kupper TS, Green DR (1984) Immunoregulation after thermal injury: sequential appearance of I-J^+, Ly-1 T suppressor inducer cells and Ly-2 T suppressor effector cells following thermal trauma in mice. J Immunol 135:3047–3053

Host-Mediated Antitumor Activity Induced by Neocarzinostatin and Its Polymer-Conjugated Derivative SMANCS in Tumor-Bearing Mice

ETSUKO MASUDA[1] and HIROSHI MAEDA[2]

Summary. Pretreatment with neocarzinostatin (NCS) or its polymer-conjugated derivative, zinostatin stimalamer (usually called SMANCS), induced Meth A tumor eradication in Balb/c mice. Tumor eradication was induced by a single administration of NCS or SMANCS between 4 weeks and 1 day before tumor transplantation. Meth A was always eradicated after transient growth. These results suggested that Meth A was eradicated by host-mediated antitumor activity induced by NCS or SMANCS. Pretreatment with these drugs also suppressed the growth of Colon 26 carcinoma and Sarcoma 180. Apo-SMANCS, which lacks cytocidal activity, did not induce antitumor activity, suggesting that cytocidal activity was required for induction of host-mediated antitumor activity. Spleen cells from Meth A-bearing mice pretreated with NCS or SMANCS showed tumor-neutralizing activity, and the effector cells were T cells, but the spleen cells did not exhibit cytotoxic activity or cytotoxic T-lymphocyte (CTL) activity. Flow cytometric analysis indicated that a decrease of B cells was prominent in spleen cells and lymph node cells after NCS or SMANCS treatment. In vivo depletion experiments using antibodies and carrageenan suggested that the effector cells for tumor eradication were asialo-GM1^{+}/Lyt2^{+}/Thy1.2^{+} and possibly asialo GM1^{-}/Lyt2^{+}/Thy1.2^{+}, and that asialo GM1^{-}/Lyt2^{+}/Thy1.2^{+} cells were essential for the generation of antitumor activity. It was also suggested that host-mediated antitumor activity was augmented not only by pretreatment but also by posttreatment. These studies suggest that NCS or SMANCS exhibits antitumor activity through both direct cytotoxicity and indirect host-mediated antitumor activity.

Key words: Biological response modifier—SMANCS (Zinostatin stimalamer)—Tumor eradication—Meth A—Effector cells—Thy1.2—Asialo GM1—Immune activation

---

[1] Molecular Medicine Research Labs, Institute for Drug Discovery Research, Yamanouchi Pharmaceutical Co., Ltd., 21 Miyukigaoka, Tsukuba, Ibaraki, 305 Japan
[2] Department of Microbiology, Kumamoto University School of Medicine, 2-2-1 Honjo, Kumamoto, 860 Japan

Introduction

Most chemotherapeutic anticancer agents exhibit intrinsic immunosuppression that is mediated by their antiproliferative action, and this is clearly detrimental to the tumor-bearing host. However, some chemotherapeutics are known to augment host-antitumor resistance. Cyclophosphamide, melphalan, busulfan, and bleomycin reportedly induce tumor eradication in animal models by augmentating host-mediated antitumor resistance [1–19]. Poly(styrene-co-maleic acid) conjugated neocarzinostatin (SMANCS), also called zinostatin stimalamer, is marketed as a chemotherapeutic agent for the treatment of liver cancer in Japan. The synthesis and pharmacological properties of SMANCS are described in detail in a chapter by H. Maeda and T. Konno (this volume), and the effect of SMANCS on the host's tumor immunity is discussed in another chapter by F. Suzuki and M. Kobayashi (this volume).

The effects of neocarzinostatin (NCS) and SMANCS on host-antitumor resistance were examined by pretreatment of mice before tumor transplantation. It was found that these drugs induced the regression and eventual eradication of Meth A (a methylcholanthrene-induced fibrosarcoma). In this chapter, the studies on antitumor resistance induced by NCS and SMANCS are reviewed.

Effect of Pretreatment with NCS and SMANCS on the Growth of Meth A Tumors

Pretreatment with NCS 1 mg/kg (Kayaku, Tokyo, Japan) or SMANCS® 2.5 mg/kg (Yamanouchi Pharmaceutical, Tokyo, Japan) administered intravenously (i.v.) 3 days before Meth A transplantation induced tumor eradication in Balb/c mice. As shown in Fig. 1, Tumor regression was always seen after transient growth. The tumor volumes of mice pretreated with NCS or SMANCS were similar to those of Meth A-bearing mice without drug treatment (tumor control mice) until about day 10, at which time tumor regression began. Meth A was eradicated in 5 of 7 mice pretreated with NCS and in 7 of 8 mice pretreated with SMANCS between days 21 and 42 after transplantation. In the remaining mice, the tumors enlarged again after transient remission. Next, the timing of pretreatment was examined. A single administration of NCS or SMANCS between 4 weeks and 1 day before tumor transplantation induced Meth A eradication with high frequency (Fig. 2; Table 1).

The facts that tumor was eradicated after transient growth, that pretreatment 4 weeks before tumor transplantation induced tumor eradication, and that cytocidal activity was not detected in vivo after intravenous injection of these drugs [20,21] strongly suggest that tumor was eradicated by host-mediated antitumor activity induced by these drugs. The optimal dose of NCS for tumor eradication was 1 mg/kg regardless of the timing of pretreatment. The optimal dose of SMANCS, however, was dependent on the timing of treatment. When

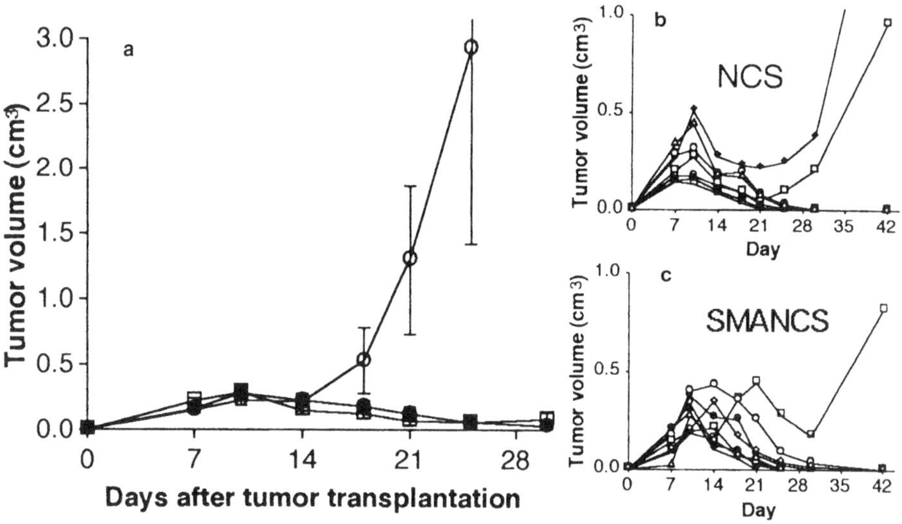

FIG. 1a–c. Kinetics of tumor growth in mice pretreated with neocarzinostatin (*NCS*) or zinostatin stimalamer (*SMANCS*). **a** Balb/c mice were administered with NCS (1 mg/kg, *open squares*), SMANCS (2.5 mg/kg, *solid circles*), or none (*open circles*); 5×10^5 Meth A cells were transplanted s.c. 3 days later. Tumor diameters were measured at indicated days, and tumor volume was calculated. Each group consisted of 7 or 8 mice; each value represents the mean tumor volume ± SE. One mouse of the control group died of tumor on day 30. **b,c** Individual tumor growth in SMANCS- or NCS-pretreated mice. (From [22])

TABLE 1. Antitumor effect induced by pretreatment with SMANCS or NCS on Meth A (From [22])

Pretreatment[a] Drug (mg/kg)		Tumor weight (g)[b] Mean ± SE	Inhibition (%)	Tumor-free mice[c]	Survival/Total[d]
SMANCS	0.5	1.79 ± 1.51	(45.1)	1	10/10
	1.0	0.15 ± 0.20**	(95.4)	4	10/10
	2.0	0.01 ± 0.02**	(99.7)	8	10/10
	4.0	0.06 ± 0.05**	(98.2)	1	9/10
NCS	0.5	0.58 ± 0.61*	(82.2)	2	10/10
	1.0	0.17 ± 0.53**	(94.8)	8	10/10
	2.0	0.01 ± 0.02**	(99.7)	2	3/10
None (Meth A only)		3.26 ± 2.48		0	9/10

* $P < .05$; ** $P < .01$ (*vs* Meth A only).

[a] Mice were administered with or without a drug (day −1) i.v. and transplanted with 8×10^5 Meth A cells (day 0).

[b] Tumor was removed and weighed on day 24.

[c] Number of tumor-free mice (day 24).

[d] Number of surviving mice (day 24). One mouse in the control group died of the tumor on day 22, and 1 mouse treated with 4 mg/kg SMANCS and 7 mice treated with 2 mg/kg NCS died of drug toxicity within 2 weeks after injection.

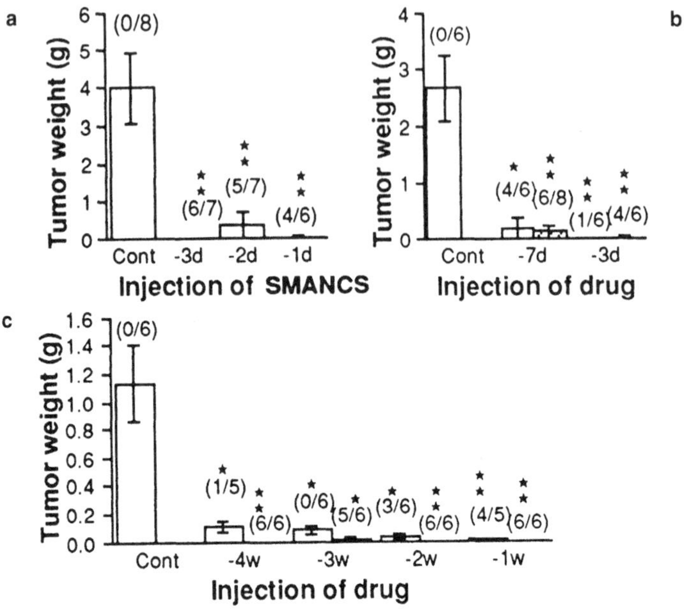

Fig. 2a–c. Effect of NCS or SMANCS pretreatment on Meth A tumor growth. NCS (1 mg/kg, *hatched bars*) or SMANCS (2.5 mg/kg, *open bars*) was injected i.v. into Balb/c mice at the indicated days before Meth A transplantation [8×10^5 cells (**a,b**), 5×10^5 cells (**c**), s.c., day 0]. Tumors were removed from mice and weighed on day 25 (**a**), day 21 (**b**), or day 24 (**c**). One of 9 control mice died of tumor on day 25 (**a**). Each value represents the mean tumor weight ± SE. *Ratios* above the bars represent the number of tumor-free mice/total. *Asterisks* indicate significant differences between control (*shaded bars*) and each experimental group (* $P < .05$; ** $P < .01$; *** $P < .001$). *Cont*, Control; *d*, days; *w*, weeks. (From [22])

SMANCS was administered between 7 days and 1 day before tumor transplantation, SMANCS 2.5 mg/kg was effective in eradicating the tumor, but when SMANCS was administered 21 days before transplantation, 2.5 mg/kg was less effective than 1 mg/kg [22]. Therefore, as seen in Fig. 2, 2.5 mg/kg of SMANCS tended to decrease the antitumor effect with time.

Among other chemotherapeutic agents that induce tumor eradication, cyclophosphamide and busulfan induce tumor eradication through pretreatment [2,13]. Although the optimal intervals between tumor transplantation and administration of these drugs are in the range of a few days to about 1 week, NCS and SMANCS can induce tumor eradication by a single injection as early as 28 days before tumor transplantation. Therefore, NCS and SMANCS have a unique profile among these agents.

After Meth A eradication induced by pretreatment with NCS or SMANCS, cured mice were challenged with Meth A cells or Colon 26 cells (an *N*-nitroso-*N*-

methyl-urethane-induced undifferentiated adenocarcinoma). Rechallenged Meth A cells were rejected without transient growth in all mice; however, Colon 26 was not rejected [22]. This suggests that tumor-specific antitumor resistance was augmented in cured mice. Meth A was not eradicated in Balb/c nu/nu mice pretreated with NCS or SMANCS [22], suggesting that T cells play an important role in tumor eradication.

Effect of Pretreatment with NCS or SMANCS on the Growth of Other Tumors

The effects of pretreatment with NCS or SMANCS on the growth of other tumors were also examined. As shown in Fig. 3, growth of Colon 26, which was derived from Balb/c like Meth A, and that of Sarcoma 180, which was derived from another strain, was inhibited by pretreatment with these drugs. Although tumor eradication was not observed in these models, it was suggested that antitu-

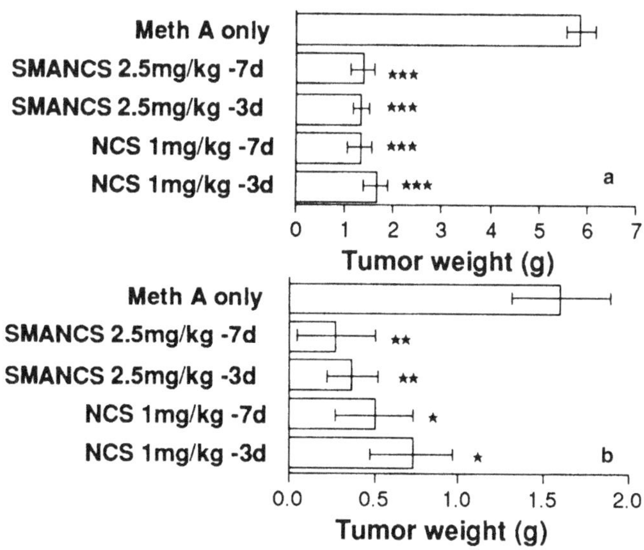

FIG. 3a,b. Effect of pretreatment with NCS or SMANCS on the growth of Colon 26 and Sarcoma 180 tumors. Mice were administered with NCS 1 mg/kg, SMANCS 2.5 mg/kg, i.v., or neither 3 or 7 days before tumor transplantation (day 0). **a** Balb/c mice were inoculated with Colon 26 (2 × 2 mm blocks), and tumor weights were determined on day 21. The control group consisted of 11 mice, and each experimental group consisted of 7 mice. Two of 11 control mice died of tumor before the end of experiment. **b** ICR mice were transplanted with Sarcoma 180 (1×10^6 cells), and tumor weights were determined on day 28. Control and experimental groups consisted of 9 and 7 mice, respectively. Each value represents the mean tumor weight ± SE. *Asterisks* indicate significant differences between control and each experimental group (*, $P < .05$; **, $P < .01$; ***, $P < .001$). (From [22])

mor resistance augmented by NCS or SMANCS might suppress not only the growth of Meth A- or Balb/c-derived tumors but also the growth of various other tumors.

Cytotoxicity of NCS and the Induction of Host-Mediated Antitumor Activity

The finding that pretreatment with NCS, as well as with SMANCS, induced Meth A tumor eradication suggests that the poly(styrene-co-maleic acid) component of SMANCS is not essential for the induction of antitumor activity. The cytotoxic action of NCS results from a nonprotein chromophore with an enediyne structure [23,24]. Apo-SMANCS was synthesized from apo-NCS, which lacks the chromophore, and poly(styrene-co-maleic acid), and was examined to determine whether there was any direct or indirect (host-mediated) antitumor activity against Meth A tumor by pre- or post-treatment. As shown in Fig. 4, SMANCS exhibited antitumor activity not only by pretreatment but also by posttreatment on 1 or 7 days after transplantation in a dose-responsive manner. The antitumor effect through posttreatment was thought to be mainly the result of the potent cytotoxic activity of SMANCS. Because neither pretreatment nor posttreatment with apo-SMANCS inhibited tumor growth, it was suggested that the cytotoxic activity of NCS or SMANCS is essential for the induction of host-mediated antitumor activity.

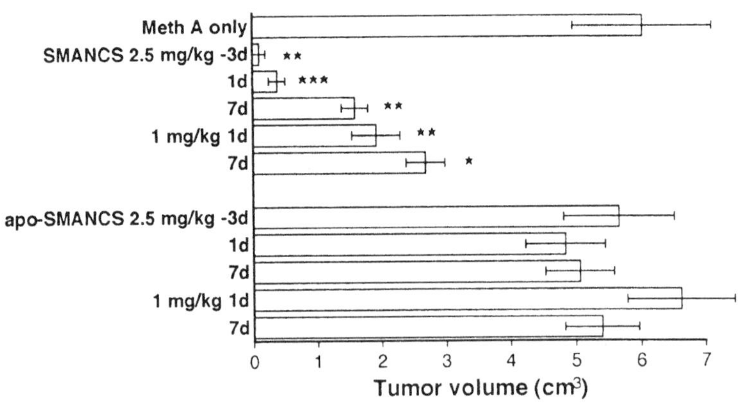

FIG. 4. Effect of pretreatment with apo-SMANCS, which is devoid of direct cytocidal activity, on Meth A growth. Mice were treated with SMANCS, apo-SMANCS i.v., none on the indicated day and transplanted with 7.5×10^5 Meth A cells (day 0). The control group and each experimental group consisted of 11 and 10 mice, respectively. Each value represents the mean tumor volume ± SE. *Asterisks* indicate significant differences between control and each experimental group (*, $P < .05$; **, $P < .01$; ***, $P < .001$)

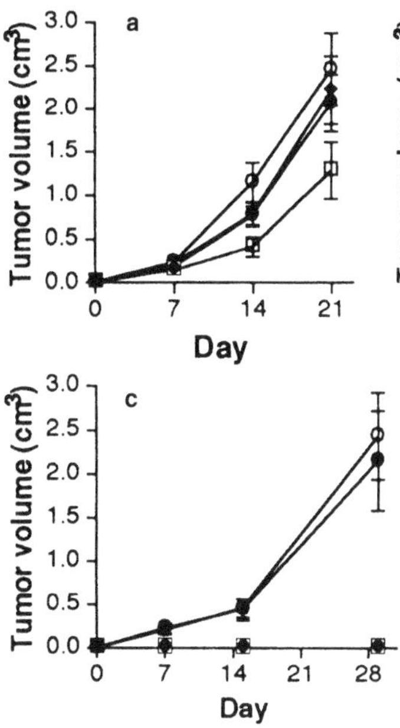

FIG. 5a–c. Tumor-neutralizing activity of splenocytes from Meth A-bearing mice pretreated with NCS or SMANCS, i.v. Mice were treated with NCS (1 mg/kg, *closed diamonds*), SMANCS (2.5 mg/kg, *open squares*), none (*closed circles*) 3 days (*d*) before Meth A transplantation (5×10^5 cells, s.c., day 0). Splenocytes were harvested on days 7 (**a**), 14 (**b**), and 32 (**c**), and inoculated into recipient Balb/c mice with Meth A (5×10^5 cells) at an E:T ratio of 50:1 (**a,b**) or 25:1 (**c**). Splenocytes from normal mice (*open circles*) were also tested as a control. Each value represents the mean tumor volume ± SE of 5–10 mice in each group. Differences between control and experimental groups were statistically significant at $P < .05$ [(**a**) *closed circles*, 7 d, *open squares*, 14 d], $P < .01$ [(**a**) *open squares*, 7 d, (**b**) *closed circles*, 7 d, (**c**) *open squares*, 7 d, *solid diamonds*, 7 d], $P < .001$ [(**b**) *open squares*, 7, 14, and 21 d, *solid triangles*, 14 and 21 d, (**c**) *open squares*, 15 and 29 d, *solid triangles*, 15 and 29 d]. (From [22])

Antitumor Effector Cells Induced by NCS or SMANCS

Generation of Tumor-Neutralizing Activity in Spleen Cells from Meth A-Bearing Mice Pretreated with NCS or SMANCS During Tumor Eradication

The tumor-neutralizing activity of spleen cells from Meth A-bearing mice pretreated with NCS or SMANCS was examined by a modified Winn assay [25], and was compared with that of cells from Meth A-bearing control mice (tumor control mice) or normal mice. The growth of Meth A transplanted with spleen cells is shown in Fig. 5. Among the spleen cells obtained 7 days after tumor transplantation (day 7 splenocytes), the only splenocytes that showed Meth A

TABLE 2. Tumor-neutralizing activity of splenic T cells from Meth A-bearing mice pretreated with SMANCS (From [22])

Experiment	Splenocytes[a]	Tumor-neutralizing activity[b] Tumor volume (cm^3) Mean ± SE	Inhibition (%)	Tumor free/Total[c]
1	Unfractionated	0.00 ± 0.00***	(100)	5/5
	Thy1.2-depleted	3.45 ± 0.51*	(32.2)	0/5
	Nylon column-passed	0.00 ± 0.00***	(100)	5/5
	None	5.09 ± 0.40		0/5
2	Unfractionated	0.00 ± 0.00***	(100)	5/5
	Nylon column-passed	0.00 ± 0.00***	(100)	5/5
	None	5.02 ± 0.61		0/7
3	Unfractionated	0.00 ± 0.00***	(100)	5/5
	Thy1.2-depleted	2.29 ± 0.36	(13.9)	0/5
	None	2.66 ± 0.34		0/5

* $P < .05$; *** $P < .001$ (*vs* None: Meth A only).

[a] Balb/c mice were administered with 2.5 mg/kg SMANCS 3 days before inoculation of 5×10^5 Meth A cells (day 0). Splenocytes were harvested on day 14 (experiments 1, 2) or day 21 (experiment 3), and treated with anti-Thy1.2 antibody and complement (Thy1.2-depleted) or nylon-wool column (nylon column-passed).

[b] Splenocytes and Meth A (5×10^5 cells) were inoculated into Balb/c nu/nu mice at effector:target (E:T) ratio of 25:1 (day 0); tumor diameters were measured on day 21 (experiments 1, 2) or day 18 (experiment 3), and tumor volume was calculated.

[c] Number of tumor-free mice (day 21).

growth inhibition were those from SMANCS-pretreated mice which showed slight inhibition. Day 14 and 32 splenocytes from NCS- or SMANCS-pretreated mice exhibited strong tumor-neutralizing activity and completely inhibited Meth A growth. In tumor control mice, tumor-neutralizing activity was observed in day 14 splenocytes, but the activity disappeared by day 32. These findings suggested that antitumor cells also appeared in tumor control mice, but that in late stage suppressor cells or suppressor factors derived from host or tumor cells appeared and suppressed the antitumor cells.

Spleen cells from NCS- or SMANCS-pretreated mice without tumor transplantation did not show tumor-neutralizing activity [22]. This suggested that tumor antigen presentation was required for generation of antitumor activity and, therefore, that nonspecific antitumor cells such as macrophages and neutrophils might not be the antitumor effector cells for Meth A eradication induced by pretreatment with NCS or SMANCS.

The tumor-neutralizing activity of T lymphocytes was examined in splenocytes from SMANCS-pretreated mice. By T-cell depletion using anti-Thy1.2 monoclonal antibody (mAb) [26] and complement, splenocytes lost most of the activity, and nylon column-passed purified T cells exhibited tumor-neutralizing activity (Table 2). From these findings, T cells were suggested to be the main effector cells for tumor neutralization.

Cytotoxic Activity of Splenocytes from NCS- or SMANCS-Pretreated Mice

The cell-mediated cytotoxic activity of splenocytes from NCS- or SMANCS-pretreated mice was examined. Meth A cells, natural killer- (NK-) sensitive YAC-1 cells, NK-insensitive or monocyte-sensitive P815 cells and EL4 cells were used as target cells. Splenocytes were harvested during the period of tumor growth, the period of tumor regression, or the period of tumor eradication, and evaluated for cytotoxic activity by 4- or 18-h ^{51}Cr-release assay at various effector:target (E:T) ratios (12.5:1 to 100:1). However, splenocytes from NCS- or SMANCS-pretreated mice showed no significant cytotoxic activity against any target cells [22].

The activity of cytotoxic T lymphocytes (CTL) following mixed-lymphocyte tumor culture (MLTC) was also examined. However, splenocytes from NCS- or SMANCS-pretreated mice, which were harvested in various periods of tumor eradication as described, did not exhibit significant killing activity on Meth A cells beyond that of normal mouse splenocytes [22].

Effects of NCS and SMANCS on the Cell Population in the Spleen and Lymph Nodes

As shown in Figs. 6 and 7, the number of spleen and inguinal lymph node cells decreased transiently when NCS 1 mg/kg or SMANCS 2.5 mg/kg was administered 3 days before tumor transplantation. The fewest spleen cells were observed on day 1 or day 0, and the cell number had recovered by day 14 [27]. Spleen and inguinal lymph node cell populations were analyzed by flow cytometry. Alteration of spleen and lymph node subsets from SMANCS-pretreated mice and tumor control mice are shown in Figs. 8 and 7, respectively. In normal Balb/c mice, spleen cells were composed of about 40% T cells, 45% B cells, 5%–10% Mac-1$^+$ cells (macrophages, etc. [28–29]), and 5%–20% asialo-GM1$^+$ cells (NK cells, etc. [30–33]). By the administration of SMANCS, percentages of L3T4-positive cells (Th/Tind) and Lyt2-positive cells (Ts/Tc) [26,34–35] were evidently increased and, as a result, the proportion of T cells reached 76.3%. Although the ratio of CD4/CD8 (L3T4/Lyt2) was not changed, B cells and Mac-1$^+$ cells were decreased by 19.8% and 3.5%, respectively. A significant change in the percentage of asialo-GM1$^+$ cells was not observed [27]. These results showed that the greatest decrease of B cells was found in spleen cell subsets. A decrease in B cells was also observed in lymph nodes from SMANCS-pretreated mice.

Almost the same findings were observed in the spleen and lymph nodes of NCS-pretreated mice, except that the decrease of Mac-1$^+$ cells was only seen on day 3 (unpublished data). On day 21, inguinal lymph nodes were enlarged in tumor control mice. Both T cells and B cells increased, but the increase of B cells was more prominent (2.5- and 10.7 fold larger in the number of those before transplantation, respectively). It was noticed that B cells, which decreased with

6

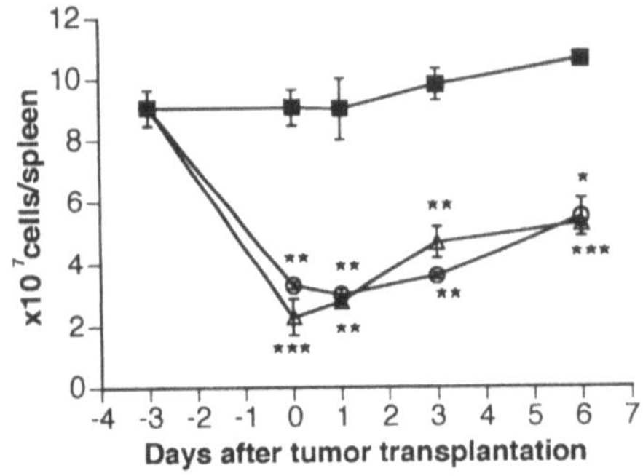

7

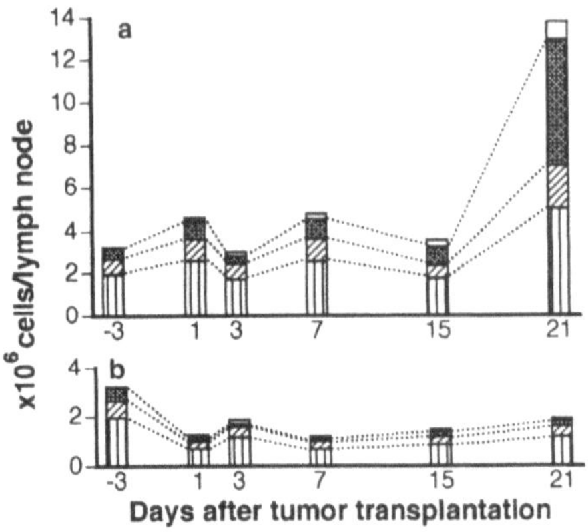

Fig. 6. Changes in numbers of spleen cells in Meth A-bearing mice pretreated with NCS or SMANCS. Mice were administered NCS (1 mg/kg, *open circles*), SMANCS (2.5 mg/kg, *open triangles*), or none (*solid squares*) 3 days before Meth A transplantation; splenocytes were harvested on the indicated days after transplantation. Each value represents the mean number of splenocytes ± SE of 3 mice in each group. *Asterisks* indicate significant differences between tumor control and each experimental group (*, $P < .05$; **, $P < .01$; ***, $P < .001$). (From [27], with permission)

Fig. 7a,b. Change of lymphocyte phenotype in lymph nodes from Meth A-bearing mice with or without SMANCS pretreatment. Mice were administered (**a**) none or (**b**) SMANCS (2.5 mg/kg) 3 days before Meth A transplantation. Inguinal lymph node cells were harvested on the indicated days, and L3T4$^+$ cells (*striped*), Lyt2$^+$ cells (*right-hatched*), and sIg$^+$ cells (*cross-hatched*) were analyzed. Each value represents the mean percentages of 3 mice. Differences between normal mice (day -3) and tumor control mice were statistically significant at $P < .05$ (Lyt2$^+$, 21 d; sIg$^+$, 21 d) and $P < .001$ (L3T4$^+$, 21 d; Lyt2$^+$, 15 d). Differences between tumor control mice and SMANCS-pretreated mice were statistically significant at $P < .05$ (L3T4$^+$, 1 d, 15 d, 21 d; Lyt2$^+$, 1 d, 21 d; sIg$^+$, 3 d, 15 d, 21 d) and $P < .01$ (Lyt2$^+$, 15 d). (From [27], with permission)

SMANCS treatment, increased in tumor-bearing mice. Antibody formation against tumor cell antigen or against cytokines having antitumor activity is thought to be one of the mechanisms by which tumor cells escape from immunological control [36–37]. It should be further investigated whether B cells are involved in suppression of the host defense against Meth A, and whether the decrease in B cells is one of the mechanisms of the antitumor effect of NCS or SMANCS.

After day 14, although splenocytes from NCS- and SMANCS-pretreated mice acquired tumor-neutralizing activity and tumor regression was observed, neither alterations in the total number of splenocytes nor alterations in the percentages of spleen cell subsets were seen. Thus, antitumor effector cells might be activated without proliferation, or the increase of a subset might be masked by the decrease of another subset because anti-L3T4 or anti-Lyt 2.2 mAb recognizes both of two different functional T-cell subsets ("Th and Tind" and "Ts and Tc," respectively) [35]. The findings that the cytotoxic activity of NCS or SMANCS was essential for the induction of tumor eradication, that the effect of these drugs on the composition of spleen and lymph node cells was only seen before tumor regression, and that a single injection of these drugs more than 3 weeks before tumor transplantation induced tumor eradication suggest that SMANCS and NCS might affect the precursor cells involved in tumor eradication and that the effect lasts for a long time.

Antitumor Effector Cells for Meth A Eradication

To identify the antitumor effector cells, in vivo depletion of T cells, NK cells, or macrophages was performed in SMANCS-pretreated mice using anti-Thy1.2

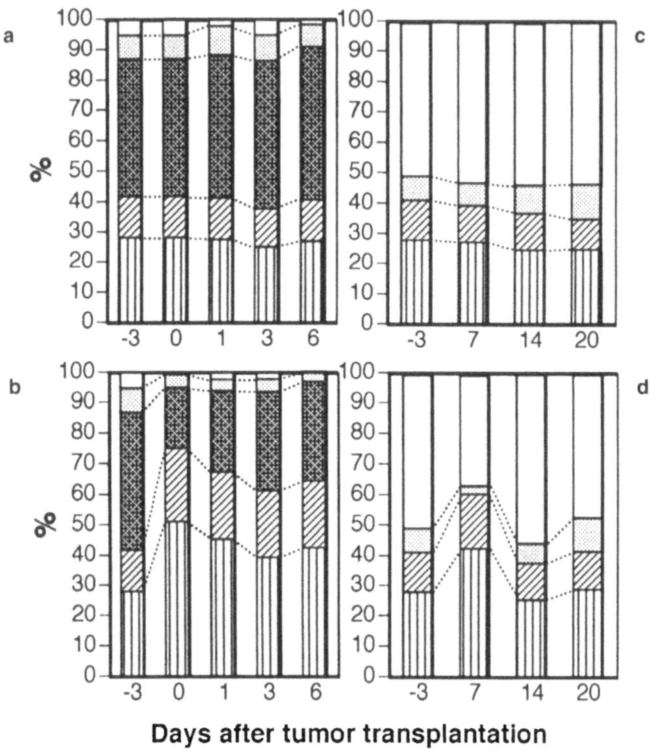

FIG. 8a–d. Change of spleen cell phenotype in Meth A-bearing mice with or without SMANCS pretreatment. Balb/c mice were administered (**a,c**) none or (**b,d**) with SMANCS 2.5 mg/kg i.v. 3 days before Meth A transplantation. Spleen cells were harvested on the indicated days, and L3T4⁺ cells (*striped*), Lyt2⁺ cells (*right-hatched*), sIg⁺ cells (*cross-hatched*), and Mac-1⁺ cells (*shaded bars*) were analyzed by flow cytometry. Each value represents the mean percentage of positive cells of 3 mice in each group. Differences between control mice and SMANCS-pretreated mice were statistically significant at $P < .05$ (L3T4⁺, 6 d; Mac-1⁺, 0 d, 3 d), $P < .01$ (L3T4⁺, 3 d, 7 d; Lyt2⁺, 3 d; sIg⁺, 1 d, 6 d; Mac-1⁺, 1 d, 6 d), and $P < .001$ (L3T4⁺, 0 d, 1 d; Lyt2⁺, 0 d, 1 d, 6 d, 7 d; sIg⁺, 0 d, 3 d; Mac-1⁺, 7 d). (From [27], with permission)

mAb [38], anti-asialo GM1 antibody (anti-as GM1) [39], and carrageenan [40], respectively. When SMANCS-pretreated mice were injected intravenously with the anti-Thy1.2 mAb on days 7, 10, and 14, tumor eradication was completely inhibited and tumor growth was similar to that of the tumor control mice. Anti-as GM1 injected intravenously with the same schedule also inhibited tumor eradication, but tumor growth was evidently suppressed. Carrageenan (i.p.) showed no effect, and the tumor was eradicated in all mice (Fig. 9). From these results, it was suggested that the antitumor effector cells were Thy1.2$^+$ T cells and that at least some of them also expressed asialo GM1 antigen.

T-cell markers of antitumor effector cells were further investigated through in vivo depletion using anti-L3T4 or anti-Lyt2 mAb. Anti-Lyt2 mAb prominently inhibited the tumor regression induced by SMANCS, but anti-L3T4 mAb had little effect (Fig. 10). This suggests that the antitumor effector cells express Lyt2 antigen. The foregoing results suggest that the antitumor effector cells for tumor eradication were asialo GM1$^+$/Lyt2$^+$/Thy1.2$^+$ and possibly asialo GM1$^-$/Lyt2$^+$/ Thy1.2$^+$.

To identify the cells involved in the generation of antitumor effector cells, in vivo depletion using antibodies and carrageenan was performed close to the day

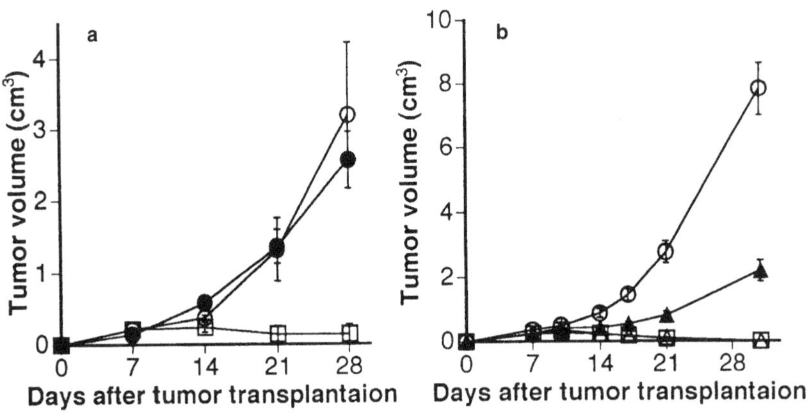

FIG. 9a,b. Effect of treatment of anti-Thy1.2 monoclonal antibody (mAb), anti-asialo GM1 antibody (anti-asGM1), and carrageenan on antitumor activity induced by pretreatment with SMANCS. Mice were administered SMANCS 2.5 mg/kg i.v. 3 days before Meth A transplantation and treated with (**a**) anti-Thy1.2 mAb (50 µl, *solid circles*) i.v. or without (*open squares*, SMANCS control) or with (**b**) anti-asGM1 (50 µl, *solid triangles*) i.v. or carrageenan (0.7 mg, *open triangles*) i.p. or without (*open squares*, SMANCS control) on days 7, 10, and 14. Tumor control mice were given Meth A transplantation only (*open circles*). Each value represents the mean tumor volume ± SE of 3–5 mice in each group. Differences between anti-Thy1.2 mAb-treated mice and SMANCS control mice were statistically significant at $P < .05$ (day 14) and $P < .01$ (day 21, day 28). Differences between anti-asGM1-treated mice and SMANCS control mice were statistically significant at $P < .01$ (day 14) or $P < .001$ (day 17, day 21, day 31). (From [27], with permission)

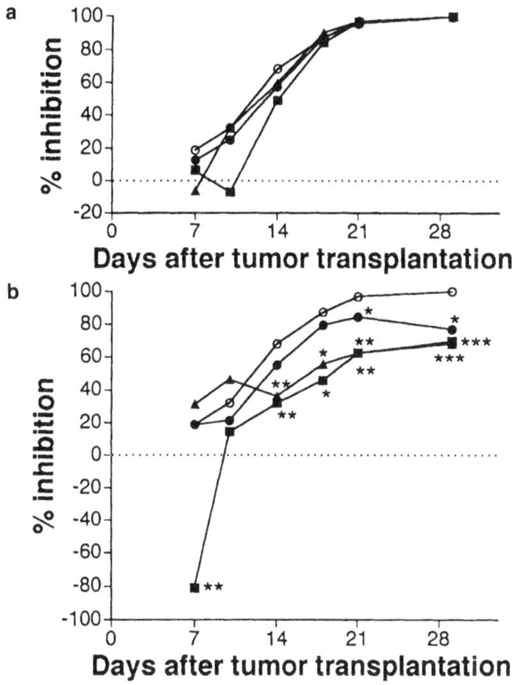

FIG. 10a,b. Effect of depletion of L3T4⁺ cells or Lyt2.2⁺ cells on antitumor activity induced by SMANCS pretreatment. Mice were administered SMANCS 2.5 mg/kg i.v. 3 days before Meth A transplantation and treated with (**a**) anti-L3T4 mAb (1 mg) or (**b**) anti-Lyt2.2 mAb 1 mg on day -2 (*closed squares*), 7 (*closed triangles*), or 14 (*closed circles*), or none (*open circles*; SMANCS control). Each value represents the percent of inhibition of tumor volume against tumor control mice (n = 5–9). *Asterisks* indicate significant differences between SMANCS control and each experimental group (*, $P < .05$; **, $P < .01$; ***, $P < .001$). (From [27], with permission)

of tumor transplantation. Anti-Thy1.2 mAb injected on days –2 and 1, or on day –1, completely abrogated the effect of pretreatment with SMANCS. However, neither anti-as GM1 injected on days –2 and 1 nor carrageenan injected on day –2 showed any effect (Table 3). These data suggest that Thy1.2⁺ T cells are important for the generation of antitumor effector cells, and that antitumor activity is induced without the help of either asialo GM1⁺ cells or macrophages.

From in vivo depletion experiments using anti-L3T4 and anti-Lyt2.2 mAbs, it was suggested that the phenotype of these T cells was Lyt2⁺ (Fig. 10). Although anti-L3T4 mAb alone on day –2 did not show a significant suppressive effect on antitumor activity induced by SMANCS, simultaneous injection of anti-L3T4 and anti-Lyt2 mAbs on day –1 augmented the inhibitory effect of anti-Lyt2 mAb [27].

TABLE 3. Effect of anti-Thy1.2 mAb or anti-asialo-GM1 antibody or carrageenan-treatment near to day of tumor transplantation on antitumor activity induced by pretreatment with SMANCS

	Treatment[a]		Tumor volume (cm^3)		
Experiment	SMANCS	Antibody	Mean ± SE	Inhibition (%)	Tumor-free/Total[b]
1	—	—	3.34 ± 0.98		0/5
	+	—	0.05 ± 0.05*	(98.5)	4/5
	+	Anti-Thy1.2 mAb −2 d, 1 d	3.02 ± 0.80	(9.6)	0/4
	+	Anti-as GM1 −1 d	0.23 ± 0.12*	(90.2)	1/4
2	—	—	2.35 ± 0.57		0/7
	+	—	0.21 ± 0.11*	(91.1)	3/6
	+	Anti-Thy1.2 mAb −1 d	2.26 ± 0.20	(3.8)	0/4
	+	Anti-Thy1.2 mAb 1 d	0.98 ± 0.40	(58.3)	0/4
3	—	—	2.23 ± 0.28		0/8
	+	—	0.05 ± 0.01***	(97.8)	8/8
	+	Carrageenan −1 d	0.02 ± 0.01***	(99.1)	3/3

mAb, Monoclonal antibody; anti-as GM1, anti-asialo GM1 antibody.
*$P < .05$; ***$P < .001$ (*vs* Meth A only).
[a] Mice were administered with or without 2.5 mg/kg SMANCS 3 days (d) before Meth A transplantation and injected with anti-Thy 1.2 mAb (i.v.), anti-asGM1 (i.v.), carrageenan (i.p.), or none on indicated days. Tumor diameters were measured on day 28 (experiment 1) or day 21 (experiments 2 and 3).
[b] Number of tumor-free mice.

These results suggested that asialo $GM1^-Lyt2^+Thy1.2^+$ cells were activated by SMANCS and tumor transplantation in an early phase followed by the generation of antitumor cells, and that $L3T4^+$ T cells were also involved in the generation of antitumor effector cells.

Effect of Therapeutic Treatment (Post-treatment) with SMANCS on Host-Mediated Antitumor Activity

NCS and SMANCS have potent cytocidal activity on various tumor cells [41–43], and the antitumor activity was previously thought to be the result of direct cytotoxicity against tumor cells when these drugs were administered after tumor transplantation. Therefore, the effect of posttreatment with SMANCS on host-mediated antitumor activity was examined. Mice were administered with SMANCS 3 days after tumor transplantation, resulting in a complete cure in 5 of 6 mice. However, when anti-Thy1.2 mAb was injected into the mice on days 10 and 13, tumor eradication was not induced in any mice, and tumor growth was only partially reversed (Fig. 11). This showed that SMANCS might augment the host-mediated antitumor activity in therapeutic administration, and that T cells take part in such antitumor activity.

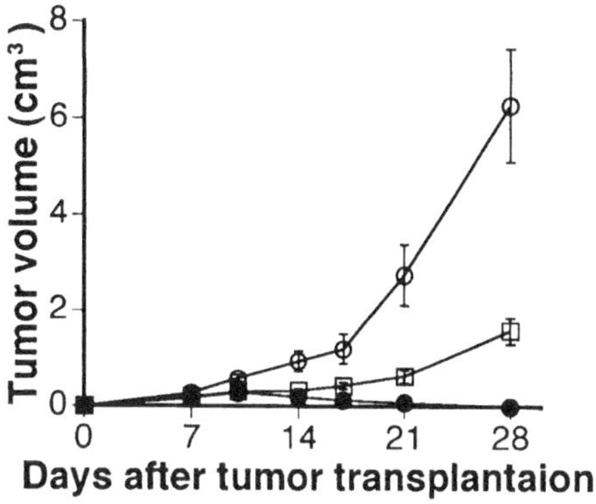

FIG. 11. Host-mediated antitumor activity induced by therapeutic administration with SMANCS. Mice were administrated SMANCS 2.5 mg/kg 3 days after Meth A transplantation, and injected with (*open squares*) or without (*solid circles*, SMANCS control) anti-Thy1.2 mAb (50 μl) on days 10 and 13. Tumor control mice were given Meth A transplantation only (*open circles*). Each value represents the mean tumor volume ± SE of 5–6 mice in each group. Differences between the anti-Thy1.2 mAb-treated group and the SMANCS control group were statistically significant at $P < .05$ (day 14, day 17) and at $P < .01$ (day 21, day 28). (From [27], with permission)

Conclusion

The effects of NCS and its derivative, SMANCS, on host antitumor defense were examined. Although these drugs transiently decrease the number of lymphocytes and macrophages, which have the potential to become antitumor effector cells, they augmented rather than suppressed host-mediated antitumor activity. This activity is unique among anticancer chemotherapeutic agents, because a single injection more than 3 weeks before tumor transplantation induced antitumor activity. It was suggested that these drugs activate asialo-$GM1^-/Lyt2^+/Thy1.2^+$ T cells after Meth A transplantation in a dose range similar to that of therapeutic administration, and that thereafter asialo $GM1^+/Lyt2^+/Thy1.2^+$ (and possibly asialo $GM1^-/Lyt2^+/Thy1.2^+$) antitumor effector cells are generated; thus, eventually the tumor was eradicated. Host-mediated antitumor activity was augmented not only by pretreatment but also by posttreatment. Consequently, it was suggested that NCS and SMANCS might exhibit antitumor activity both through direct cytotoxic activity and through indirect host-mediated antitumor activity.

References

1. Hengst JCD, Mokyr MB, Dray S (1980) Importance of timing in cyclophosphamide therapy of MOPC-315 tumor-bearing mice. Cancer Res 40:2135–2141

2. Ray PK, Raychaudhuri S (1981) Low-dose cyclophosphamide inhibition of transplantable fibrosarcoma growth by augmentation of the host immune response. J Natl Cancer Inst 67:1341–1345
3. Hengst JCD, Mokyr MB, Dray S (1981) Cooperation between cyclophosphamide tumoricidal activity and host antitumor immunity in the cure of mice bearing large MOPC-315 tumors. Cancer Res 41:2163–2167
4. El-Sady E, Parker D, Turk JL (1986) The effect of cyclophosphamide in vivo on the expression of lymphocyte markers, detected by monoclonal antibodies, in the rat. Int J Immunopharmacol 8:961–966
5. Awwad M, North RJ (1989) Cyclophosphamide-induced immunologically mediated regression of a cyclophosphamide-resistant murine tumor: a consequence of eliminating precursor L3T4^{+} suppressor T-cells. Cancer Res 49:1649–1654
6. Čulo F, Klapan I, Kolak T (1993) The influence of cyclophosphamide on antitumor immunity in mice bearing late-stage tumors. Cancer Immunol Immunother 36:115–122
7. Ben-Efraim S, Bocian RC, Mokyr MB, Dray S (1983) Increase in the effectiveness of melphalan therapy with progression of MOPC-315 plasmacytoma tumor growth. Cancer Immunol Immunother 15:101–107
8. Mokyr MB, Barker E, Weiskirch LM, Takesue BY, Pyle JM (1989) Importance of Lyt2^{+} T-cells in the curative effectiveness of a low dose of melphalan for mice bearing a large MOPC-315 tumor. Cancer Res 49:4597–4606
9. Shoval S, Ophir R, Ben-Efraim S (1989) Deficiency in immunocompetence of mice cured from large MOPC-315 plasmacytomas by melphalan therapy. Cancer Immunol Immunother 29:279–287
10. Takesue BY, Pyle JM, Mokyr MB (1990) Importance of tumor-specific cytotoxic CD8^{+} T-cells in eradication of a large subcutaneous MOPC-315 tumor following low-dose melphalan therapy. Cancer Res 50:7641–7649
11. Michael R, Mokyr MB (1993) Characterization of the exogenous interleukin-2 requirements for the generation of enhanced antitumor cytotoxicity by thymocytes from low-dose melphalan-treated MOPC-315 tumor bearers. Cancer Immunol Immunother 36:37–44
12. Gorelik L, Prokhorova A, Mokyr MB (1994) Low-dose melphalan-induced shift in the production of a Th2-type cytokine to a Th1-type cytokine in mice bearing a large MOPC-315 tumor. Cancer Immunol Immunother 39:117–126
13. Sendo F, Miyake T, Fuyama S, Arai S (1978) Spontaneous regression of syngeneic transplanted tumor in rats pretreated with the antileukemia drug busulfan. J Natl Cancer Inst 60:385–389
14. Mizushima Y, Sendo F, Miyake T, Kobayashi H (1981) Augmentation of specific cell-mediated immune responses to tumor cells in tumor-bearing rats pretreated with the antileukemia drug busulfan. J Natl Cancer Inst 66:659–665
15. Mizushima Y, Sendo F, Takeichi N, Hosohawa M, Kobayashi H (1981) Enhancement of antitumor transplantation resistance in rats by appropriately timed administration of busulfan. Cancer Res 41:2917–2921
16. Morikawa K, Hosokawa M, Hamada J, Sugawara M, Kobayashi H (1985) Host-mediated therapeutic effects produced by appropriately timed administration of bleomycin on a rat fibrosarcoma. Cancer Res 45:1502–1506
17. Morikawa K, Hosokawa M, Hamada J (1986) Possible participation of tumoricidal macrophages in the therapeutic effect of bleomycin on a transplantable rat fibrosarcoma. Cancer Res 46:684–688

18. Hosokawa M, Xu Z, Morikawa K, Hamada J, Kobayashi H (1988) Host immune responses to tumor cells augmented by bleomycin and their therapeutic effects on rat fibrosarcoma. Mol Biother 1:74–80
19. Xu Z, Hosokawa M, Morikawa K, Hatakeyama M, Kobayashi H (1988) Overcoming suppression of antitumor immune reactivity in tumor-bearing rats by treatment with bleomycin. Cancer Res 48:6658–6663
20. Fujita H, Nakayama N, Sawabe T, Kimura K (1970) In vivo distribution and inactivation of neocarzinostatin. Jpn J Antibiot 23:471–478
21. Takeshita J, Maeda H, Kanamaru R (1982) In vitro mode of action, pharmacokinetics, and organ specificity of poly(maleic acid-styrene)-conjugated neocarzinostatin, SMANCS. Gann 73:278–284
22. Masuda E, Maeda H (1995) Antitumor resistance induced by zinostatin stimalamer (SMANCS), a polymer-conjugated neocarzinostatin (NCS) derivative I. Meth A tumor eradication and tumor-neutralizing activity in mice pretreated with SMANCS or NCS. Cancer Immunol Immunother 40:329–338
23. Koide Y, Ishii F, Hasuda K, Koyama Y, Edo K, Katamine S, Kitame F, Ishida N (1980) Isolation of a non-protein component and a protein component from neocarzinostatin (NCS) and their biological activities. J Antibiot (Tokyo) 33:342–346
24. Sato K, Akaike T, Suga M, Ando M Maeda H (1994) Generation of free radicals from neocarzinostatin mediated by NADPH/cytochrome P-450 reductase via activation of enediyne chromophore. Biochem Biophys Res Commun 205:1716–1723
25. Winn HJ (1961) Immune mechanisms in homotransplantation. II. Quantitative assay of the immunologic activity of lymphoid cells stimulated by tumor homografts. J Immunol 86:228–239
26. Ledbetter JA, Herzenberg LA (1979) Xenogeneic monoclonal antibodies to mouse lymphoid differentiation antigens. Immunol Rev 47:63–90
27. Masuda E, Maeda H (1996) Changes in cellular components of spleen and lymph node cells and the effector cells responsible for Meth A tumor eradication induced by zinostatin stimalamer. Cancer Res 56:1868–1873
28. Springer T, Galfré G, Secher DS, Milstein C (1979) Mac-1: a macrophage differentiation antigen identified by monoclonal antibody. Eur J Immunol 9:301–306
29. Ault KA, Springer TA (1981) Cross-reaction of a rat-anti-mouse phagocyte-specific monoclonal antibody (anti-Mac-1) with human monocytes and natural killer cells. J Immunol 126:359–364
30. Kasai M, Iwamori M, Nagai Y, Okumura K, Tada T (1980) A glycolipid on the surface of mouse natural killer cells. Eur J Immunol 10:175–180
31. Akagawa K, Tokunaga T (1982) Appearance of a cell surface antigen associated with the activation of peritoneal macrophages in mice. Microbiol Immunol 26:831–842
32. Herrick CA, Carucci JA, Kleiner GI, Durkin HG (1993) Regulation of hapten-specific memory IgE responses induced in vitro: requirement for both Thy-1^+ AsGM1^+ and Thy-1^+ As GM1^- cells and for IFN-α and IL-4. J Immunol 151:3440–3449
33. Mercurio AM, Schwarting GA, Robbins PW (1984) Glycolipids of the mouse peritoneal macrophage: alterations in amount and surface exposure of specific glycolipid species occur in response to inflammation and tumoricidal activation. J Exp Med 160:1114–1125
34. Dialynas DP, Quan ZS, Wall KA, Pierres A, Quintáns J, Loken MR, Pierres M, Fitch FW (1983) Characterization of the murine T cell surface molecule, designated L3T4,

identified by monoclonal antibody GK1.5: similarity of L3T4 to the human Leu-3/T4 molecule. J Immunol 131:2445–2451
35. Ledbetter JA, Evans RL, Lipinski M, Cunningham-Rundles C, Good RA, Herzenberg LA (1981) Evolutionary conservation of surface molecules that distinguish T lymphocyte helper/inducer and cytotoxic/suppressor subpopulations in mouse and man. J Exp Med 153:310–323
36. Hellström KE, Hellström I (1974) Lymphocyte-mediated cytotoxicity and blocking serum activity to tumor antigens. Adv Immunol 18:209–277
37. Barbuto JAM, Grimes WJ, Hersh EM (1995) Antibodies to tumor necrosis factor: a component of B cell immune responses with a role in tumor/host interaction. Cancer Immunol Immunother 40:31–36
38. Opitz HG, Opitz U, Hewlett G, Schlumberger HD (1982) A new model for investigations of T-cell functions in mice: differential immunosuppressive effects of two monoclonal anti-Thy1.2 antibodies. Immunobiology 160:438–453
39. Kasai M, Yoneda T, Habu S, Maruyama Y, Okumura K, Tokunaga T (1981) In vivo effect of anti-asialo GM1 antibody on natural killer activity. Nature 291:334–335
40. Thomson AW, Fowler EF (1981) Carrageenan: a review of its effects on the immune system. Agents Actions 11:265–273
41. Maeda H, Ichimura H, Satoh H, Ohtsuki K (1978) Evaluation of succinyl neocarzinostatin in vivo. J Antibiot (Tokyo) 31:468–472
42. Kallistratos G, Fasske E (1978) Treatment of experimental sarcoma in rats with neocarzinostatin. Folia Biochim Biol Graeca 15:1–14
43. Numasaki Y, Takahashi K, Masuda E, Nohara C, Maeda H (1991) Antitumor effects of a new antitumor agent, zinostatin stimalamer (YM881)—effects on experimental tumors in vitro and in vivo (in Japanese). Jpn J Cancer Chemother 18:2289–2294

Pharmacological Uniqueness and Clinical Effects of Neocarzinostatin

Hiroshi Maeda

Summary. In vivo distribution after i.v. injection, a brief summary of the pharmacokinetic properties, proteolytic degradation, and toxicity of neocarzinostatin (NCS) are described. The uniqueness of NCS, compared with other drugs, is its very rapid urinary clearance, short plasma half-life, high level of accumulation in the urinary bladder, and reabsorption from the bladder into the circulating blood. The clinical efficacy of NCS against bladder cancer, brain tumor, cerebrospinal carcinomatosis, leukemia, and three solid tumors is discussed. Therapeutic tactics using NCS based on the pharmacokinetic one-compartment model for meningeal carcinomatosis and leukemia, and the two-compartment model for the treatment of brain tumor infused via the carotid artery, are described, and the rationales for the use of NCS is also discussed.

Key words: Neocarzinostatin—Distribution—Pharmacokinetic computer simulation—Cancer therapy—Bladder cancer—Brain tumor—Leukemia

Introduction

Neocarzinostatin (NCS) is the first antitumor protein drug as well as enediyene-containing drug used clinically, having been approved by the Ministry of Health and Welfare of Japan in 1971. Its approved use is for leukemia and cancers of digestive organs, and lately cancers of the bladder have been added. As described elsewhere in this volume, NCS exhibits extremely potent activity in vitro as well as in vivo. The minimal inhibitory concentration is lower than 0.1 μg/ml against susceptible microbes such as *Micrococcus luteus*. It is interesting that *Shigella flexneri* 2a-11 is more sensitive than *M. luteus* (0.05 μg/ml). Although NCS is

---

Department of Microbiology, Kumamoto University School of Medicine, 2-2-1 Honjo, Kumamoto, 860 Japan active against *M. luteus*, *Staphylococcus aureus*, *Shigella flexneri* 2a-11, and *Mycoplasma* [1], it is usually inactive against *E. coli* and other gram-negative bacteria. Mutants of *E. coli* became sensitive to NCS [2], perhaps similarly to the *Shigella* strain. NCS also was shown to exhibit bacteriostatic action against some

TABLE 1. Distribution of radioactivity of [^{14}C]succinyl neocarzinostatin in organs or tissues of rats (From [4], with permission)

Specimens	After 20 min Mean dpm/min ± SE	After 60 min Mean dpm/min ± SE
Kidney	151 570 ± 15 661	99 514 ± 6 333
Bladder	108 759 ± 28 756	133 423 ± 29 556
Ovary–uterus	14 502 ± 5 878	3 873[a] —
Bone marrow	12 291 ± 673	5 394 ± 1 002
Blood plasma	10 246 ± 639	5 541 ± 198
Lung	8 261 ± 2 664	1 785 ± 390
Skin	7 243 ± 767	4 278 ± 353
Tongue	7 076 ± 611	5 158 ± 907
Stomach	5 558 ± 706	2 693 ± 471
Muscle	4 289 ± 1 302	1 482 ± 206
Large intestine	4 041 ± 938	1 487 ± 266
Lymph nodes	3 884 ± 499	1 604 ± 762
Small intestine	3 336 ± 1 033	939 ± 77
Testicles	— —	1 350[b] ± 62
Adrenal gland	3 305 ± 445	6 684 ± 2 469
Pancreas	2 339 ± 417	1 223 ± 113
Liver	2 290 ± 466	1 575 ± 114
Submaxillary–salivary gland	2 279 ± 106	1 375 ± 100
Bone	2 251 ± 1 115	1 026 ± 79
Heart	2 179 ± 246	1 032 ± 206
Thymus	2 072 ± 505	927 ± 139
Eye	1 880 ± 317	599 ± 194
Spleen	1 209 ± 126	827 ± 71
Brain	96 ± 28	108 ± 8
Cellular fraction in blood[c]	951 ± 129	1 001 ± 195
Hypophysis[d]	4 050 ± 680	3 822 ± 689
Urine[e] (from 1-g bladder)	5 469 ± 1 254	6 110 ± 1 230

Data are the mean of four rats at 20 min and 60 min, respectively; disintegrations per minute (dpm) counts were expressed as per gram (wet weight) of organ. The body weight of each rat was adjusted to 100 g by calculation.
[a] Sample from one rat.
[b] Sample from three rats.
[c] Cellular components were washed with phosphate buffered saline (PBS) two times by centrifugation (1500 rpm × 5 min).
[d] Diffusible counts from 1-g bladder assigned arbitrarily as urine.
[e] The weight of the hypophysis was about 5 mg, and thus the error of 20% was unavoidable.

of *Mycobacteria* at or above 10 μg/ml [3]. A mutant of *E. coli* susceptible to actinomycin D was found to be also susceptible to NCS (Y. Ono, unpublished observation).

An early in vivo study using an experimental murine tumor, e.g., an ascitic form of S-180, showed activity as low as 0.01 μg/ml. It was later found that NCS was one of the most remarkable drugs against L-1210 leukemia in mice; i.e., its early L-1210 rating was surpassed only by actinobolin and phosphorodiamidic acid among 43 or more anticancer agents tested in the late 1960s.

Subsequent pharmacological study showed that it was degraded very rapidly in the blood and in vivo, as is discussed later [4]. During the course of my distribution study of NCS in vivo, two characteristics of NCS were found to be unique and very different from many other conventional drugs: one is that its degradation can be suppressed by protease inhibitors [5,6], and the other is that NCS accumulates at an exceedingly high concentration in the bladder and the kidney when injected i.v. [4], or at an extremely high concentration in the lymphatic nodules when it is injected s.c. [7]. These unique findings of NCS initially prompted me to devise a derivative of NCS to target to the lymphatic system because this is the most frequent site of tumor metastasis, and most frequently, therapeutic failure is the result of lymphatic metastasis. Thus, eradication of cancer metastasis in the lymphatics is the key for successful cancer therapy.

To make the drug more lymphotropic, two factors were considered. One was lipophilicity and the other larger molecular size; such molecules are known to be recovered into the lymphatic system, the site of metastasis [8]. By a chemical modification study to clarify the structure–activity relationship of NCS [9,10], the bis-*N*,*N*′-(Ala^1, Lys^{20}) succinyl derivative of NCS was found to be biologically superior to NCS [11], although poly(styrene-maleic acid)-conjugated neocarzinostatin (SMANCS) [12] was found to be much superior in various respects in later studies, as described in the chapter by Maeda and Konno in this volume.

In such circumstances I came across an advertisement for poly(styrene-co-maleic acid) *n*-butylester (SMA) as the basic material of car and floor wax. SMA can confer the two factors just mentioned for lymphotropic delivery. I also had sufficient knowledge of the chemical modification of NCS in which the two free amino groups (at Ala 1 and Lys 20) [9,10] and some carboxyl groups can all be modified without loss of activity. That is, succinylated NCS is two- or threefold superior to NCS in the aspect of antitumor activity, but that is not enough for a large-scale project of general clinical use [11]. SMA conjugation was initiated under such circumstances, and study on the SMA–NCS conjugate (SMANCS) was first reported at the International Protein Symposium in Taipei in 1978 [12]. The following chapter discusses more on SMANCS.

The last review article on the clinical effects of NCS appeared in 1981 [13]. Since then, enthusiasm for NCS has diminished because many new anticancer agents, such as 5-fluorouridine (5-FU), antracyclines, and *cis*-platinum have gained more attention in the clinic. However, the uniqueness of NCS remains remarkable when compared with other anticancer agents, and it is time to re-evaluate and apply NCS clinically in a more rational manner than with mere enthusiasm or denial. Therefore, a large part of this article was taken from the previous one [13]; also, more data on tissue distribution, pharmacokinetics, and computer simulation models have been added.

In Vivo Distribution and Pharmacokinetics of NCS

The earlier distribution study of NCS showed that its biodistribution/accumulation was high in the pancreas and the kidney. However, without appropriate controls, the assay based on antimicrobial activity in the pancreas seemed to yield false-positives. A later detailed distribution study using a ^{14}C-labeled derivative of NCS injected i.v. in rats [4] showed extremely high accumulation in the kidney and bladder (Table 1). Furthermore, extensive degradation of NCS occurred in the first 60 min after i.v. injection (Fig. 1) [4–7,14–17]. It was also noteworthy that accumulation of NCS in the ovary–uterus and bone marrow within 20 min was

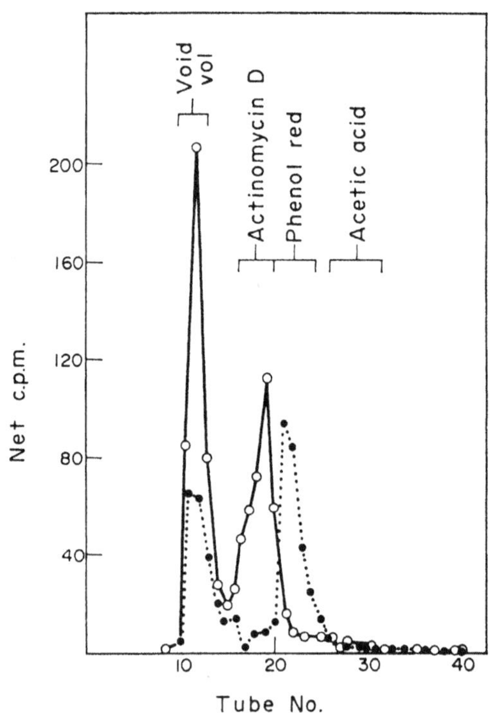

FIG. 1. Breakdown of [^{14}C]succinyl neocarzinostatin (NCS) in vivo (rat) as revealed by Sephadex G-15. *Open circle/solid line* and *solid circle/dotted line*, plasma samples obtained 20 min and 60 min, respectively, after injection of the drug. Note the decrease of the first peak near the void volume after 60 min and the increase in amount and the size shift to the low molecular weight region of the second peak, thus indicating an extensive degradation of the protein. *c.p.m.*, counts per minute

higher than the level in blood, lung, skin, or intestine, which explains that the primary side effect of NCS appears as bone marrow suppression and then renal toxicity. In any event, the drug level in the kidney and the urinary bladder far exceeds, that of the bone marrow or blood plasma, more than 10 to 30 fold, or 50 fold that of the liver.

These data prompted us to examine the vesicorenal recirculation mechanism and thus the pronounced accumulation in the bladder. When we injected [^{14}C]NCS into the bladder, it was recovered in the blood, the kidney, and in the urine at the ureter [14]; that is, NCS in the urine is reabsorbed into the bladder and recirculated into the kidney and then to the urine (Figs. 2 and 3). Data from ureterostomized rabbits showed that about 50% of the NCS was derived from the urinary route and the rest via the blood vessels. Therefore, intravesical administration for the treatment of bladder cancer became reasonable, and NCS can be given intravesically as well as via the i.v. route. It is inactivated by blood plasma, and inactivated rapidly by liver homogenate, and slowly by cerebrospinal fluid or the lung; it is first activated, then inactivated, in urine [4,18].

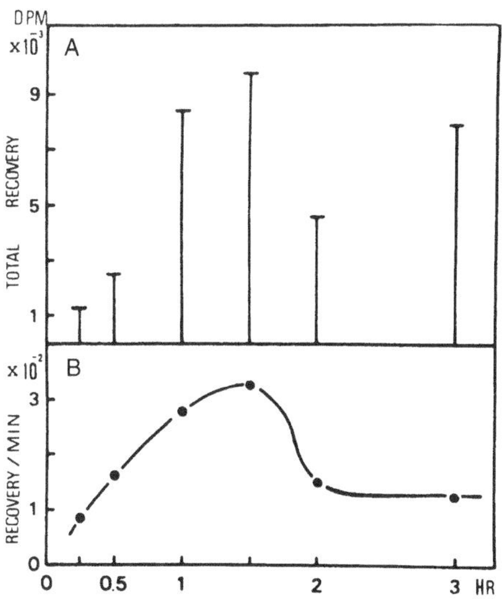

2

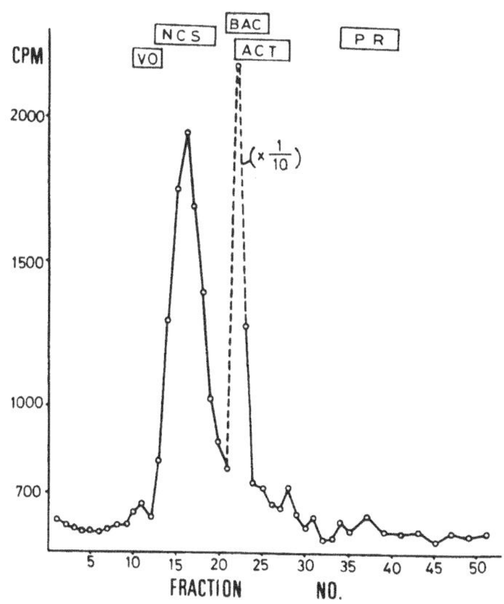

3

FIG. 2A,B. Urinary recovery of [^{14}C]NCS from the ureter after intrabladder administration. The recovery is measured by radioactivity (*dpm*, disintegrations per minute). **A** Total activity recovered during each collection period. **B** Rate of excretion (activity excreted per minute) of samples taken at the indicated times

FIG. 3. Analysis of molecular size distribution of absorbed and metabolized [^{14}C]NCS from urine. The sample was obtained from the ureter or ureterostomized rabbits between 60 and 90 min after intrabladder administration of the drug and analyzed on a Sephadex G-50 column. The markers for various molecular sizes are shown at the top. *VO*, Void volume (molecular weight, $>1.5 \times 10^4$); *NCS*, pure neocarzinostatin (11 000); *BAC*, bacitracin (1460); *ACT*, actinomycin D (1255); *PR*, phenol red (354). The notation $\times 1/10$ indicates the fractions were actually 10 times higher than shown

Toxicology of Neocarzinostatin

The lethal dose 50% (LD_{50}) of NCS is shown in Table 2. Toxicity of NCS in animals varies with the species. However, it is affected more by the route of administration than by the difference in species; i.e., and ministration p.o. causes practically no toxicity, while the i.v. route causes the most severe toxicity. The low toxicity of the oral and s.c. routes may be explained by proteolytic and other hydrolytic mechanisms of inactivation. Succinylated NCS [9,10], an interesting derivative, has been shown to be more stable than NCS in the serum [11]. Its toxicity is increased about twofold in rats, while its antitumor activity is increased three- to fivefold [11]. In humans, the maximum tolerable dose of NCS when given by drip infusion (6 mg/m^2 over 24 h on 5 consecutive days) was reported to be 30–50 mg [19] (Tables 3 and 4). Most of the patients with cancer of the bladder

TABLE 2. Toxicity of neocarzinostain (From [13], with permission)

Animal	Route of administration (single injection)	LD_{50}, mg/kg[a]
Mouse	i.v.	1.61
	i.p.	1.94
	s.c.	7.25
	p.o.	~1000.0
Rat	i.v.	1.13
	i.p.	0.94
	s.c.	5.04
	p.o.	≥300
Rabbit	i.v.	0.96
Dog	i.v.	<10.0[b]
	i.p.	0.5–1.0

[a] Based on observation period of 30 days.
[b] Daily consecutive lethal dose: <1.25 mg/kg per day.

TABLE 3. Hematological toxicity of neocarzinostatin

Protocol	Schedule of one course: Sakamoto et al. [20]: Bolus, 1 mg i.v. daily ×7, followed by 1-week intermission	Schedule of one course: Ohnuma et al. [19]: Continuous infusion, i.v., 6 mg/24 h, ×5 consecutive days
No. of courses (total dose, mg)	4 (24)	1 (30)
Total dose per day (mg)	1	6
WBC (per μl)	3300 (median value)	<4000, 24/33[a]; <2000, 10/33[a]
Platelet counts (per μl × 10^6)	2.8 (median value)	<0.1, 26/32[a]; <0.5, 13/32[a]
Fall of hemoglobin (>1.5 per dl)	Insignificant	16/30

WBC, White blood cells.
[a] Number of patients in group/Total number of patients in the study.

TABLE 4. General side effects of neocarzinostatin in humans

Side effects	Kamimura [40]: i.v. infusion, 2–4 mg/day, 4–6 days (%)	Masaoka et al. [41]: i.v. infusion, 2–4 mg/day, 4–6 days (%)	Sakamoto et al. [20,33]: 1 mg bolus, i.v. (in 20 s) (%)	Ohnuma et al. [19]: 6 mg i.v. infusion, continuous for 24 h, for 5 successive days (%)
Anorexia or stomatitis	50.0	—	87.5	19.2
Nausea	42.3	33.0	57.5	46.2
Vomiting	11.5	—	2.5	30.7
Dullness	11.5	16.6	—	—
Fever or chills	9.6	12.5	0	24.2
Cutaneous eruption	3.8	8.3	5	9.6
Cardiac stimulation	3.8	—	—	—
Hepatic dysfunction	17.3	Insignificant (GPT, ALP, LDH)	0	23.1
Hemorrhage	1.9	—	0	—
Shock	1.9	—	—	—
Phlebitis or cellulitis at i.v. site	—	—	2.5	21.1
No. of cases	55	20	40	52

GPT, Glutamic pyruvic transaminase; ALP, alkaline phosphatase; LDH, lactic acid dehydrogenase.

or other organs showed low white blood cell (WBC) counts after four courses of treatment (to about 3000/mm^3) in which one course consisted of a bolus i.v. injection (in 20 s) of 1 mg/60 kg per day for 7 successive days followed by a 1-week intermission [20]. The primary side effect was bone marrow suppression, followed by decrease of WBC and, to a lesser extent, of platelets [21,22]. An American report showed more severe thrombocytopenia than a fall in the hemoglobin level [23]. The phase I study by Ohnuma et al. [19] indicated azotemia [blood urea nitrogen (BUN) > 25 mg/dl: 14/51], hypophosphatemia (<2.5 mg/dl), amino aciduria, and hyponatremia, all of which may be associated with renal toxicity as observed previously [24].

A high lethal-dose-toxicity study of NCS in dogs showed hemorrhagic enterocolitis and depression of bone marrow and lymphoid tissue, with resulting death from sepsis, necrotic tonsillitis, and phlegmonous paw abscesses. The toxicity and antitumor activity were considered to be radiomimetic [24]. The nontoxic doses in dogs, using a single dose, a weekly dose for 2–5 weeks, or 28 daily doses, were 1.25, 0.625, and 0.16 mg/kg, respectively [24]. All sublethal toxicity so far reported was reversible.

Antigenicity and Immunogenecity of NCS

Because NCS is a proteinanceous substance, one can envisage the possibility of eliciting an antibody against NCS in vivo, which in turn could result in an adverse side effect. Previous studies have shown that NCS, as it is, does not generally produce antibodies in rabbits [25,26]. Moreover, NCS is an immunosuppressive agent. However, one can produce an NCS antibody by conjugation with poly-L-lysine followed by sensitization with Freund's complete adjuvant [26] or after a lethal dose [24]. We have examined more than 100 sera from patients with gastric and bladder cancer who were treated with NCS. The results showed no indication of anti-NCS antibody formation, except in one ambiguous positive case (unpublished). These results may agree with the observation that NCS is degraded and excreted extremely rapidly, and incidentally fails to produce antibody.

A few patients exhibited an immediate type of reaction on first-time injection with NCS. These immediate reactions, including fever, chills, rigor, hypertension, and mental confusion, were dose limiting for the 2-h infusion schedule, occurring at about 3 mg/m^2 and higher [19]. The reactions are unlikely to be associated with antibody formation because of the insufficient time period. These patients, when treated with corticosteroid, acetaminophen, and diphenhydramine, improved in 2 h [19]. However, in Japanese studies of more than 2800 patients with cancers of the digestive organs who were under extensive NCS treatment, there was only a single case of shock that may or may not have been associated with NCS [27].

In conclusion, NCS is not generally immunogenic, although it is advisable to prepare for such a possibility in the clinical setting.

Clinical Effects of Neocarzinostatin

Bladder Cancer

Based on the distribution study as just described, it became apparent that NCS would be a useful agent for cancer of the urinary bladder. Sakamoto et al., at Kumamoto University, initiated human trials and found a beneficial effect of NCS [20,23]. A standard course of treatment consisted of a single daily i.v. injection of 1 mg of NCS in 20 s for 7 consecutive days, followed by an intermission of 1 week. Most patients withstood three or four courses before their WBC dropped to about 3000/mm^3. Leukopenia (WBC $\leqq$ 3000/mm^3) was the main reason for cessation of NCS administration.

The results are summarized in Table 5. About 79% of the patients showed a positive response, while the tumor stabilized in 15.6% and tumor growth continued in 5.3%. All patients underwent transurethral resectioning (TUR) after chemotherapy, which was much easier to perform because of the diminished tumor size. In postoperative maintenance therapy, intravesical instillation of NCS or mitomycin C was used. It is noteworthy that the recurrence rate was much lower when NCS was combined with TUR (Table 6). The low rate of relapse (about one-fourth to one-fifth of historical controls) may be attributed to the efficient targeting (high concentration of NCS in the bladder tissue) and the great antitumor effect as well as the highly lymphotropic nature of NCS, which may reduce micrometastasis through the lymphatic system.

TABLE 5. Systemic i.v. administration of neocarzinostatin (NCS) and response to NCS on bladder cancer (From [33], with permission)

Total NCS (courses tolerated)	No. of patients (%)	Anticancer effect (%)				
		+++	++	+	±	–
1	3 (7.9)	0	2 (66.7)	0	1 (33.3)	0
2	6 (15.8)	0	2 (33.3)	2 (33.3)	2 (33.3)	0
3	13 (34.2)	0	5 (38.5)	4 (30.8)	2 (15.4)	2 (15.4)
4	16 (42.2)	2 (12.5)	10 (62.5)	3 (18.8)	1 (6.3)	0
Total	38 (100)	2 (5.3)	19 (50.0)	9 (23.7)	6 (15.7)	2 (5.3)
			79.0%			
				94.7%		

TABLE 6. Rate of recurrence: comparison with historical control on bladder cancer

Type of therapy	Observation period (yr) and recurrence rate (%)					
	<1/2	<1	<2	<3	<4	<5
Historical control:		67[b]				
Transurethral	16.3[a]	25.4[a]				
resection (TUR) only	60.5[c]	69[c]	92.3[c]			
NCS + TUR	2.5	5.0	15.0	24.3	27.3	36.0
No. of cases applicable[d]	1/40	2/4	6/40	9/27	9/33	9/25

[a] Wescott [28].
[b] Hiramatsu et al. [29].
[c] Saitoh et al. [30].
[d] Maeda [13].

Other than by systemic administration, NCS can be administered via the internal iliac artery semiselectively. The result was very marked: 100% tumor disappearances (10/10) with no side effects [31]. Asahi et al. [32] compared the results of four different regimens of NCS administration: (a) intravesical instillation (30 mg/day, 3 consecutive days, two courses in 2 weeks); (b) a small dose (1 mg daily by i.v. infusion for 7 days, one course); (c) a large dose (4 mg/day, twice per week, in combination with bleomycin, 25 mg i.v., for 5 consecutive days, plus citarabine, 100 mg i.v., on day 5; three courses in 3 weeks); and (d) a small i.v. dose of NCS (1 mg daily in combination with adriamycin given intravesically for a few weeks). They found only (b) (3/6) and (d) (7/10) to be effective. Later, Sakamoto et al. [33] reported the combination of NCS given i.v. and mitomycin C via the internal iliac artery and found satisfactory results.

A standard procedure used by Sakamoto et al. is a combination of NCS followed by surgical resectioning (TUR) in which NCS is given i.v. at 1 mg/day for 1–4 weeks until leukopenia (WBC < 3000 mm^3) is observed. A most pronounced effect is a great reduction of the frequency of recurrence (at 2 years, 80%–90% in the group of TUR alone versus 15% in the group of NCS + TUR), and improved long-term survival rate. An antitumor effect on bladder cancer was observed in 75%; complete remission was seen in 5%; partial response (size less than 50% of the initial size) in 47.5%; marginal response in 22.5%; tumor stabilization in 20%; and tumor growth in 5%.

Later Yoshida and Ikegami [34] retrospectively compared the efficacy of NCS and other antitumor agents such as mitomycin C. By routes other than the arterial NCS injection just mentioned (that showed 100% response), NCS i.v. (1 mg/day) with mitomycin C by the intravesical route showed significant reduction in recurrence rate (20.7% in 5 years versus 31.8% with NCS i.v. alone) as well as a high response rate (>80%). Mitomycin C given via the internal iliac artery (recurrence, 30% in 5 years) could be as effective as NCS given i.v. Because of the procedural simplicity (i.e., via i.v. route) of administration of NCS,

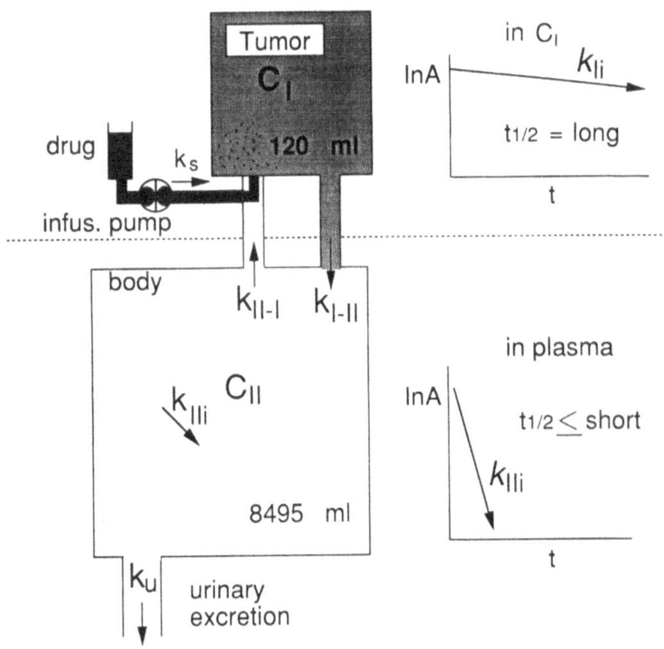

FIG. 4. Two-compartment model for the intraarterial infusion of drugs for the treatment of brain tumor. C_I, Drug concentration in the cerebral blood of a hemisphere that has a distribution volume of 65 ml; C_{II}, drug concentration in the systemic blood circulation outside the cerebral blood circulation, which has an apparent volume of drug distribution of 8495 ml; $k_{II\text{-}I}$ and $k_{I\text{-}II}$, transfer rate constants from compartments II to I and I to II, respectively; k_u urinary excretion rate constant; k_{Ii} and k_{IIi}, inactivation rate constants in compartments I and II, where $t_{1/2}$ is near or less than 3 s; k_s, infusion rate constant. *lnA*, drug concentration is in biologic activity (log). (From [18], with permission)

TABLE 7. Pharmacokinetic two-compartment simulation model for patients with brain tumor: infusion velocity via internal carotid artery and designed intracerebral concentrations of NCS (From [18], with permission)

Cerebral arterial concentration of drug (μg/min), C_{1d}	Infusion constant, k_i (μg/min)	Infusion time required (min) per amount of drug 0.25 mg	0.5 mg	1.0 mg	1.5 mg
0.0001	15.6	16.0	32.1	64.1	96.2
0.0002	18.8	13.3	26.6	53.2	79.8
0.0005	24.2	10.3	20.7	41.3	62.0
0.001	29.2	8.6	17.1	34.2	51.4
0.002	35.5	7.0	14.1	28.2	42.3
0.004	43.3	5.8	11.5	23.1	34.6
0.008	53.3	4.7	9.4	18.8	28.1
0.016	66.5	3.8	7.5	15.0	22.6
0.032	84.7	3.0	5.9	11.8	17.7
0.064	111.1	2.3	4.5	9.0	13.5
0.128	152.1	1.6	3.3	6.6	9.0

Data based on the human model of body weight of 60 kg. For calculation, see [18].

NCS i.v. in combination with mitomycin C intravesically should be better than mitomycin C via the iliac artery route. Further, NCS i.v. is better than adriamycin because the toxicity of NCS is reversible, while the cardiac toxicity of adriamycin could be fatal, and no drug resistance for NCS is reported in contrast to the high incidence of that of adriamycin [35].

In conclusion, NCS appears to be most beneficial for patients with bladder cancer.

Brain Tumor

The uniqueness of NCS permits a rational treatment of brain tumors. First, glioblastoma cells in culture were found very sensitive compared with normal glia cells [18]. Second, intracarotid arterial infusion at a well-regulated infusion velocity can make only the drug concentration of the brain compartment significantly high [18,36], because the blood–brain barrier at a tumor site is no longer functioning properly and a tumor exhibits a more enhanced vascular permeability than normal blood vessels, as discussed (17).

We have published a two-compartment model for the treatment of brain tumor and simulated the dosing regimen (Fig. 4). and Table 7 shows examples of the results of the computer simulation [18]. The tactic is theoretically novel and sound, but it was not pursued thoroughly clinically.

When NCS is slowly infused via the carotid artery at a given velocity with sufficient drug concentration in the cerebral compartment, as described, NCS can leak into the tumor compartment from the tumor blood vessel because the tumor

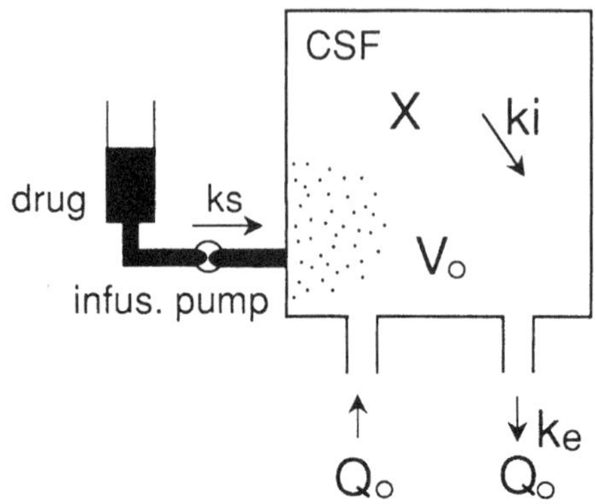

FIG. 5. One-compartment model in which the *drug* is infused into the cerebrospinal compartment by infusion pump (*infus. pump*). V_o, volume of cerebrospinal fluid (*CSF*); X, drug concentration at time (t), k_i, inactivation rate constant; k_e, elimination rate constant, which is replaced by circulating blood (Q_o). (From [37], with permission)

TABLE 8. Pharmacokinetic one-compartment simulation model for patients with brain tumor: continuous cerebrospinal infusion and drug level in cerebrospinal fluid (CSF) (From [37], with permission)

Concentration of NCS attained in CSF	Infusion rate constant, k_s	Infusion time required (min) per amount of drug			
(μg/ml)	(μg/min)	0.25 mg	0.5 mg	1.0 mg	1.5 mg
0.001	3.70	67.8	135.6	270.3	405.4
0.002	4.44	56.3	112.6	225.2	337.8
0.004	5.31	47.1	94.2	188.3	282.5
0.008	6.36	39.3	78.6	157.2	235.9
0.016	7.62	32.8	65.6	131.2	196.9
0.032	9.14	27.4	54.8	109.4	164.1
0.064	10.98	22.7	45.5	91.0	136.6
0.128	13.15	19.0	38.0	76.0	114.0

Data based on the human model of body weight of 60 kg. See [37] for details.

neovasculature has a highly leaky character and the blood–brain barrier is no longer intact. Thus, only the tumor interstitium, but not the normal tissue, will pick up the drug. The drug then will return into the general circulation and be diluted to a nontoxic level of drug concentration and then excreted into the urine. The diagram in Fig. 4 shows this compartment model; compartment I is the cerebral one and II is the blood pool of the body [18]. By similar reasoning, infusion into the subarachnoid space for meningial carcinomatosis has been found effective (Fig. 5).

The very labile nature of NCS at dilute concentration in blood plasma but not in the cerebrospinal fluid makes NCS ideal to apply for a single-compartment model, and a good rationale for the treatment of cerebrospinal metastasis or meningeal carcinomatosis is deduced from the result of simulation of the single-compartment model [37] (Fig. 5). When NCS moves into the second compartment, its activity or toxicity will be inactivated rapidly as well as diluted to the non-toxicity concentration [18,37]; the drug level of the spinal compartment can be maintained by intrathecal perfusion or an Ommaya V-P shunt. Table 8 summarizes the simulation study of the one-compartment model. A preliminary clinical result is encouraging (Table 9) [38,39].

Leukemia

The very first clinical effect of NCS in leukemia was reported for mylogeneous leukemia by Hiraki et al. at Okayama University [22]. Subsequently, they and other groups published various regimens in combination with other drugs, mostly for adult leukemia. Their earlier studies for adult leukemia patients treated with NCS alone (2 or 3 mg/day for 4 days followed by 7–10 days of intermission) showed complete remission of 5/11 (45.5%), while 1/11 (9.4%) was in partial remission, and 5/11 (45.5%) showed no effect. A later report with an increased

TABLE 9. Treatment of malignant glioma by radiation followed by intracarotid infusion of NCS (From [36], with permission)

Response[a]			
Complete response:	11.5%	Overall response: 69.2%	
Partial response:	38.5%		
Marginal response:	19.2%		
No change:	23.1%		
Disease progressed:	7.7%		

Survival rate	Overall (%)[b]	<5 mg NCS	>5 mg NCS
1 year	84.6	—	—
2 years	54.2	—	—
3 years	33.0	12.5%	50%

Side effect: Ocular toxicity[c]: 35.3%
No toxicity in WBC, RBC, PLT counts.
No liver or kidney toxicity

[a] Total dose of NCS: 5 mg or more is preferred; 0.5 or 1 mg per dose was infused in 30–40 min via the carotid artery (if the catheter can be located beyond the ocular artery, 1 mg of NCS can be infused). NCS was administered every 4 days. Lineac radiation: total doses were 40–60 Gy; 1.5–2 Gy was used five times per week.
[b] Performance status: all cases showed Karnofsky scale 100%.
[c] Toxicity appeared if the catheter tip failed to be located beyond the origin of the ophthalmic artery. WBC, White blood cells; RBC, red blood cells; PLT, platelets.

TABLE 10. Results of treatment of acute leukemia with NCS in combination with other drugs (From [40], with permission)

	ALL		ANLL		Total	
Protocol[a]	No. of cases	Complete remission	No. of cases	Complete remission	No. of cases	Complete remission
NVMP	15	13 (86.7%)	6	4 (66.7%)	21	17 (81.0%)
NDMP	0	0	15	11 (73.3%)	15	11 (73.3%)
NAMP	3	2 (66.7%)	4	3 (75.0%)	7	5 (71.4%)
NADP	—	—	18	16 (88.9%)	18	16 (88.9%)
NAVP	—	—	2	2	2	2
Total	18	15 (83.3%)	45	36 (80.0%)	63	51 (81.0%)

ALL, Acute lymphocytic leukemia; ANLL, acute nonlymphoblastic leukemia.
[a] N, Neocarzinostatin, 0.04 mg/kg; V, vincristine, 0.04 mg/kg; M, 6-mercaptopurine riboside, 6.0 mg/kg; P, prednisolone, 1.2 mg/kg; D, daunomycin, 0.6 mg/kg; A, cytosine arabinoside, 1.2 mg/kg.
One course of treatment consisted of 4 successive days of i.v. dosing, followed by 4–10 days intermission. Vincristine was administered on the first day only. Two to three courses were usually carried out.

number of leukemic patients indicated the beneficial effect of NCS, particularly in combination with other antineoplastic agents (Table 10), was promising: 81% complete remission (n = 63) [40]. Among these patients, 88.9% exhibited complete remission for acute nonlymphocytic leukemia when NCS, cytosine arabinoside, daunomycin, and prednisolone were used. The data of Masaoka et al. [41] indicated that NCS brought a rapid initial decrease in the number of leukemic cells but with a relatively short duration of complete remission.

The results of recent preliminary trials suggested a combination chemotherapy for relapsed adult leukemia including NCS, cytosine arabinoside, 6-mercaptopurine, and prednisolone in which daunomycin replaced by NCS showed more promise [42,43]. This regimen avoided liver toxicity as well as coronary-cardiac toxicity, especially in elderly patients. The results of this combination with NCS showed a rate of complete remission (66.7%) comparable to that with daunomycin (61%–86%). The median duration of complete remission was 16.5 months, which compared well to the similar combination with daunomycin in the place of NCS, as reported by Clarkson et al. [44] and Uzuka et al. [45]. In any case, this author questions whether the infusion velocity of NCS in these studies was sufficiently fast to attain an effective drug level in the plasma or slow enough to avoid systemic toxicity.

The inconsistent results of NCS for leukemias as well as other tumors may be caused by the different infusion velocity or other factors such as

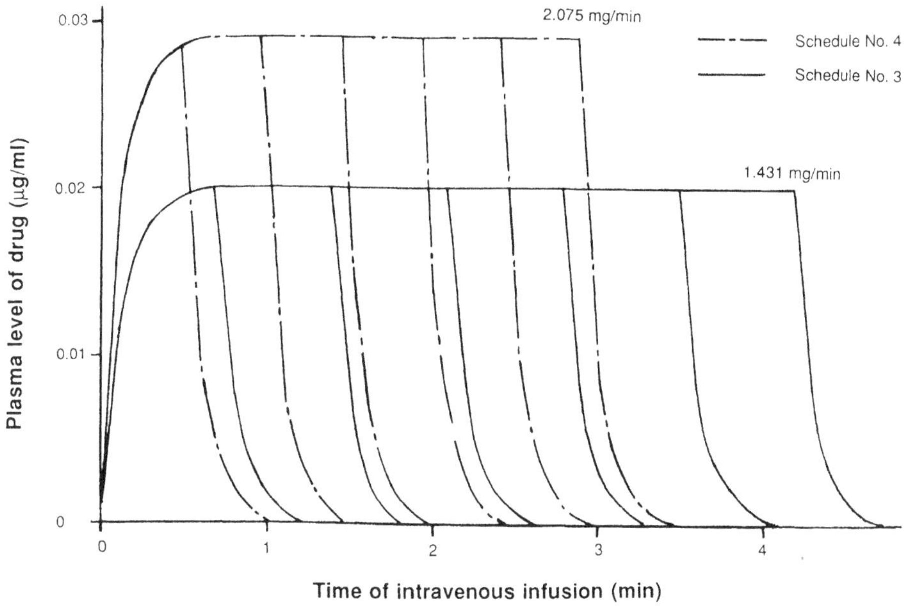

FIG. 6. Plasma level of NCS determined by computer simulation based on one-compartment model for a human with body weight of 60 kg. The infusion velocity of NCS is shown on each curve; infusion is started at time zero. Schedules no. 3 and 4 refer to Table 11. Each line dropping off from the plateau level is the completion point of NCS dosing at 1, 2, 3, 4, 5, and 6 mg. See [46] for details; cf. Fig. 5. in [37]

protection from light. Because NCS has an extremely short half-life in plasma, it must be infused within a relatively short period of time to attain a desired plasma concentration. We have simulated a single-compartment model (Figs. 5–7) considering all possible pharmacokinetic parameters and obtained a result which shows that a 2-mg dose of NCS, for instance, must be infused within 2 min to attain a plasma concentration of 0.02 μg/ml [an arbitrary minimal inhibitory concentration (MIC) for leukemic cells] for a man with a body weight of 60 kg (Table 11) [46]. However, standard antileukemia protocol does not include NCS. With the advent of antiemetic drugs and with the possible use of glanurocyte-stimulating factor (GCSF), as well as the advanced theoretical knowledge of pharmacokinetics, it is time to reevaluate this drug for the treatment of leukemia.

Gastric and Other Cancers in Digestive Organs Including the Pancreas and the Liver

Because gastric cancer is the most common cancer in Japan, the efficacy of NCS was tested first on this cancer and has been explored thus most extensively ever

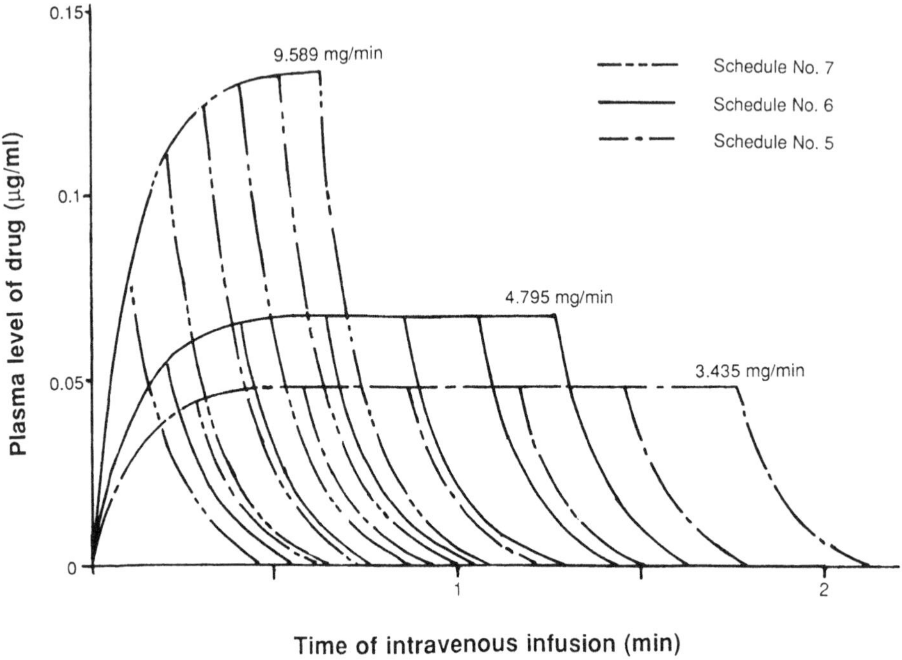

FIG. 7. Plasma level of NCS determined by computer simulation based on one-compartment model (same as in Fig. 6; cf. legend of Fig. 5). In this model, a higher plasma level than that in Fig. 5 was the aim. See [46] for details; cf. Table 11

TABLE 11. Dose regimen for i.v. infusion of NCS (From [46], with permission)

Schedule no.	Dose velocity (mg/min)	Attainable plasma level (µg/ml)	Time (min) required for dosing per amount of drug					
			1 mg	2 mg	3 mg	4 mg	5 mg	6 mg
1	0.3578	0.005	2.80	5.59	8.38	11.18	13.97	16.77
2	0.7156	0.01	1.40	2.80	4.19	5.59	6.99	8.38
3	1.431	0.02	0.70	1.40	2.10	2.80	3.49	4.19
4	2.075	0.029	0.48	0.96	1.45	1.93	2.41	2.89
5	3.435	0.048	0.29	0.58	0.87	1.16	1.46	1.75
6	4.795	0.067	0.21	0.42	0.63	0.83	1.04	1.25
7	9.589	0.134	0.10	0.21	0.31	0.42	0.52	0.63

Conditions for the model: Body wt: 60 kg, apparent volume of distribution: 8.56 l, inactivation and clearance constant:
$K = K_u + K_i + K_m \doteqdot 8.36$/min
K_i: determined at 0.2µg/ml of NCS in vitro.

since [47,48]. Of the more than 2000 patients treated, NCS alone, or in combination with 5-FU, has been administered primarily by i.v. drip infusion; the effect of NCS was significant when patients tolerated a dose of more than 20 mg. NCS in combination with 5-FU was also recommended as beneficial. However, most patients subjected to NCS treatment were at stage III and IV; thus, the response rate as well as improvement in the Karnofsky rating (I-C) was rather low. The drug could be proven to be more promising if it were applied to early-stage cancer and then evaluated for longer term of survival and relapse rate because of possible antimetastatic activity [7]. If NCS were applied locally it would result in high drug concentration at the regional lymph nodes, thus antilymphatic metastasis would be expected [49]. This seems to be the cose in bladder cancer [20]. In this respect, NCS may be more useful if it is used under angiotensin II-induced hypertension because hypertension would facilitate drug delivery to the tumor, as discussed in the chapter on SMANCS (Chapter by H. Maeda and T. Konno, this volume).

After the early observation of relatively high levels of NCS in the pancreas [50], Ishii et al. [51] initiated the use of NCS against pancreatic cancer. Of 88 cases treated, 11 (12.4%) were found responsive. This is relatively low in response rate although pancreatic cancer is one of the most difficult cancers to cure. Other reports showed subjective improvement of 16.7% and objective improvement (primarily reduction in tumor size) of 50% [52]. A recent large-scale trial with NCS alone [27] showed similar results (in Karnofsky's rating; 1-A, -B, -C: 13.7%; O-A, -B, -C: 37.1%), with a total evaluable case number of 124. Preliminary clinical trials in a number of American institutions suggested a similar degree of efficacy, although the numbers of patients were small [21].

Issell et al. [21] suggested the possible efficacy of NCS against liver cancer based on their phase I study with a limited number of patients. Subsequently, Falkson et al. [53] treated 30 evaluable patients with histologically confirmed primary liver cancer and found 7 patients with partial response and 9 patients with no change; in 14 patients, however, there was progressive disease. This may indicate that hepatocellular carcinoma is relatively sensitive to NCS, which may be consistent with the results of SMANCS discussed later. Side effects were similar to those described by others.

Skin Cancer

For the treatment of warts, antineoplastic agents such as bleomycins and 5-FU are occasionally used. Kumasaka and Miyazawa [54] applied NCS for warts in situ at a dose of 13–63 μg by s.c. injection three to five times. Of the 24 cases treated, 75% showed either disappearance of the warts or considerable regression in their size. No side effect was observed. This treatment was extended to malignant tumors of the skin [55]. In this case only s.c. injection in situ showed excellent results, while i.v. injection elicited no response. As the systemic toxicity of NCS given s.c. is quite low, a more extensive investigation should be encouraged.

Other Cancers

There are only a limited number of tumors other than those described here for which the use of NCS may be benefial, and its efficacy or rationale in these tumors remains to be established. Among them are lymphoma (1/2, effective/treated cases, in [21]), although with a merit of lymphotropicity [7] and high susceptibility in cultured lymphoblastoid cells [56]; colon cancer (0/8 in [21], 1/1 in [50], and 5/35 as Karnofsky's rating of 1-A, -B, -C in [27]; lung cancer (1/7 in [21] and [57]; esophagus cancer (1/1 in [47]); maxillary antrum carcinoma [58]; cervical and ovarian cancer [59]; and cancers of the tongue (1/1 in [47]).

Immunosuppressive Effect

As expected, NCS is an immunosuppressive agent. We have previously reported that normal peripheral lymphocytes, which do not proliferate, are more resistant to NCS than the transformed lymphoblastoids or leukemia/lymphoma cells [56,60]. Yamashita et al. [61] confirmed that NCS had no cytocidal effect on small (normal) lymphocytes as judged by cell viability or oxygen consumption measurements. A study was conducted to clarify the effect of NCS on the reactivity of parental lymphocytes to a major histocompatability complex of F_1 hybrid cells in rats. For this purpose, the graft-versus-host reaction (GVHR) was tested primarily by measuring the magnitude of popliteal lymph node enlargement and the incidence of runting syndrome. The result showed that preincubation of parental lymphocytes (graft) with NCS at 1 μg/ml for 10 min at 37°C inhibited the enlargement of popliteal lymph nodes of the recipients. Treatment of the donors' lymphocytes prevented the runting syndrome when pretreated donors lymphocytes were injected i.v. into F_1 hybrid rats [61]. This result encourages further investigation of this unique GVHR suppression of NCS with respect to bone marrow transfusion in which such immunosuppression is the most critical requisite.

Tan et al. [57] tested NCS in comparison with azathiopurine and oxisuran for the suppression of skin allografts, for inhibition of blastogenesis in mixed-lymphocyte culture, and in response to phytohemagglutinin, as well as in antibody formation in rats. The results suggest a strong immunosuppressive activity of NCS.

Our unpublished results on purging cancer (neuroblastoma) cells in the autologous bone marrow transfusion model in mice showed that stem cells are relatively resistant, compared to dividing cancer cells. Consequently, reduction of at least two log units in cancer cells was observed. In any event, this area needs more research.

Concluding Remarks

NCS is unique, particularly in its distribution in vivo and its pharmacokinetics. Its pronounced distribution in the bladder is explained by urinary reabsorption from the urine as well as vascular access through the circulating blood, while the renal excretion of NCS into the urine is extremely rapid. This constitutes the rationale for the treatment of bladder cancer. The second unique quality is its extremely rapid clearance ($t_{1/2}$, ~2 min in mice), which is attributed to rapid renal clearance and proteolytic degradation. These characteristics became advantageous for the use of NCS against brain cancer; intracarotid arterial infusion is based on the two-compartment model, and thecal (meningeal) or spinal perfusion is used against meningospinal carcinomatosis.

The pharmacokinetic model for leukemia needs to be reevaluated in a clinical setting in which there is a good rationale to use NCS. The efficacy of NCS against other cancers in various organs has been described as having some limited success, but enthusiasm has declined. The author believes that NCS has a good rationale for the treatment of cancers of the bladder and brain (cerebral/meningeal carcinomatosis) and of leukemia. With the advent of antiemetics and GCSF as well as application of angiotensin II-induced hypertension chemotherapy, more rigorous reevaluation of the efficacy of NCS needs to be pursued, which will prove the clinical value of NCS per se.

References

1. Ishida N, Miyazaki K, Kumagai K, Rikimasu M (1965) Neocarzinostatin, an antitumor antibiotic of high molecular weight. J Antibiot (Tokyo) 18:68–76
2. Ohishi T, Yamagata H, Udaka S (1979) Mutants of *Escherichia coli* hypersensitive to an antitumor protein: neocarzinostatin. Agri Biol Chem 43:371–378
3. Kallistratos G, Kazda J, Freerksen R (1978) Bacteriostatic properties of neocarzinostatin. Folia Biochim Biol Graeca 15:15–19
4. Maeda H, Yamamoto N, Yamashita A (1976) Fate and distribution study of [^{14}C]succinyl neocarzinostatin in rats. Eur J Cancer 12:865–870
5. Maeda H, Takeshita J (1975) Degradation of neocarzinostatin by blood sera in vitro and its inhibition by diisopropylfluorophosphate and *N*-ethylmaleimide. Gann 66:523–527
6. Maeda H, Takeshita J (1976) Inhibitors of proteolytic enzymes prevent the inactivation by blood of protein antibiotic neocarzinostatin and its succinyl derivative. J Antibiot (Tokyo) 29:111–112
7. Maeda H, Takeshita J, Yamashita A (1980) Lymphotropic accumulation of an antitumor antibiotic protein, neocarzinostatin. Eur J Cancer 16:723–731
8. Courtice FC (1963) The origin of lipoprotein in lymph. In: Myersen HS (ed) Lymph and the lymphatic system. Thomas, Springfield, pp 89–126
9. Maeda H (1974) Preparation of succinyl neocarzinostatin. Antimicrob Agents Chemother 5:354–355
10. Maeda H (1974) Chemical and biological characterization of succinyl neocarzinostatin. J Antibiot (Tokyo) 27:303–311

11. Maeda H, Ichimura H, Satoh H, Ohtsuki K (1978) Evaluation of succinyl neocarzinostatin in vivo. J Antibiot (Tokyo) 31:468–472
12. Maeda H, Takeshita J, Kanamaru R (1978) A lymphotropic derivative of neocarzinostatin with antimetastatic property: conversion into a derivative soluble in water and solvents with improved pharmacological property. In: Proceedings of the International Symposium on Protein, March 1978, Taipei, Taiwan
13. Maeda H (1981) Neocarzinostatin in cancer chemotherapy (review). Anticancer Res 1:175–186
14. Maeda H, Sakamoto S, Ogata T (1979) Mechanism of accumulation of the antitumor protein antibiotic neocarzinostatin in bladder tissue: intravenous administration, urinary excretion, and absorption into bladder tissue. Antimicrob Agents Chemother 11:941–945
15. Lowenthal IS, Parker LM, Greenblatt DJ, Brown BL, Samy TSA (1979) Pharmacokinetic analysis of neocarzinostatin in normal and tumor-bearing rodents. Cancer Res 39:1547–1551
16. Maeda H, Matsumura Y, Oda T, Sasamoto K (1986) Cancer selective macromolecular therapeusis: tailoring of an antitumor protein drug. In: Feeney RE, Whitaker JR (eds) Protein tailoring for food and medical uses. Dekker, New York, pp 353–381
17. Matsumura Y, Maeda H (1986) A new concept for macromolecular therapeutics in cancer chemotherapy; mechanism of tumoritropic accumulation of proteins and the antitumor agent SMANCS. Cancer Res 46:6387–6392
18. Maeda H, Sano Y, Takeshita J, Iwai Z, Kosaka H, Marubayashi T, Matsukado Y (1981) A pharmacokinetic simulation model for chemotherapy of brain tumor with an antitumor protein antibiotic, neocarzinostatin: theoretical considerations behind a two-compartment model for continuous infusion via an internal carotid artery. Cancer Chemother Pharmacol 5:243–249
19. Ohnuma T, Nogeire C, Cuttner J, Holland JF (1978) Phase I study with neocarzinostatin: tolerance to two-hour infusion and continuous infusion. Cancer 42:1670–1679
20. Sakamoto S, Ogata J, Ikegami K, Maeda H (1980) Chemotherapy for bladder cancer with neocarzinostatin: evaluation of systemic administration. Eur J Cancer 16:103–113
21. Issell BF, Prestayke AW, Comis RL, Crook ST (1979) Zinostain (neocarzinostatin). Cancer Treat Rev 6:239–249
22. Hiraki K, Kamimura O, Takahashi I, Nagao T, Kitajima K, Irino S (1973) Neocarzinostatin: une approche nouvelle dans la chemiotheprapie des leucemies aigues. Nouv Rev Fr Hematol 13:445–451
23. Sakamoto S, Ogta J, Ikegami K, Maeda H (1978) Effects of systemic administration of neocarzinostatin, a new protein antibiotic, on human bladder cancer. Cancer Treat Rep 62:453–455
24. Schaeppi U, Menninger F, Fleishman RW, Bodgen AE, Shein PS, Cooney DA (1974) Toxicity of neocarzinostatin (NSC-69856): an antitumor antibiotic with radiomimetic and antigenic characteristics. Cancer Chemother Rep Part 3 5:43–52
25. Toki H (1976) Absence of anti-neocarzinostatin (NCS) antibody production in leukemia patients treated with NCS. Acta Med Okayama 30:135–142
26. Sakamoto S, Maeda H, Matsumoto T, Ogata J (1978) Experimental and clinical studies on the formation of antibodies to neocarzinostatin, a new protein antibiotic. Cancer Treat Res 62:2063–2070
27. Anonymous (1980) Summary of national cumulative data on neocarzinostatin treatment for cancers in the digestive organs in Japan (in Japanese). Yamanouchi Pharmaceutical Co, Tokyo Japan

28. Wescott JW (1966) The prophylactic use of Thio-TEPA in transitional cell carcinoma of the bladder. J Urol 96:913–918
29. Hiramatsu T, Okajima E, Motomiya Y, Iriya K, Kondo T, Hirao Y, Matsushima S (1973) Clinical studies on tumor of the urinary bladder. II. Clinico-statistical tumor of the urinary bladder (in Japanese). Jpn J Urol 64:287–194
30. Saitoh K, Kubota Y, Takai S (1978) The recurrence after conservative therapy of bladder cancer (in Japanese). Jpn J Urol 69:373–380
31. Matsumoto K, Seto K (1975) A trial of subselective intraarterial continuous infusion with neocarzinostatin in urinary bladder cancer (in Japanese). In: Proceedings, 34th Annual Meeting of the Japanese Cancer Association, Osaka, Oct 1–3, p 85
32. Asahi T, Ike N, Takamoto J, Takanashi T, Suyama B, Fujita Y, Ohmori H, Matsumura Y, Nishi M (1979) The treatment of the bladder cancer by neocarzinostatin (in Japanese). Nishinihon J Urol 41:91–94
33. Sakamoto S, Ogata J, Maeda H (1981) Chemotherapy of bladder cancer: effect of continuous infusion of mitomycin C via internal iliac artery and systemic administration of neocarzinostatin (in Japanese). Nishinihon J Urol 43:223–234
34. Yoshida M, Ikegami K (1984) Studies of prophylactic effects of chemotherapy on the recurrence of bladder carcinoma. Nishinihon J Urol 46:1087–1091
35. Miyamoto Y, Oda T, Maeda H (1990) Comparison of the cytotoxic effects of the high- and low-molecular-weight anticancer agents on multidrug-resistant Chinese hamster ovary cells in vitro. Cancer Res 50:1571–1575
36. Uemura S, Matsukado Y, Sonoda H, Kuratsu J, Ohtsuka T, Yoshioka S, Yoshida A, Kochi M, Marubayashi T (1986) Treatment of malignant glioma with neocarzinostatin (in Japanese). Neurol Med-Chir (Tokyo) 26:304–310
37. Maeda H, Matsukado Y, Iwai Z, Uemura S, Kuratsu J, Takeshita J, Sano Y (1982) Pharmacokinetic one-compartment model using neocarzinostatin as a prototype drug and its clinical application to chemotherapy for brain tumor. Part I. Theory and computer simulation for cerebrospinal infusion (in Japanese). Jpn J Cancer Chemother 11:1042–1045
38. Uemura S, Matsukado Y, Fujioka K, Kuratsu J, Sonoda H, Yano T, Ohtsuka T, Yoshida S (1985) Treatment of meningeal carcinomatosis: NCS perfusion therapy in the CSF pathway (in Japanese). Jpn J Cancer Chemother 12:1794–1800
39. Matsukado Y, Uemura S, Kuratsu J (1980) Subarachnoid dissemination of the brain tumor cells (in Japanese). Neurol Surg 8:1113–1123
40. Kamimura O (1977) Studies on the chemotherapy in acute leukemia (in Japanese). Okayama Igakkai Zasshi (Okayama Med J) 89:701–711
41. Masaoka T, Nakamura H, Hasegawa Y, Shibata H, Tatsumi N, Ueda T, Takubo K, Koshitake J, Senda N, Kinoshita K, Miyake M, Tsubaki T (1974) Treatment of acute leukemia with neocarzinostatin (NCS) (in Japanese). Rinsho Ketsueki (Clin Hematol) 15:1309–1316
42. Tokioka M (1981) Induction chemotherapy for acute leukemia. Part 1. Clinical trial of NCMP two-step regimen for treatment of acute non-lymphocytic leukemia (in Japanese). Okayama Igakkai Zasshi (Okayama Med J) 93:1–13
43. Tokioka M (1981) Induction chemotherapy for acute leukemia. Part 2. Clinical trial of 5 drug combination chemotherapy (NCDVP, Ncyclo-cDVP) for treatment of refractory acute leukemia (in Japanese). Okayama Igakkai Zasshi (Okayama Med J) 93:14–24
44. Clarkson BS, Dowling MD, Gee TS, Cunningham IB, Burchenal JH (1975) Treatment of acute leukemia in adults. Cancer 36:775–795

45. Uzuka Y, Liong SK, Yamagata S (1976) Treatment of acute nonlymphoblastic leukemia using intermittent combination chemotherapy with daunomycin, cytosine arabinoside, 6-mercaptopurine and prednisolone-DCMP two-step therapy. Tohoku J Exp Med 118(suppl):217–225
46. Maeda H, Takeshita J, Iwai Z, Yamanami Y, Sakamoto S (1979) Computer simulation of dose regimen of neocarzinostatin infusion for the complete remission of leukemia based on a single compartment model (in Japanese). Jpn J Cancer Chemother 6: 95–103
47. Takahashi M, Toriyama K, Maeda H, Kikuchi M, Kumagai K, Ishida N (1969) Clinical trials of new antitumor polypeptide, neocarzinostatin. Tohoku J Exp Med 98:273–280
48. Aneha Y, Kikuchi K, Sugano H (1974) The effect of high molecular weight antitumor agent, neocarzinostatin, on the gastric cancer (in Japanese). Nihon Rinsho (Jpn Clin) 32:870–874
49. Hase Y, Kondo T, Nakagawa H, Matsumoto T, Maekawa T, Sasakibara N (1987) Intraoperative administration of neocarzinostatin in the duodenal subserosa to the regional lymph nodes in pancreatoduodenal cancer (in Japanese). Jpn J Cancer Chemother 14:3119–3124
50. Fujita H, Nakayama N, Sawabe T, Kimura K (1970) In vivo distribution and inactivation of neocarzinostatin. Jpn J Antibiot 23:471–478
51. Ishii K, Nakamura K (1974) Cooperative studies on chemotherapy for the pancreatic cancer by neocarzinostatin (NCS) (in Japanese). Jpn J Cancer Chemother 1:433–442
52. Kinami Y, Yazaki H, Koyama F, Noguchi M, Konishi K, Furukawa I, Nishida Y, Miyazaki I (1975) The treatment of pancreatic carcinoma with neocarzinostatin (in Japanese). J Jpn Soc Cancer Ther 10:502–508
53. Falkson G, Von Hoff D, Klaassen DDU, Plessis H, Van Dermerwe CF, Van Dermerwe AM, Carbone PP (1980) A phase II study of neocarzinostatin (NSC 157365) in malignant hepatoma. Cancer Chemother Pharmacol 4:33–36
54. Kumasaka T, Miyazawa T (1973) The effect of new antitumor antibiotic neocarzinostatin on warts (in Japanese). Rinsho Hifuka (Clin Dermatol) 27:163–168
55. Ishihara K, Komatsu N, Yanagida H (1975) The treatment of skin tumors with neocarzinostatin (NCS) (in Japanese). Rinsho Hifuka (Clin Dermatol) 29:155–160
56. Suzuki F, Okuno Y, Maeda M, Maeda H (1987) Tumor growth inhibitory effects of SMANCS, a poly(styrene maleic acid) conjugated neocarzinostatin, on various tissue cultured cells (in Japanese). Jpn J Cancer Chemother 14:3305–3312
57. Tan M, Nishihira T, Tsutsumi E, Kasai M (1976) The effect of oxisuran on differential inhibition of cell-mediated immunity. Tohoku J Exp Med 119:333–337
58. Takeda C, Washizu K, Ono S, Ebihara S, Suzuki K, Fujii M, Baba K (1974) Intraarterial infusion of neocarzinostatin combined with radiotherapy as a treatment method for carcinoma of the maxillary antrum (in Japanese). Jpn J Cancer Clin 20:937–943
59. Takada M, Koizumi K, Ochiai M, Kamemori H (1974) Clinical studies on neocarzinostatin in gynecological malignant tumors (in Japanese). Jpn J Cancer Chemother 1:801–812
60. Maeda H, Matsumoto M (1979) Cytotoxic effect of neocarzinostatin on human lymphoid cells. Tohoku J Exp Med 128:313–323
61. Yamashita A, Kitawaki Y, Miyamoto M, Maeda H (1979) Suppression of graft-versus-host reaction by preincubation of the graft with an antitumor protein, neocarzinostatin. Immunopharmacology 1:255–265

Metamorphosis of Neocarzinostatin to SMANCS: Chemistry, Biology, Pharmacology, and Clinical Effect of the First Prototype Anticancer Polymer Therapeutic

Hiroshi Maeda[1] and Toshimitsu Konno[2]

Summary. A polymer conjugated derivative of neocarzinostatin was prepared by cross-linking with two chains of poly(styrene-co-maleic acid half-butylate) (SMA) to neocarzinostatin (NCS), one at N-terminal alanine and the other at lysine 20. The conjugate was designated SMANCS. SMANCS exhibits a molecular weight of about 16000, and its oily formulation with Lipiodol (lipid contrast agent) became possible because of the hydrophobic property of SMA. When administered via the tumor-feeding artery, SMANCS in Lipiodol remained predominantly in tumor tissue: a concentration more than 2000 fold higher in the tumor than in the blood plasma. SMANCS showed increased stability in blood, and the immunogenecity was much lower than that of the parental NCS.

A predominant tumor-selective targeting mechanism was found not only for SMANCS administered intravenously as an aqueous formulation, but also was found later in many other macromolecular/polymer drugs as a common phenomenon if they were biocompatible. This mechanism is now called the enhanced permeability and retention (EPR) effect of macromolecules and lipids. These macromolecules circulate much longer in the blood and permeate solid tumors. Most solid tumors have a high vascular density and are defective in their anatomical architecture. Further, solid tumors produce multiple types of vascular permeability factors (e.g., bradykinin, nitric oxide, vascular permeability factors) that facilitate the vascular permeability of macromolecules to tumor interstitium. In addition, there is much less efficient clearance of these macromolecules from a tumor by the lymphatic system, thus they remain in the tumor for a long time. SMANCS also was shown to accumulate effectively in normal lymph nodes, thus it is considered beneficial as an anti-lymphatic-metastatic agent. In vitro studies showed that SMANCS is more effectively internalized into tumor cells and cleaves DNA.

---

[1]Department of Microbiology, Kumamoto University School of Medicine, 2-2-1 Honjo, Kumamoto, 860 Japan
[2]First Department Surgery, Kumamoto University School of Medicine, 2-2-1 Honjo, Kumamoto, 860 Japan

The unprecedented clinical effect of SMANCS for human hepatoma is now well established; SMANCS became an approved drug in 1993 in Japan. The survival rate of hepatoma can be improved with SMANCS beyond 5 years; otherwise these patients usually die within 6 months. If hepatoma patients have no liver cirrhosis or metastasis to other organs, and the tumor is confined within one segment, the chances of 5-year survival could be as high as 90%. Future applications of SMANCS/Lipiodol with ascertained clinical benefit include treatment of renal cell carcinoma via the renal artery, metastatic hepatoma via the hepatic artery, and pleural and ascitic carcinomatoses using intracavitary application. Treatment of tumors of the lung, stomach, pancreas, and gallbladder, as well as lymphoma and melanoma with SMANCS/Lipiodol shows promise and awaits further evaluation. Major side effects are mild fever (50%), which usually lasts for a few days, and dull pain (about 30%), which lasts for a half-hour. Leucocytosis and immunological activation are seen frequently and they are beneficial effects. Hematological suppression and toxicity to the liver and kidneys are not usually observed.

Key words: SMANCS—Polymer drug—Tumor targeting—Liver cancer—Oily formulation—Lipiodol—Slow release—Clinical effect

Introduction

As described elsewhere, neocarzinostatin (NCS) possesses extremely potent biological activity. Untargeted drug delivery therefore might cause severe side effects in organs and tissues with rapid cell divisions such as bone marrow and intestinal epithelium. In fact, its molecular weight of 12kDa is actually very small and indeed it is freely diffusible in and out of blood vessels; for example, its renal clearance in mice is very rapid ($t_{1/2}$ = 1.9min), as has been described. Thus, the general toxicity of NCS is similar to other low molecular weight anticancer drugs that affect many organs indiscriminately. However, only when it is injected subcutaneously, its recovery route or clearance from the site of injection does show a significant difference from results with low molecular weight drugs such as mitomycin C; i.e., it showed high accumulation in the regional lymph nodes but not in distant lymph nodes, blood, or other organs [1].

It is well known that the major failure of cancer treatment is mostly metastasis of cancer via the lymphatics. Thus, one of us (H.M.) undertook a series of experiments to modify NCS to make it more lymphotropic so as to target the drug to the lymphatic system, which is the major route of cancer metastasis. The efforts resulted in the invention of SMANCS, which is NCS conjugated with a copolymer of styrene-maleic acid *n*-butyl ester [2,3]. To understand the mechanism of lymphotropism, a knowledge of lymphology was needed; i.e., macromolecular and lipophilic material are recovered via the lymphatics.

When the right polymer with these properties was looked for, the poly(styrene-co-maleic acid half-butyl ester), abbreviated SMA, was found.

During the course of the study on SMANCS, however, the most interesting observation was that it exhibits exceptionally high tumor-accumulating efficacy in addition to lymphotropicity, for which it was designed [4,5], and remarkably high lipophilicity, which made its lipid formulation (in Lipiodol) possible. Lipiodol is a contrast media prepared by iodination and esterification of poppyseed oil. This Lipiodol-formulated SMANCS, which was eventually applied to arterial injection in the clinic, gives the highest tumor-to-blood ratio of any drug known, as is described here. Thus, it is like a tumor-seeking magic missile, with the result that its systemic side effects, such as bone marrow toxicity, are minimal and its antitumor effect maximal.

A preclinical study, a clinical pilot study, and then a multiinstitutional clinical-phase I/II study of SMANCS, led by us in Kumamoto in collaboration with Yamanouchi Pharmaceutical Co. Ltd., Kuraray Co. Ltd., and Kayaku Laboratories for approval by the regulatory agency, have been pursued since 1984. In the meantime, it was approved by the Japanese Government at the end of August 1993, and it has been marketed since February 1994 for use against hepatoma. Clinical data generated by our group are unprecedented for the treatment of hepatoma with SMANCS, as otherwise patients with this cancer die within a few months. Efficacy for other tumor types is also similarly promising, as is discussed in this chapter.

Maeda et al. found eventually that the basic mechanism of tumor-selective targeting is based on the unique characteristics of tumor blood vessels, called the EPR (enhanced permeability and retention) effect of solid tumor, which is also discussed later in this chapter. Many laboratories throughout the world have been observing this EPR effect, and the rational strategy of drug design, looking for polymer therapeutics for cancer treatment, is pursued with obvious advantages discussed herein.

Chemistry of SMANCS: Synthesis and Characteristics

Synthesis of SMANCS

SMA poly(styrene-co-maleic acid) or its various alkyl esters have been used as floor- or car-waxing materials, having a amphipathic character that is soluble in water and some organic solvents. From studies of chemical modification of NCS, it was known that modification of free amino groups does not affect its biological activity. As described in an earlier chapter of this book (see chapter by Kuromizu and Maeda), there are only two free amino groups in NCS, one at N-terminal alanine and other at lysine 20. Thus, these residues are available for polymer conjugation without the loss of biological activity (see chapter by Kuromizu and Maeda). A typical coupling reaction of NCS with SMA ([poly[styrene-co-maleic

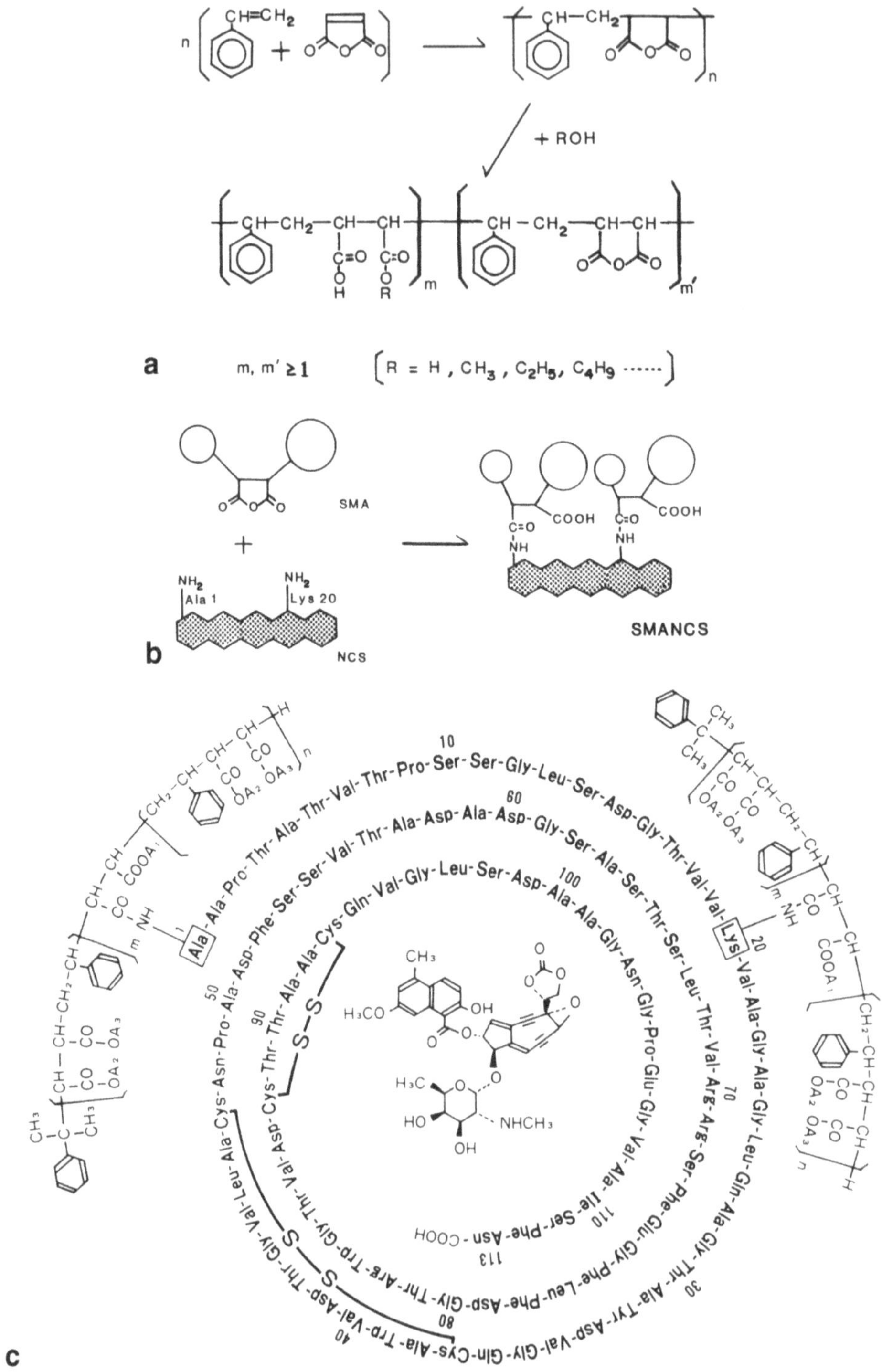

FIG. 1a–c. Scheme of synthesis of (**a**) copoly(styrene-maleic acid/anhydride/ester) (*SMA*), (**b**) its conjugation of SMA with neocarzinostatin (NCS) to synthesize SMANCS, and (**c**) the chemical structure of SMANCS. Enediyene chromophore is covered by protein portion

2
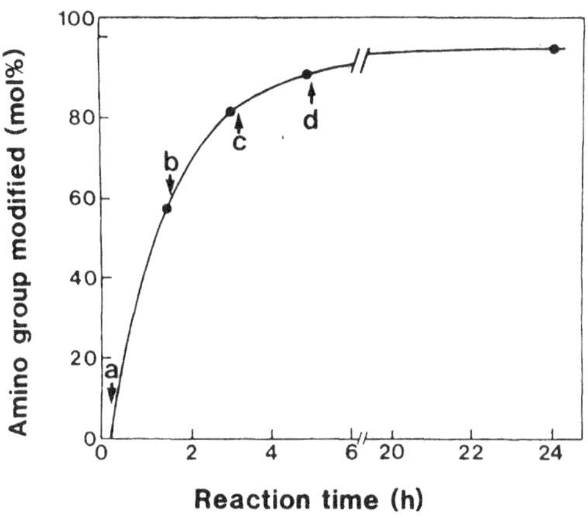

3
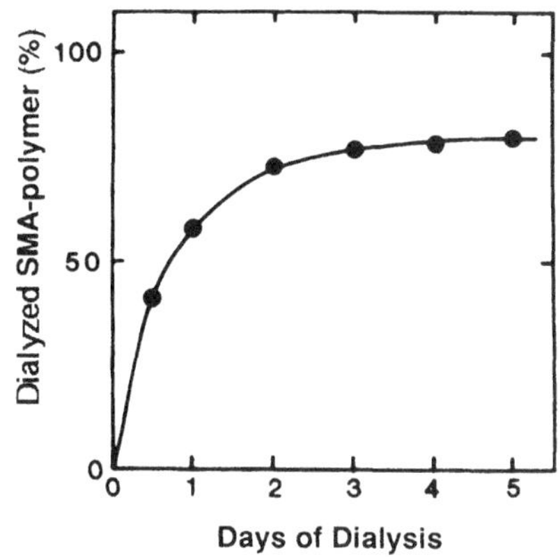

FIG. 2. (*left*) Synthesis of SMANCS: reaction of NCS with partially hydrolyzed (p-H)-SMA. To a solution of NCS (0.2g) in 0.8 *M* $NaHCO_3$ (5 ml) at 4°C in the dark was added powdery p-H-SMA at the times indicated by the *arrows* (*a*, 0.3 g; *b*, 0.3 g; *c*, 0.2 g; *d*, 0.1 g). The reaction mixture was stored in a refrigerator after the complete dissolution of the last addition (*d*). At 24 hr after the beginning of the reaction, the amino group modified was 97.3 mol%. (From [3], with permission)

FIG. 3. (*right*) Time-course of dialysis of the reaction mixture of Fig. 2. Dialysis was carried out under a pressure of about 1–2 atm in 10 m*M* ammonium bicarbonate at 4°C. The amount of styrene maleic acid (*SMA*) copolymer removed was quantified by absorption at 260 nm and the volume of the dialysis solution. (From [4], with permission)

acid)]partial maleic anhydride/partial half-*n*-butyl ester), as shown diagrammatically in Fig. 1, was carried out at 4°C in the dark by stepwise addition of the SMA derivative to the solution of NCS in 0.8 *M* $NaHCO_3$ at pH 8.5, and pH was maintained by addition of 0.5 *M* Na_2CO_3 while the solution was stirred (see legend, Fig. 1). The reaction of the amino group of NCS was quantified by the trinitrobenzene sulfonic acid method [3]. The overall chemical structure is shown in Fig. 1c.

Figure 2 illustrates the time-course of modification of amino groups of NCS with SMA [3,4]. Usually, a 10–20 molar excess amount of SMA was added to the amino group (mol/mol) of NCS [2,3]. SMA, possessing 30%–40% maleic anhydride, was prepared by alcoholysis with *n*-butanol (Fig. 1a). Among the variety of different alkyl esters examined, including methyl-, ethyl-, *n*-propyl-, *n*-butyl, *n*-pentyl-, ethyl cellosolve-, etc., *n*-butyl ester was found most ideal in view of solubility both in water and in solvents [3], protein binding [7], and macrophage activation [7,10]. The size of polymer for conjugation was settled at about 1600 daltons (Da) from the original size of about 6000 Da [2]. Although the original larger size was better in biological or pharmacological data, the smaller copolymer exhibited better purity properties. Thus, the SMA copolymer of about 1600 Da was used henceforth, and the conjugate referred to as SMANCS hereafter is composed of half-*n*-butylate SMA of about 16 kDa.

TABLE 1. Analysis and solubility of various partially hydrolyzed/alcoholyzed styrene-maleic acid (SMA) copolymer (From [3], with permission)

Type of SMA	M_w	M_w/M_n	M_{an}[a]	Solubility					
				n-Hexane	Toluene	Acetone	Dioxane	H_2O[b]	0.5 *M* $NaHCO_3$
p-H-SMA[c]	1760	1.20	48.3	Insoluble	Swell	Soluble	Soluble	Insoluble	Soluble
p-Bu-SMA[d]	2270	1.17	30.0	Insoluble	Swell	Soluble	Soluble	Insoluble	Soluble
p-EtCe-SMA[e]	1700	1.18	25.3	Insoluble	Swell	Soluble	Soluble	Insoluble	Soluble

M_w, weight average molecular weight; M_n, number average molecular wt.; M_{an}, residual content (%) of maleic anhydride.
[a] Residual content of maleic anhydride residue.
[b] pH 5.0.
[c] Partially hydrolyzed SMA.
[d] Partially half-butyl-esterified SMA.
[e] Partially half-ethylcellosolve-esterified SMA.

^{1}H- and ^{13}C-NMR studies showed 1:1 ratio in the proton signal area, and the degree of polymerization was 7–8 [3]. A separate experiment using nuclear magnetic resonance (NMR) showed that the styrene and maleate groups alternate (Kuraray Co., unpublished data). Free SMA polymers seem to form micellar aggregates in a dose-dependent manner (at >1 μ*M* even in 0.05 *M* Tris HCl buffer at 25°C, while SMANCS does not [3].

Characteristics of different SMA and their NCS conjugates are summarized in Table 1. These are easily purified from the reaction mixture by extensive dialysis in weakly alkaline volatile buffer (10 m*M*) ammonium bicarbonate, pH about 8) (see Fig. 3), followed by lyophilization and gel filtration chromatography on Sephadex G-75 using 5 m*M* NH_4HCO_3 as a carrier.

Removal of SMA components by precipitation from the reaction mixture by lowering the pH was unsuccessful because all the components are insoluble below pH 7.0. Cation-exchange chromatography such as on CM-cellulose was not applicable because the components are insoluble in an acidic medium. DEAE-cellulose column chromatography showed little separation of the conjugates (SMANCS) because of the anionic similarity of each component. Precipitation with either ammonium sulfate or sodium sulfate proved unsuccessful.

Molecular Weight Estimation: Sodium Dodecyl Sulfate-Polyacrylamide Gel Electrophoresis and Vapor Pressure Osmometry

Polyacrylamide (15%) gel electrophoresis in 0.1 *M* sodium dodecyl sulfate (SDS-PAGE) at pH 8.6 (Fig. 4) showed the apparent molecular weight to be 17000 (Table 2). Vapor pressure osmometery with a VPO instrument (Kanauer GmbH, Berlin, Germany) was done at 45°C with ethyl acetate as the solvent for SMA and its derivatives [3]. Molecular weight based on chemical analysis either by increase in C (%) content from SMA or by decrease in nitrogen/protein content (%) from SMA conjugation was compared with that of NCS. The

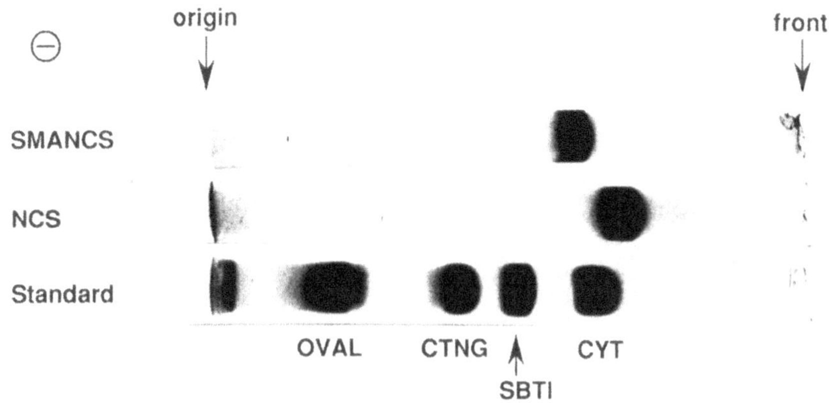

FIG. 4. Disk gel electrophoresis in the presence of sodium dodecyl sulfate at 0.2% in 0.025 *M* Tris-glycine buffer containing 2 m*M* ethylenediaminetetraacetate (EDTA) at pH 8.3. Standard proteins and their molecular weight (in parentheses) are as follows: *OVAL*, ovalbumin (48000); *CTNG*, chymotrypsinogen (24000); *SBTI*, soybean trypsin inhibitor (20000); *CYT*, cytochrome *c* (14000); *NCS*, neocarzinostatin (12000). (From [3], with permission)

TABLE 2. Molecular weights of SMANCS determined by various methods (From [3], with permission)

SMANCS	Molecular weight estimation based on: Nitrogen content	Protein content	HPLC SMA[a]	HPLC Protein[b]	$NaDodSO_4$-PAGE[c]	FP[d]	Theoretical[e]
H-SMANCS	16455	17113			17000		15261
Bu-SMANCS	17003	18430	12200	30000	17000	17000	16281
EtCe-SMANCS	16162	21062			17000		15101

[a] Using SMA as standard.
[b] Using protein as standard.
[c] Polyacrylamide gel electrophoresis in sodium dodecylsulfate.
[d] Fluorescence polarization [3].
[e] See [3]. Note that parental neocarzinostatin has a molecular weight of 11741.

theoretical value was also calculated. All these data are in fairly good agreement (Table 2).

Protein Binding of SMA and SMANCS

Because of the aromatic ring, we anticipated that SMANCS may bind with albumin. When SMANCS with a molecular size of about 16–17 kDa could bind with serum albumin, it would behave like a much larger protein of about 80000. It is known that the apparent molecular size of a substance in blood plasma is critical for the renal clearance rate if the molecules have high biocompatibility

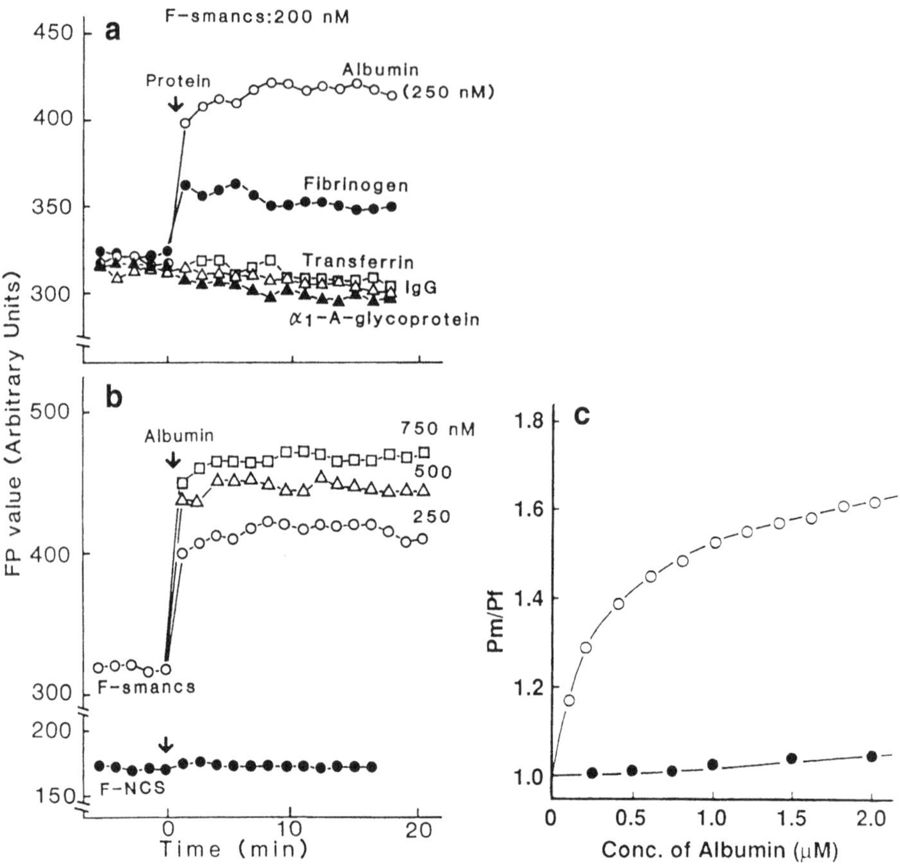

FIG. 5. **a** Binding of F-SMANCS to various proteins. Proteins (0.25 μ*M*) were added to F-SMANCS (0.2 μ*M*) in 0.15 *M* sodium phosphate buffer (pH 7.4) at 37°C (*arrows*). Fluorescence polarization (*FP*) values were determined before and after the addition of human serum albmin (*open circles*), fibrinogen (*closed circles*), transferrin (*open triangles*), α_1-acid glycoprotein (*solid triangles*), or IgG (immunoglobulin, *open squares*), all at 0.25 μ*M*, respectively. **b** Time-course of binding of F-SMANCS or F-NCS to different doses of human serum albumin. F-SMANCS (0.2 μ*M*) was incubated with 0.25 μ*M* (*open circles*), 0.5 μ*M* (*open triangles*), or 0.75 μ*M* (*open squares*) of human serum albumin at 37°C (pH 7.4). F-NCS (0.2 μ*M*) was incubated with 0.75 μ*M* human serum albumin (*closed circles*) at similar conditions. FP values were determined similarly. **c** Dose–response binding curves for F-SMANCS and F-NCS to human serum albumin. F-SMANCS (0.2 μ*M*) (*open circles*) was incubated with human serum albumin (0.25–2.0 μ*M*) at 37°C (pH 7.4). F-NCS (0.2 μ*M*) (*closed circles*) was also incubated with human serum albumin (0.25–2.0 μ*M*) at similar conditions FP values were determined before (P_f) and after (P_m) the addition of human serum albumin, and normalized polarization values (P_m/P_f) were calculated. (From [7], with permission)

and are free from adsorption to the vascular bed surface or the liver sinusoidal structure, which are heavily charged with negative charge of COO^-, and SO_4^- of heparan sulfate/chondrosulfate, hyauronic acid, etc. That is, a substance with a molecular size larger than the renal threshold (>40 kDa) would remain in circulation for a long time if it were neutral or acidic. In fact, we knew from the in vivo pharmacokinetic study of SMANCS, as described later, that its plasma half-life is about ten times longer than that of parental NCS, while retaining the DNA-cleaving character [5] as well as a much improved anticancer effect because of the much higher tumor accumulation [4–6]. Thus, it is interesting to know if there is any protein binding of SMANCS.

We studied this subject primarily using the fluorescence polarization method, which can be interpreted as molecular rotational or Brownian motion by fluorescence emission that takes place during the time frame of the fluorescence relaxation of the order of *n* sec [7–9]. NCS was labeled with fluorescin isothiocyanate (FITC) as described [8,9]. Because there is no free amino group in SMANCS to react with FITC, it is necessary to attach L-lysine by using carbodiimide to introduce free amino groups [7,8].

The fluorescence polarization value (FP value) was determined with the MAC-2 polarization spectrofluorophotometer (Japan Immunoresearch Laboratories, Takasaki) as described previously [8,9]. A cuvette containing F-NCS or F-SMANCS (200–1000 n*M* each) in 0.15 *M* sodium phosphate buffer at pH 7.4 was placed in the fluoropolarimeter and equilibrated to 37.0°C. The FP value of F-NCS or F-SMANCS in the free state, P_f, was determined before the addition of various serum proteins. Then, after the addition of the proteins to each cuvette (final concentration, 180–750 n*M*), the FP value P_m was determined.

As shown in Fig. 5a, albumin is the only main protein to affect the FP value of F-SMANCS, and to a lesser extent fibrinogen, showing the binding of SMANCS to albumin. A clear dose–response and saturation profile is seen with albumin binding (Fig. 5b,c). It should be also noted that F-NCS does not bind any plasma proteins. This binding of F-SMANCS was blocked by nonlabeled SMANCS; the apparent binding constant was calculated to be $2.19 \times 10^6 M^{-1}$, and 3.2 moles of F-SMANCS could bind to human serum albumin via the SMA portion [7]. The longer SMA chain and higher alkyl group ($(n\text{-butyl})_6 > (n\text{-butyl})_4$ > ethyl > free carboxylate) showed a tighter binding constant in this order. It appears from the competition experiments using known standard albumin ligands that warfarin or diazepan shares the same binding site as SMA. Bilirubin may also partially share the binding site [7].

Membrane Perturbation by Various SMA Derivatives and Macrophage Activation

As described [10], the SMA portion confers potential hydrophobicity to the conjugate, although the type of alkyl ester residues on maleic acid makes a great difference. At higher concentrations of SMA, it exhibits moderately high hemolytic activity that will be, however, greatly lowered in the presence of a 0.1%

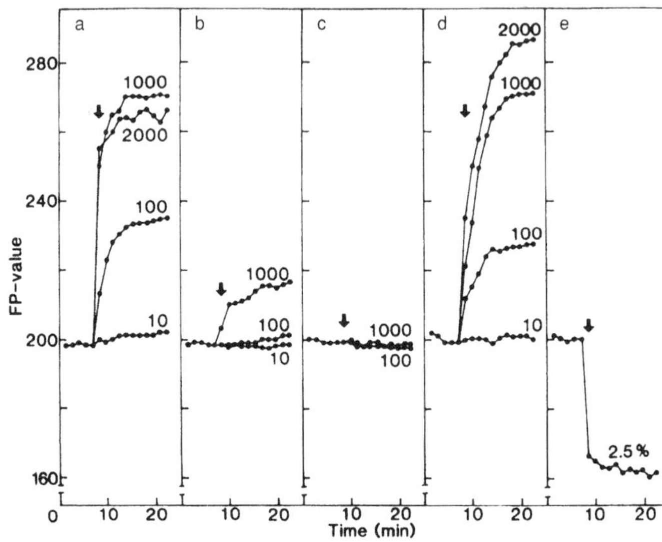

FIG. 6a–e. Effect of various SMA derivatives and *n*-butanol on macrophage membrane fluidity. FP values of diphenylhexatriene, as fluorescent probe for lipid layer bound to macrophages, were measured at the indicated doses (mg/ml or %) at 25°C, added at the *arrows*. **a** Butyl-SMA. **b** Ethyl-SMA. **c** Hydrolyzed SMA. **d** SMABSA. **e** *n*-Butanol. (From [10], with permission)

TABLE 3. Hemolytic activity of various SMA derivatives in the presence or absence of 1 mg/ml of bovine serum albumin (BSA) (From [10], with permission)

Sample	Molecular weight	Minimum concentration for hemolysis (μ*M*) Without BSA	With BSA (1 mg/ml)
Hydrolyzed SMA	2500	500	2000
Ethyl-SMA	2500	63	1000
Butyl-SMA	2500	1.56	125
SMANCS[a]	17000	37 (74)[b]	590 (1180)[b]
SMA-BSA[c]	100000	1.6 (16)[b]	50 (500)[b]

[a] SMANCS contained 2 mol butyl-SMA polymers/mol NCS.
[b] The values in parentheses are expressed as concentrations of butyl-SMA.
[c] SMA-BSA contained about 10 mol butyl-SMA/mol BSA.

or higher concentration of albumin (Table 3). Figure 6 shows an interesting trend: membrane perturbation of macrophages is greatly facilitated in the order of butyl-SMA > ethyl-SMA > H-SMA (free carboxylate), in which 1,6-diphenyl-1,3,5-hexatriene as a fluorescent probe was used to label the macrophage membrane [10].

The next experiment conducted was to determine how macrophages are activated by various SMA derivatives in terms of the release of H_2O_2, superoxide anion ($O_2\cdot^-$), and lysozyme. The results are consistent with the foregoing results as well as the release of H_2O_2, etc. (Figs. 7 and 8). It is also noted that these dose responses are bell shaped. The optimal concentrations to stimulate the adherent macrophage to release H_2O_2 are 50, 5.9, 10, 1.0, and 0.3 μ*M* for butyl-SMA, SMANCS, SMA-albumin, concanavalin A, and wheat germ agglutinin, respectively [10]. This means that SMANCS can activate macrophages, although how this is significant in vivo needs to be clarified. In this regard, immunological stimulation by SMANCS is discussed elsewhere (see chapters by F. Suzuki and M. Kobayashi, and by E. Matsuda and H. Maeda, respectively), such as activation of T cells and natural killer (NK) cells and induction of interferons, tumor necrosis factor (TNF), etc. may be mediated through the interaction of SMANCS though the membrane.

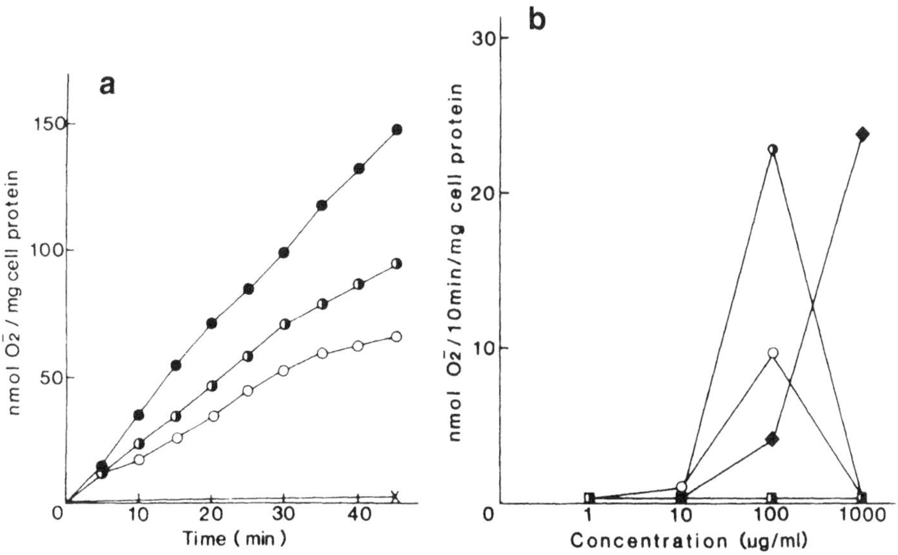

FIG. 7a,b. O_2^- release from macrophages after the addition of PMA and SMA derivatives. **a** Time-course. Macrophages were exposed to 0.01 μg/ml of PMA (*closed circles*), 100 μg/ml of SMANCS (*half-closed circles*), or 10 μg/ml of SMA (*open circles*), and the resultant release of O_2^- was quantified at various times thereafter. ×, Control; no stimulant. Data are plotted as nanomoles of O_2^- secreted per milligram of macrophages protein against time. **b** Dose–response relationship. The release of O_2^- from adherent macrophages in response to various doses of SMANCS (*half-closed circles*), SMABSA (*closed diamonds*), SMA (*open circles*), or NCS (*half-opened squares*) was quantified. The data are plotted as nanomoles of O_2^- released in 10 min per milligram of macrophage protein. (From [10], with permission)

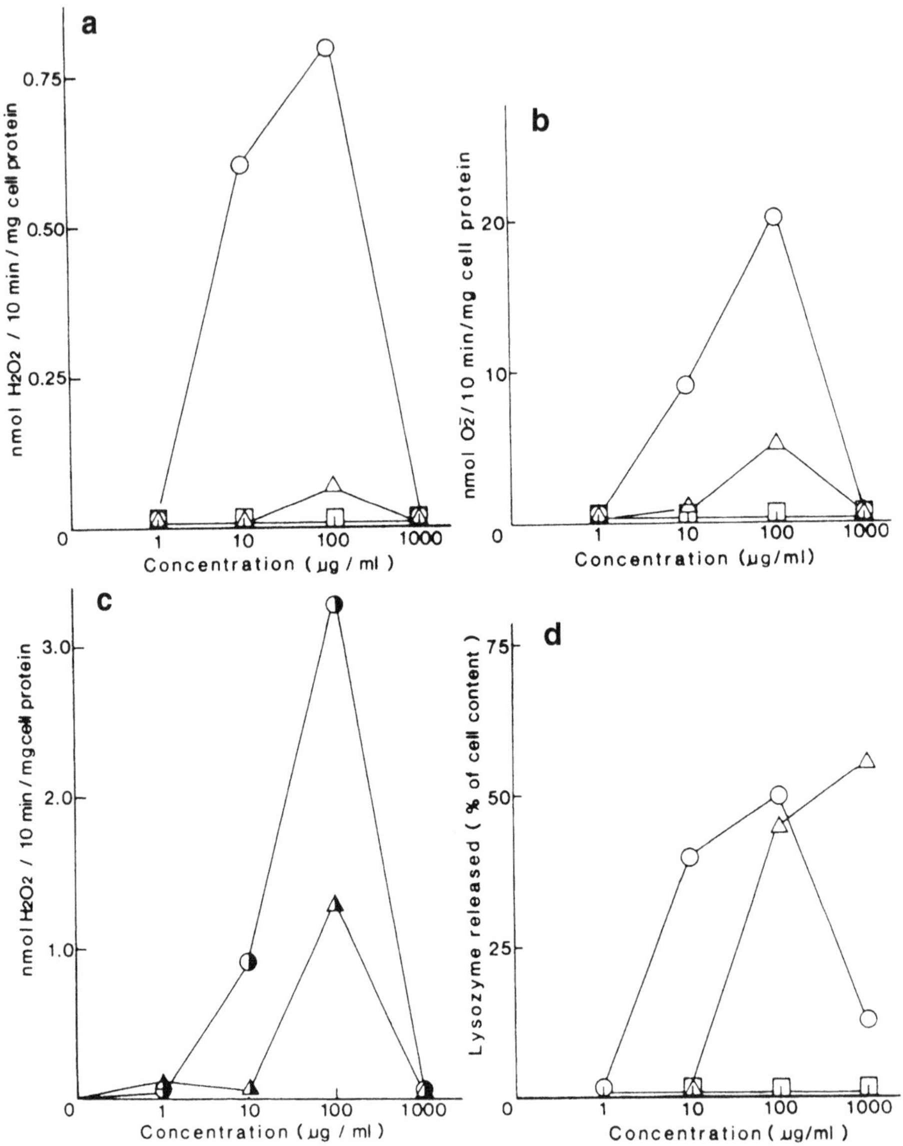

FIG. 8a–d. Release of H_2O_2, O_2^-, and lysozyme from macrophage after the addition of various SMA derivatives. **a** Dose response in H_2O_2 release. **b** Dose response in O_2^- release. The secretion of H_2O_2 or O_2^- from adherent macrophages in response to various doses of butyl-SMA (*open circles*), ethyl-SMA (*open triangles*), or hydrolyzed SMA (*squares*) was quantified. **c** Dose–response relationship of butyl- (*half-closed circles*) and ethyl- (*half-closed triangles*) SMANCS in H_2O_2 release. Results are expressed as in Fig. 7. **d** Dose–response relationship of various SMA polymers to lysozyme release. Adherent macrophages were incubated with various doses of butyl-SMA (*open circles*), ethyl-SMA (*open triangles*), or hydrolyzed SMA (*open squares*) for 1 hr at 37°C; lysozyme activity in supernatant fluid was then assayed as described in the text. Lysozyme release is expressed as percentage of total cellular lysozyme content. (From [10], with permission)

Subcellular Action of NCS and SMANCS: Enhanced Internalization of SMANCS

The molecular mechanism of action of NCS in bacteria and mammalian tumor cells is known to be the arrest of DNA synthesis by direct DNA strand scission involving its prosthetic group, as described elsewhere in this monograph (see chapter by K. Ohtsuki and Y. Ono). In addition, Maeda et al. previously reported the subcellular fate of NCS and clarified its penetration into the cytosol and the nucleus of cancer cells by utilizing fluorescent and ^{14}C-labeled NCS [11–14].

In view of the higher lipophilicity of SMANCS, it was anticipated that SMANCS would have higher membrane affinity and could be internalized. Thus, the binding and cytotoxicity of NCS, SMANCS, and various derivatives of SMA copolymers to HeLa cells and WISH cells were studied in detail and to correlate the role of the SMA copolymer portion of SMANCS in its binding to the cell surface and the internalization process [15].

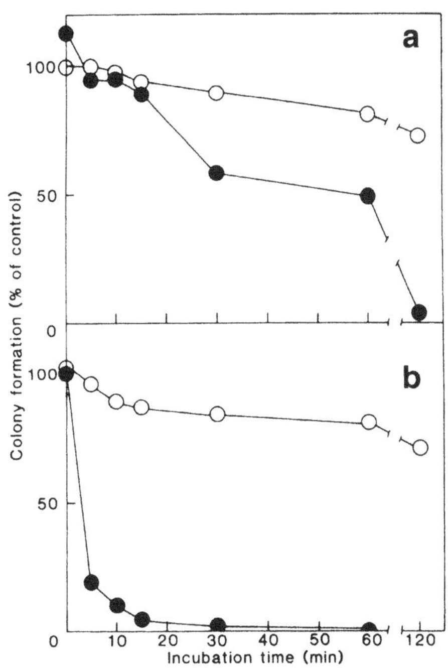

FIG. 9a,b. Toxicity of NCS and SMANCS as measured by their ability to inhibit colony formation of HeLa cells. HeLa cells (250–300 cells in 35-mm-diameter tissue culture dishes) were incubated with 30 n*M* NCS (**a**) or SMANCS (**b**) for various times at 0°C (*open circles*) or 37°C (*closed circles*). At the end of each incubation, the cells were washed with phosphate-buffered saline (PBS), and 2 ml of growth medium was added to the culture for 7–8 days. The number of colonies was determined as described in the text. (From [15], with permission)

The binding of SMANCS and NCS to cultured cells was investigated using drugs labeled with FITC that retained biological activity. The binding of these drugs to HeLa cells was dependent on time and temperature, with 2-fold more drug being bound at 37°C than at 0°C. In the presence of a 100-fold molar excess of unlabeled NCS, the binding of SMANCS or NCS to HeLa cells was similarly inhibited. Therefore, it was suggested that SMANCS binds to NCS-binding sites (receptor) of the cell surface. However, the amount of cell-bound SMANCS was increased about 20 fold compared with that of NCS. Scatchard plot analyses of the binding of these drugs to HeLa and WISH cells indicated that this increase was caused by increased affinity to the receptor resulting from polymer (SMA) conjugation to the drug rather than by an increase in the number of drug-binding sites at the cell surface (Table 4). Furthermore, when the cytotoxicity of these drugs to HeLa cells was compared, SMANCS needed only 5 min to achieve 50% inhibition of the control (Fig. 9). In contrast, at the same dose NCS required more than 90 min to achieve the same toxic effect. More rapid internalization of SMANCS than NCS was also elucidated under fluorescence microscopy at 37°C. There was no intracellular incorporation of these drugs below 20°C, indicating an endocytotic mechanism for internalization of NCS and SMANCS.

These results indicated that an increased lipophilicity of SMANCS appears to be responsible for its increased cell-surface affinity, internalization rate, and toxic effect. Concordant to this interpretation was that among various copoly(styrene-

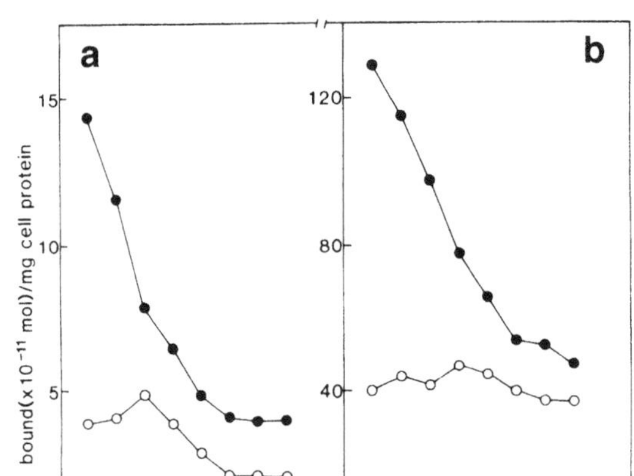

FIG. 10a,b. pH Dependence of the binding of fluorescein isothiocyanate- (FITC-) labeled drugs to cells. HeLa cells (5×10^5 cells/well) were incubated with 100 µg/ml of FITC-NCS (**a**) or FITC-SMANCS (**b**) for 2 h at 0°C(*open circles*) or 37°C (*closed circles*) in 0.5 m of KRP (Krebs-Ringer phosphate buffer) at the indicated pH values. After incubation, the amounts of cell-bound FITC drugs were determined as described in the text. (From [16], with permission)

TABLE 4. Cell binding constant and numbers of binding sites of NCS and SMANCS at 37°C (From [15], with permission)

Cell	Drug	K_a ($\times 10^{-5} M^{-1}$)[a]	No. of binding sites per cell($\times 10^{-7}$)
HeLA	NCS	0.05	3.20
	SMANCS	1.34	4.48
WISH	NCS	0.11	7.15
	SMANCS	1.69	3.48

[a] Ka, association constant.

maleic acid) esters, a more hydrophobic derivative showed more internalization (butyl > ethyl > carboxylate). These results suggest the general applicability of this tactic of utilizing protein tailoring with lipophilic residues to augment the subcellular activity of functional proteins.

When the effect of environmental pH on the binding and cytotoxicity of the antitumor proteins NCS and SMANCS to cultured cells was studied, it was found that at 37°C the binding of these drugs to HeLa cells was pH dependent: the amount of cell-bound drugs increased with an increase in the acidity of the medium, corresponding to that of tumor tissue (Fig. 10). The pH-dependent change in the binding of the drugs was not as evident at 0°C [16]. The cytotoxic action of these drugs was much more rapid at acidic pH compared with that at neutral or slightly alkaline pH. Furthermore, FITC-labeled drugs could be utilized to probe the microenvironmental pH in Meth-A cells in which the drug was located by the ratio of fluorescent intensities excited at 450 and 490nm, respectively. The microenvironmental location of the cell-bound FITC-labeled drugs became acidic from a neutral environment with progress of incubation time at 37°C. It required about 20min or longer to achieve the acidic environment, but this did not occur at 0°C. Inasmuch as these drugs directly attack DNA, these results suggest that NCS and SMANCS are translocated across the membrane of acidic vesicles into the cytosol after endocytotic uptake. This hypothesis is also supported by the finding that NH_4Cl and chloroquine protected the HeLa cells against the cytotoxicity of the drugs. Taken together, our results show that the acidic pH of tumor tissue is preferable for effective binding and internalization into the cytosol of NCS and SMANCS [16].

This idea, that both NCS and SMANCS can enter the cells for their action, is also consistent with a notion that they liberate superoxide and other free radicals, which have extremely short half-lives; thus an extracellular site of action is too remote from the real molecular site of action. This notion of enhanced internalization into cells agrees with the result of intestinal uptake in that uptake into the circulating blood is further augmented by lipid formulation [17]. The highly hydrophilic small protein of NCS is virtually not absorbed from the intestine [17]. Application of this principle, i.e., hydrophobic polymer conjugation and its lipid formulation for oral delivery and enhanced intestinal absorption, should be applicable to other small proteins and peptides.

Pharmacological Properties of SMANCS

Elongation of Plasma Half-Life and Tumor-Selective Delivery Based on the EPR Effect

Because of albumin binding, which results in an apparent large molecular size and anionic nature, the plasma half-life of SMANCS becomes 19 min in mice compared to the 1.9 min of NCS [18]. The human clinical data obtained in the phase I study of NCS in the United States and SMANCS in Japan (Fig. 11) show that SMANCS exhibits prolonged plasma half-life because of its decreased clearance rate, nonspecific uptake by the vascular endothelium or the reticuloendothelial system, and decreased inactivation rate by plasma protease and other proteases, as seen from the animal data [4–6,18]. This last notion, of the lowered inactivation rate of SMANCS, was verified by incubating NCS or SMANCS with blood plasma or whole blood (Fig. 12) [4,5]. The conjugate was found to be much more stable than native NCS in either blood or serum. It was demonstrated earlier that incubation of SMANCS with cell homogenates, which contain various cathepsins and other proteolytic enzymes, will liberate NCS [5].

Maeda and Matsumura have carried out extensive studies with regard to tumor accumulation of various proteins and this macromolecular drug, SMANCS, and their extravasation or highly enhanced permeability in solid tumors was observed [18]. When various ^{51}Cr-labeled proteins, conjugated with chelating agent (diethyltriaminepentaacetic acid, DTPA) and chelating

11

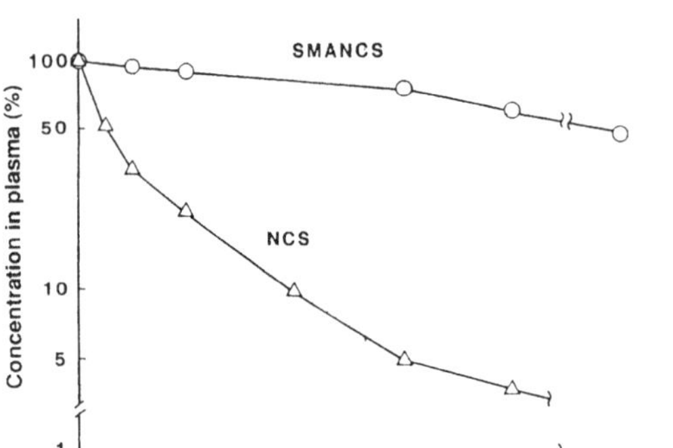

12

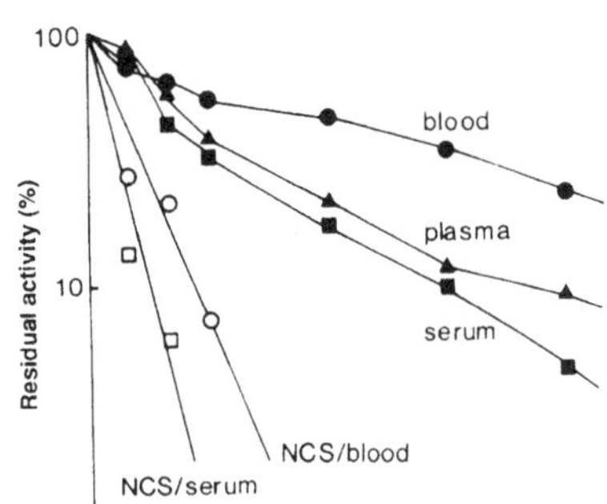

FIG. 11. (*left*) Plasma concentration of NCS and SMANCS in humans after i.v. bolus injection was determined by radioimmunoassay of NCS and by immunological assay for SMANCS. (From [18a], with permission)

FIG. 12. (*right*) Inactivation of SMANCS and NCS by blood plasma and by serum. *Closed symbols*, SMANCS; *open symbols*, NCS. (From [4], with permission)

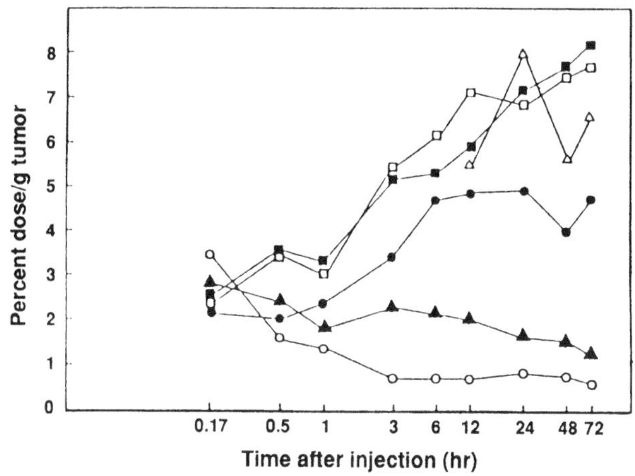

FIG. 13. Intratumor accumulation of [^{51}Cr]-labeled proteins and SMANCS in S-180 tumor in mice. *Open circles*, NCS (M_r, 12000); *closed circles*, SMANCS (M_r, 16000); *closed triangles*, ovomucoid (M_r, 29000); *open squares*, BSA (M_r, 69000); *closed squares*, mouse serum albumin; *open triangles*, mouse IgG (M_r, 160000). Radioactive proteins were injected i.v. at time zero. Values are based on radioactivity. (From [18], with permission)

radioemitting radionuclei (^{51}Cr), were injected intravenously (i.v.) into the tumor-bearing mice, it was found that larger proteins (>48kDa) tended to accumulate in tumor tissue much more effectively than a low molecular weight protein such as NCS (12kDa) or ovomucoid (29kDa) over a prolonged time period (Fig. 13). It was also visually demonstrated by injecting Evans blue dye (i.v.) (Fig. 14). Namely, macromolecules or proteins of a size above a certain threshold (>50kDa) could accumulate more selectively in tumor than normal tissues without using any tumor-seeking device such as an antibody-directed tumor antigen. This gradually led Maeda and colleagues to a new concept of the EPR effect (see below).

Enhanced Permeability and Retention Effect of Macromolecules and Lipid in Solid Tumors: Mechanism of Tumor-Selective Drug Targeting

This passive and spontaneous tumor targeting (Figs. 13 and 14) is attributed to the unique character of most solid tumors, and the term EPR (enhanced permeability and retention) effect of solid tumor was proposed [4,18,18a,19,21]. The EPR effect was also found to be valid for lipids [22,24–26]. Both macromolecules and lipids show accelerated accumulation in tumors, and they remain for a much longer time without being cleared from the site (tumor) where they are delivered [4,18,18a,19,22,24,25]. The uniqueness of tumor blood vessels and that of the

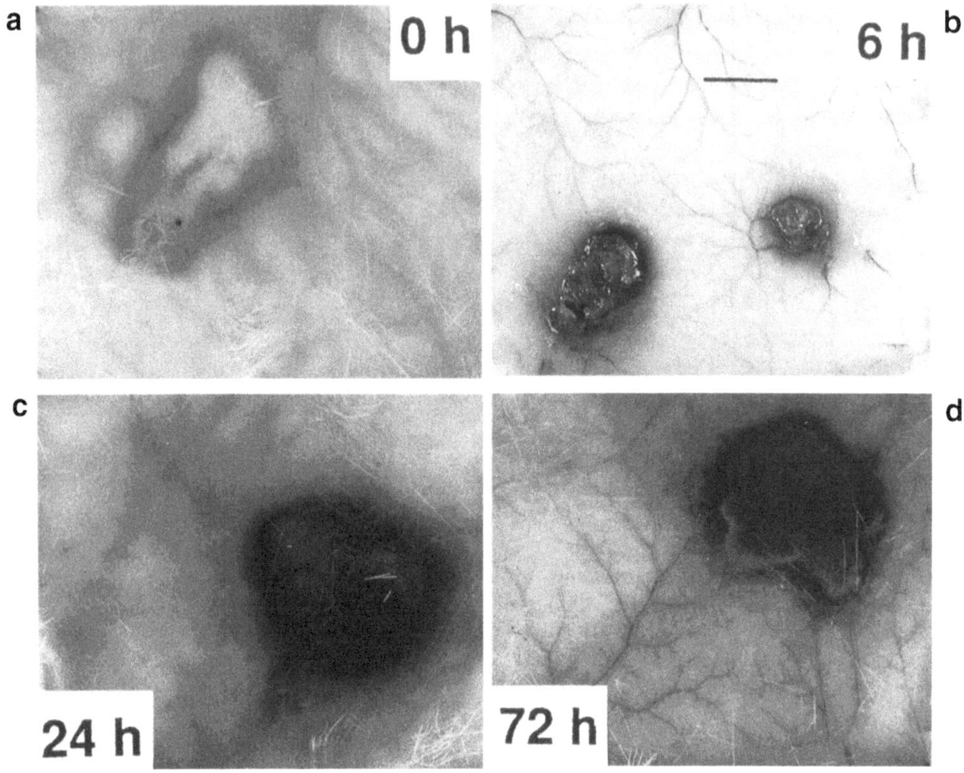

FIG. 14a–d. Accumulation of Evans blue–albumin complex in tumor tissue and normal skin in tumor-bearing mice. Tumor S-180 was injected into the skin. **a** to **d** provide a macroscopic picture of the tumor in the skin taken at 0, 6, 24, and 72 h, respectively, after i.v. injection of Evans blue. The *dark area* indicates the tumor, which became progressively bluer after injection. (From [18], with permission.) (*For color reproduction see color insert in the frontmatter*)

transport system for various substances resides in (a) hypervasculature as a result of extensive angiogenesis; (b) defective vascular architecture, i.e., that mostly lacking smooth muscle layer and response to vascular signal mediators such as angiotensin II; (c) enhanced production of various vascular signals or mediators such as NO, bradykinin, prostaglandin, various cytokines from the cancer cells, and infiltrated lymphocytes/macrophages, etc., to facilitate extravascular leakage; and (d) suppressed lymphatic clearance from the tumor tissue, perhaps partly the result of the less well developed lymphatic system compared to the well-developed blood vessels in tumor tissue.

Tumor uptake of the Evans blue–albumin complex (70 kDa) in mice is shown in Fig. 14; the quantified amount of dye–albumin complex that leaked out of circulating blood to the interstitial tumor bed is shown in Fig. 15. The intratumor concentration of the dye became higher than the plasma concentration at 12h after the dye injection or later, and after about 100h the ratio of dye concentration in S-180 tumor to normal tissue (where the tumor is implanted) in mice could be 30. The tumor/normal tissue ratio at 6–12h is more than 5. Thus, it should be realized one can direct tumor delivery by making the molecular size of drugs larger than 50kDa.

When SMANCS was injected i.v. in rabbits and the intratumor concentration was also quantified, the amount of drug in the tumor is much higher (about 30- to 100 fold) than in the normal tissue (liver) where the tumor was injected [4,19]. The data in Fig. 16 show accumulation of SMANCS in VX-2 tumor, comparing the values with those obtained for parent compound NCS and low molecular weight mitomycin C (Fig. 16a). It should be noted that normal liver accumulated much less than tumor (Fig. 16b). Tumors, whether implanted (S-180, Meth A, L-1210, P388, Ehrlich, or B-16 in mice; AH66 or Walker in rats; VX-2 in rabbits) or DMBA- (dimethylbenz[a]anthracene-) induced rat mammary tumors, did show good response to SMANCS when given intraperitoneally or intravenously [3,4,6,23].

As discussed in the *Introduction*, the distribution of SMANCS and NCS in the lymph nodes as well as in various organs was studied in rats, and it was found that accumulation in the lymph nodes is very pronounced (Fig. 17) [4–6,19]. Only the kidney and the bladder, other than the tumor and lymph nodes, have a high accumulation, which may indicate the route of excretion.

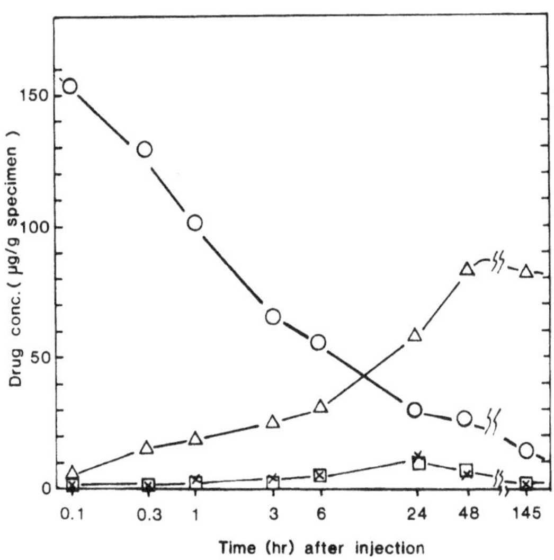

FIG. 15. Quantification of the concentration of the Evans blue–albumin complex at different times for plasma (*circles*), normal skin (*squares*), muscle (×), and tumor (*triangles*). (From [18], with permission)

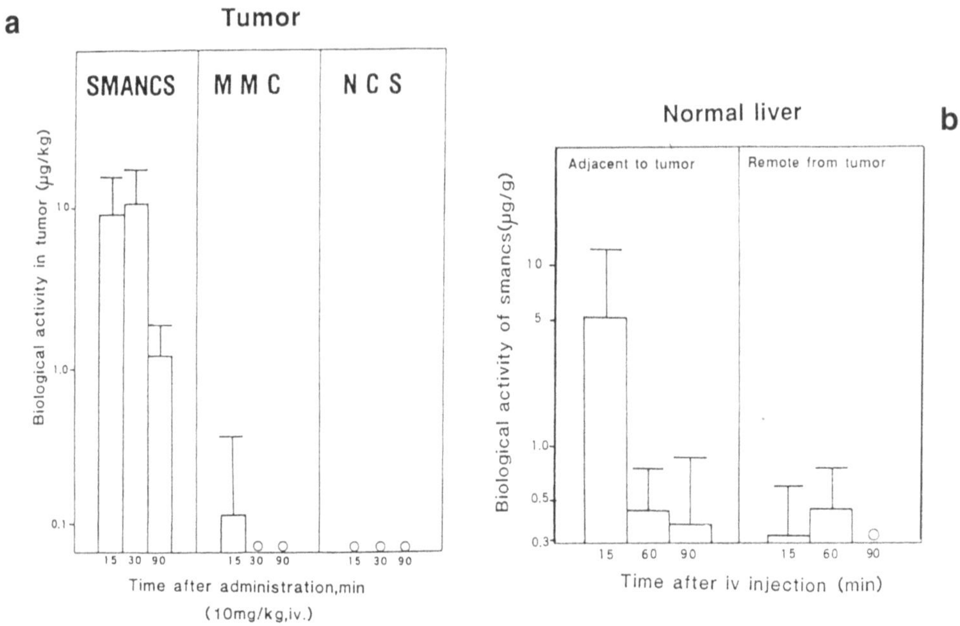

FIG. 16. Intratumor accumulation of various antitumor agents in tumor (**a**) and normal liver (**b**). SMANCS exhibits a much higher and more prolonged tumor concentration than mitomycin C (*MMC*) and NCS. All drugs were given as an i.v. bolus at 10 mg/kg to rabbits bearing VX-2 tumor in the liver (from [4], with permission). Assay was by antibacterial activity (*Micrococcus luteus*), which was standardized by separate standard curves and experiments

Lipid Formulation of SMANCS: Stability, Tumor Targeting, Antitumor Effect, and Antimetastatic Effects

Soon after its synthesis, the general biochemical and biological characteristics of SMANCS were clarified. Subsequently, Konno and Maeda found that it can be solubilized in a lipid lymphographic agent [Lipiodol®][1] in which most of the conventional water-soluble anticancer agents are not usable. Konno and co-workers had found previously that when Lipiodol® was injected into the ligated artery, secured by laparatomy, it remained selectively in the tumor. However, there was no ideal anticancer drug at that point. Maeda anticipated that SMANCS would be soluble because of its high lipophilicity (as shown in Table 1) and could be a good candidate [3]. Following accounts described here are very interesting aspects of the oily-formulated drug that are applicable for other oily drugs in general.

[1] Lipiodol is an ethylester of iodinated poppyseed oil that is prepared and marketed by Laboratoire Guerbet, France, and which has a density of 1.37 g/cm^3

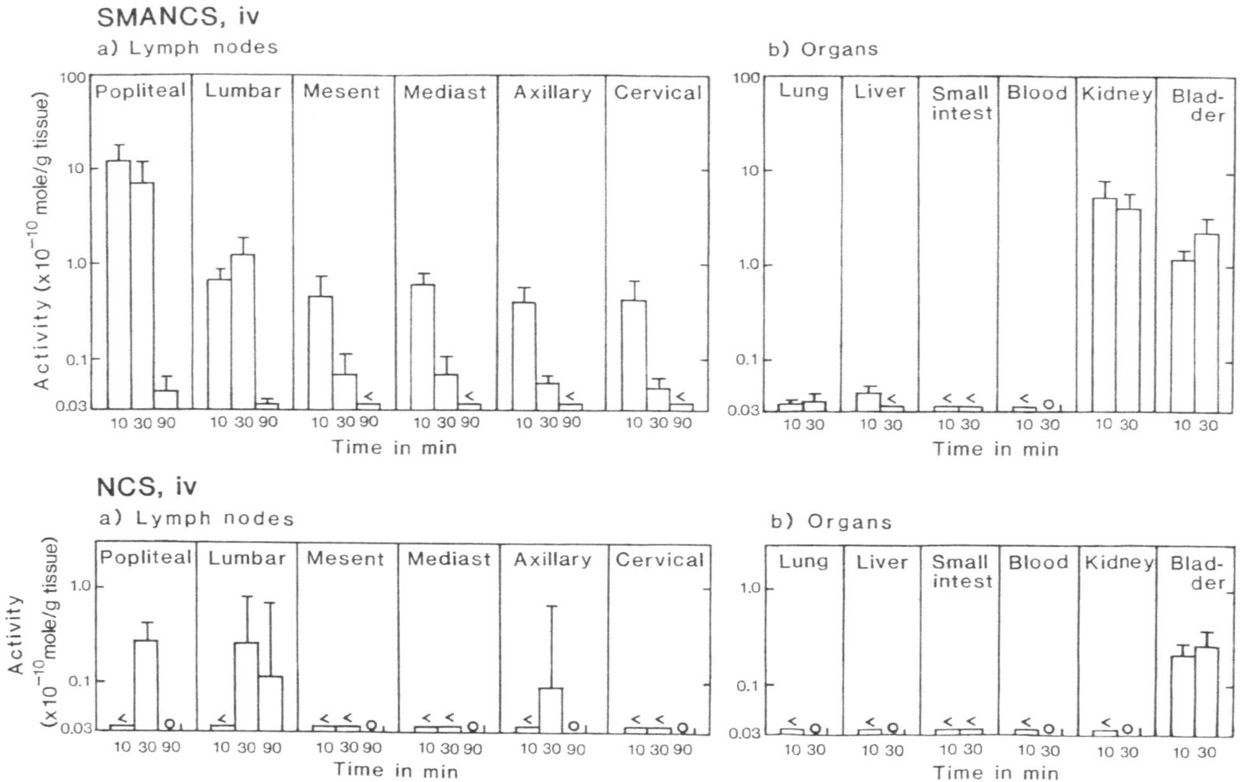

FIG. 17. Drug accumulation of SMANCS (*top*, ***a*** and ***b***) and NCS (*bottom*, ***a*** and ***b***) in various lymph nodes and organs after i.v. injection in rats. Dose levels were as in Fig. 16. A *circle* indicates value below detection limit (0.03 μg/g). (From [4], with permission)

Stability of SMANCS in Lipids and Slow Release from Lipids

As shown in Fig. 18, the stability of SMANCS is much improved in the lipid milieu compared to the water phase [20]. SMANCS is more stable than NCS either in water or in lipids. When SMANCS is suspended in Lipiodol®, its stability

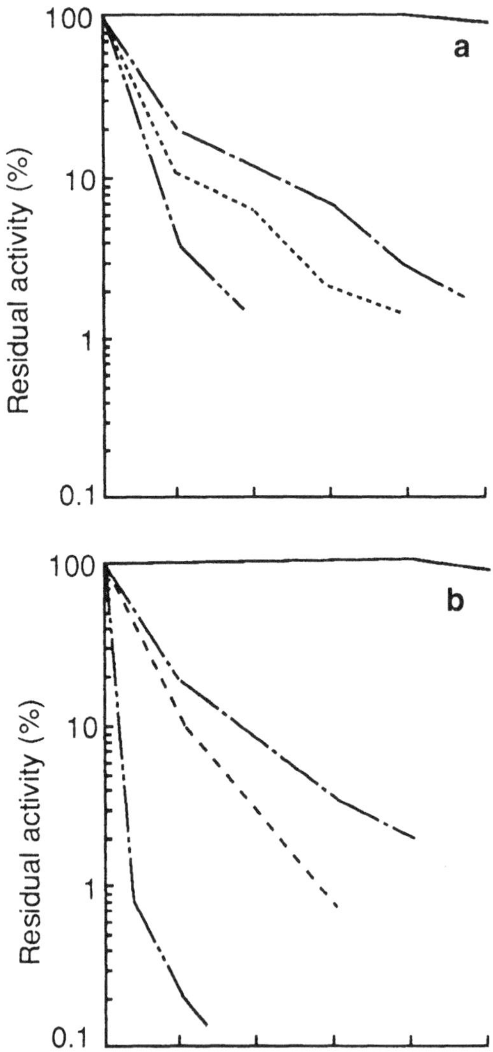

FIG. 18a–c. Inactivation time-course at various temperatures. (**a**) SMANCS/7% sodium bicarbonate solution, pH 8.4; (**b**) NCS/7% sodium bicarbonate solution; (**c**) SMANCS/Lipiodol, 1 mg/ml in bicarbonate or Lipiodol. (From [20], with permission)

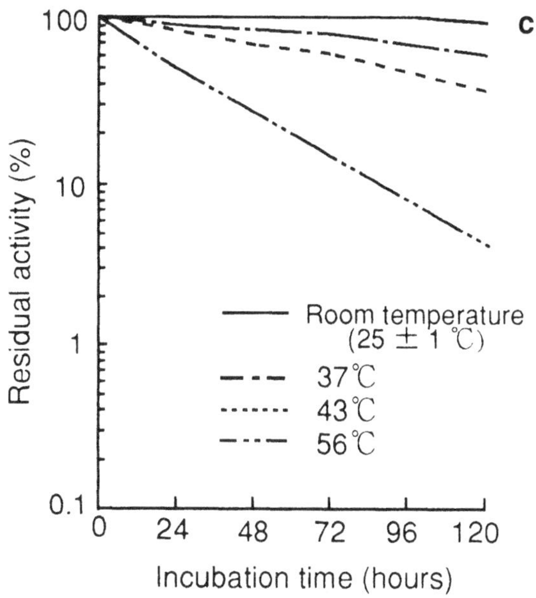

FIG. 18a–c. *Continued*

in the dark at ambient room temperature is greatly improved; more than 80% of the activity remained after 3 months.

Arterial injection of any anticancer agents into the tumor-feeding artery is the most direct way of tumor delivery, mainly because of the so-called first path effect. There is, however, only a little advantage for this method of drug administration because low molecular weight substances will be readily washed out from the target tumor from which a momentarily high drug concentration can diffuse into the circulating blood quickly. Macromolecules and lipids are quite different in this respect. Once macromolecules and lipids reach the interstitial space, they cannot traverse freely back into the circulating blood, as was the case for soluble macromolecules; this is described as the EPR effect, which permits prolonged retention. Lipids are also retained for a long period of time [4,21,22,24,25].

In Fig. 19, retention of SMANCS in Lipiodol was quantified with or without shaking at 1 Hz to simulate heart rate. In this experiment, SMANCS/Lipiodol (partial suspension and partial solution in Lipiodol) was placed to make direct contact with blood or serum in test tubes. SMANCS/Lipiodol, being of high density (1.37 g/cm^3), remained in the bottom layer in the test tube while serum stayed in the upper phase. Most of the conventional drugs, such as mitomycin C, adriamycin, and 5-fluorouracil, are water soluble and readily transferred into the aqueous phase and thus not retained in Lipiodol. Furthermore, conventional drugs dissolved in aqueous high-density contrast medium (Urographin®) adjusted to d = 1.37, for instance, cannot be retained more than 10 min, while

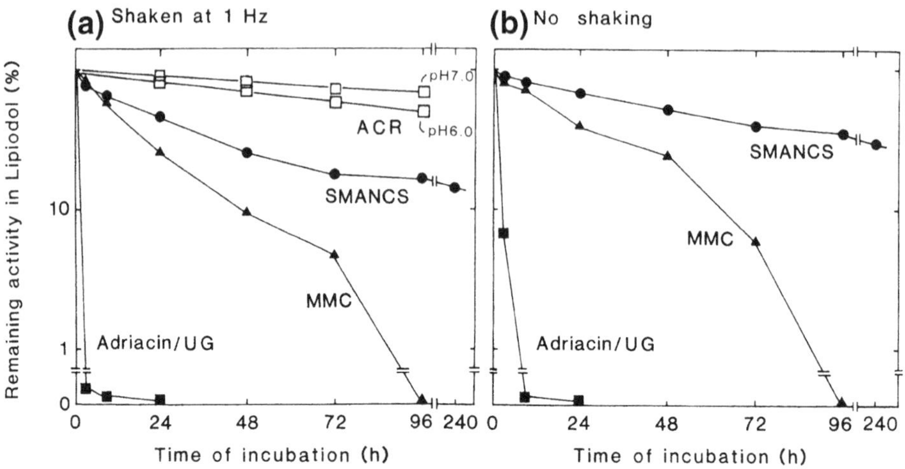

FIG. 19a,b. Transfer of biological activity of anticancer agents from Lipiodol to blood serum (see [4] for details) with (**a**) or without (**b**) shaking. Experiments were carried out in test tubes at 37°C. Remaining activity in Lipiodol is shown. (From [25a], with permission)

SMANCS could be retained at more than 20% in Lipiodol even after more than 100h which is in contact with serum (Fig. 19). The usual concentration (1.0mg/ml) of SMANCS is 10000 times higher than that which inhibits tumor cell growth; thus even a 10% level of drug concentration in the tissue is more than adequate to kill tumor cells.

Tumor Targeting by Arterial Injection of Oily-Formulated SMANCS

When SMANCS in Lipiodol (SMANCS/Lipiodol) is injected into the tumor-feeding artery, drug delivery to the tumor is found to be the most effective; it is selective and almost exclusive to tumor tissue [24]. We have verified this by using rabbit with VX-2 carcinoma implanted in the liver; the [^{14}C]-Lipiodol moiety was injected arterially, and its distribution in various tissues including the cancer tissue and blood plasma and organs was examined (Table 5). The drug was found in the tumor at more than 1000-fold compared to that in blood plasma [24]. This targeting efficiency is found to be much higher than any of the so-called missile drugs using a monoclonal antibody targeted to the tumor antigen. As explained in the section on EPR effect, the high drug concentration is retained for a long time. This result was valid not only when Lipiodol used as carrier but also when using other vegetable oils, triglycerides, and fatty acids [25].

Because of the high electron density of this carrier (Lipiodol), we could actually visualize how it is selectively localized in the tumor under soft X-ray

TABLE 5. Distribution of ^{14}C after the injection of [^{14}C]Lipiodol via the proper hepatic artery in rabbit bearing tumor in the liver (injection volume, 0.2 ml) (From [24], with permission)

Sample	Radioactivity (dpm/g × 10^{-3}) 15 min (n = 3)	3 days (n = 2)	7 days (n = 1)
Tumor	444.92	97.18	37.03
Liver[a]	200.67	9.01	4.13
Liver[b]	29.24	3.23	0.75
Small intestine	0.48	2.13	1.60
Lung	2.51	0.85	0.41
Kidney	0.70	1.03	0.60
Stomach	7.78	0.53	0.48
Heart	1.34	1.72	—[c]
Large intestine	0.20	1.06	—
Spleen	1.92	3.28	0.55
Bladder	0.19	1.31	—
Brain	0.1	0.83	—
Muscle (hind leg)	0.1	0.46	—
Skin (hind leg)	0.1	1.42	—
Mesenteric lymph node	0.21	2.21	—
Cervical lymph node	0.28	1.61	—
Thymus	0.22	0.93	—
Plasma	0.37	0.03	—
Blood cell	0.64	1.57	—
Bone marrow	0.15	2.91	—
Urine (fresh)	0.18	1.09	—
Urine (stored)	—	1.14	—
Bile	38.18	2.23	—

[a] Nontumorous portion adjacent to tumor.
[b] Nontumorous portion distant from tumor.
[c] —, Not measured.

[23,24,26]. Figure 20 shows the time-course of a soft X-ray picture of tumor in rabbit liver. As seen in Fig. 20, Lipiodol was retained more than 1 week in the tumor while normal liver tissue cleared this carrier within a day or so. In human liver cancer, it was actually quantified even after 4 months; the concentration was high enough to kill the tumor cell or inhibit tumor growth (Table 6).

When the localization of Lipiodol at microscopic level was examined in tumor tissue, it was found in the interstitial tissue and in the vascular wall (endothelium), and in some cases was found completely filling the blood vessel. This was confirmed by radioautography and by lipid staining (Sudan III) (Fig. 21) [24].

Drug concentration in the tumor of these rabbits was quantified after arterial administration and found to be consistently high (Tables 5 and 6). In metastatic tumors in the liver in patients who were injected with SMANCS/Lipiodol

TABLE 6. Biological activity of SMANCS in resected liver after arterial injection of SMANCS/Lipiodol

Drug administration in patients with metastatic hepatoma				Biological activity of SMANCS (μg/g tissue)		
					Normal liver	
Patient	Dose of SMANCS	Artery of drug administration	Period between the last administration and hepatectomy[a]	Tumor area	Adjacent to tumor	Distant from tumor
1	5 mg	Common hepatic	2 wk	(a)[b] 10	5	0
				(b) 3	0	
2	4 mg	Proper hepatic	2 mon	(c) 33	7	0
				(d) 9		0

[a] The assays were carried out at the time of resection after homogenization using agar plate method (from [25a,26], with permission).
[b] Samples from different sites in the tumor.

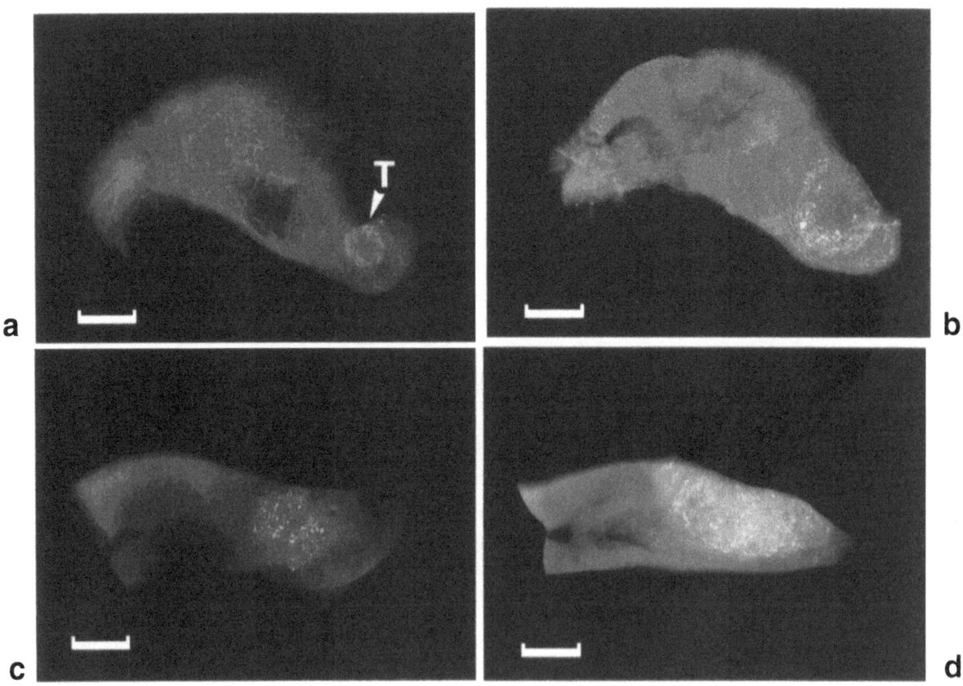

FIG. 20a–d. Softex film of slices of rabbit liver; *a* to *d* from 15 min, 1 day, 3 days, and 7 days, respectively, after injection of SMANCS/Lipiodol via the proper hepatic artery. **a** Deposition of Lipiodol in the tumor (*T, arrowhead*) and normal liver arterial lumina is seen. **b** Deposition in normal liver arterial lumina was less than that in the tumor; a more selective tumor deposition is seen. **c** Lipiodol deposition in normal liver arterial lumina was much less than that at 1 day, but that of the tumor persisted. **d** Deposition of Lipiodol was confirmed only in the tumor tissue. Lipiodol disappeared completely from the normal liver arterial lumina. *Bars*, 1 × 1 cm. (From [24], with permission)

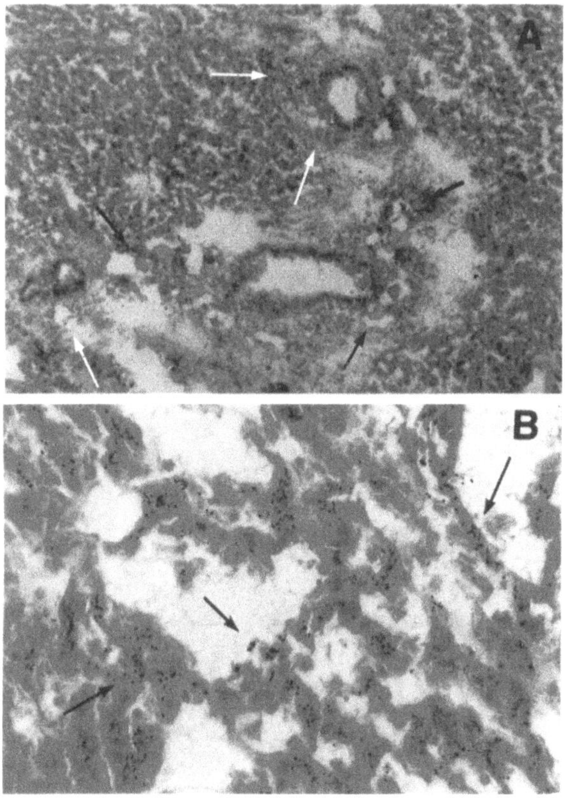

FIG. 21A,B. Microscopic examination of distribution of [^{14}C]-Lipiodol moiety in VX-2 carcinoma in rabbit liver. **A** Sudan III lipid staining in which cluster of lipid is seen in and out of the blood vessels (*black arrows*), although less intense radioactive microparticles are also seen in the interstitial space (*white arrows*). **B** Radioautograph shows *black dots* of Lipiodol throughout tumor tissue (*arrows* and elsewhere). (From [24], with permission.) (*For color reproduction see color insert in the frontmatter*)

intraarterially, its concentration was similarly high while that in normal liver was undetectable (Table 6) [26].

Antitumor Effect of SMANCS/Lipiodol Given via the Tumor-Feeding Artery: Effect Against Experimental Metastatic Liver Tumor

As described in the following section, the therapeutic effect of SMANCS against primary liver cancer (hepatocellular carcinoma) is unprecedented. To extend this idea we prepared a model of metastatic liver cancer by injecting VX-2 carcinoma (about 2.5×10^6 cells) into the portal vein of rabbits [27]. Tumors are allowed to

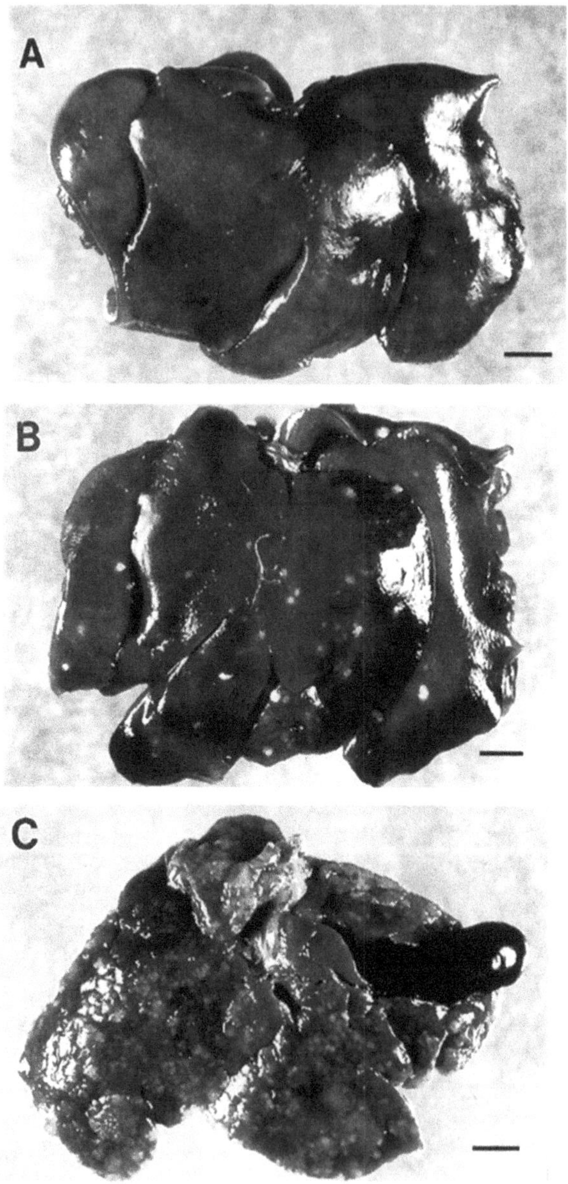

FIG. 22A–C. Suppression of metastatic liver tumor by SMANCS/Lipiodol against experimental metastatic liver tumor (VX-2) in rabbits. Tumor cells were inoculated via the portal vein, and the drugs were injected into the portal vein just before tumor inoculation. All liver specimens were obtained 12 days after inoculation. **A** SMANCS/Lipiodol, 0.4 mg/0.4*ml/kg. **B** SMANCS in 5% glucose (0.4 mg/0.4 ml/kg) followed by free Lipiodol as in **A**. **C** No drug control; tumor nodules number more than 500. (See Table 7.) *Bars*, 1 cm. (From [27], with permission.) (*For color reproduction see color insertin frontmatter*)

grow for about a week or two, and numerous tumor nodules are observed by laparotomy. Therapeutic effect against this metastatic tumor model was examined by injecting SMANCS/Lipiodol (about 0.4 in 0.4 ml/kg) into the mesenteric vein in rabbits. As controls, one group without any drug, and another group using SMANCS in 5% glucose followed by Lipidol, and the last group with SMANCS in Lipiodol were used for comparison (Fig. 22).

As shown in Table 7, significant suppression (less than 1/100) of tumor nodules was observed in the SMANCS/Lipiodol group. This result indicates a possible therapeutic application of this method in the clinical setting to control metastatic liver cancer by injecting SMANCS/Lipiodol into the portal vein before or during resection of colon cancer when the patient is under laparotomy. During the operational procedure it is believed that the colon tumor is being seeded most frequently into the liver, and hence metastatic foci develop. This notion is being investigated at Monash University (Professor O'Brein, and Dr. Christophi et al. of the Department of Surgery, in collaboration with us) using a rat model. The preliminary results show similar results obtained by us (unpublished data).

Although the portal vein is the route of blood-borne metastasis from the colon, a high incidence is known to occur via the lymphatic route. The route of the feeding artery to these metastatic liver cancers or lung cancers or other metastatic lymph nodules is a matter of controversy. Metastatic liver cancer and metastatic lung cancer are now known to be fed by the hepatic artery and by the bronchial artery, respectively. The metastatic lymph nodes were also found to be fed by the regional artery, not by the lymphatic vessels, when the metastatic nodule is larger than 5 mm, in our data [28]. Therefore, the EPR principle will be applicable to all metastatic tumor nodules. In any event, the direct arterial injection of an oily formulation (SMANCS/Lipiodol) is expected to be the best choice.

Table 7. Effect of SMANCS/Lipiodol against metastatic liver cancer and drug concentration after injection into the portal vein (From [27], with permission)

Drug/formulation	No. of rabbits	No. of range of tumor nodules per rabbit liver: corresponding no. of rabbits (%)[a]				Drug concentration (μg/ml)			
		0	1–50	51–500	>500	Specimen/organ	5 h	24 h	72 h
SMANCS/Lipiodol[b] (0.4 mg/kg)	18	8 (44.4)	6 (33.3)	4 (22.2)	0 (0)	Liver	130	70	41
						Lung	1.5	1.5	<1.5
						Kidney	9.6	7.5	30.0
SMANCS/5% glucose followed by Lipiodol[c]	15	0 (0)	2 (13.3)	5 (33.3)	8 (53.3)	Liver	14.1	2.7	2.6
						Lung	1.1	0	0
						Kidney	8.3	0	0
No drug, tumor only	12	0 (0)	0 (0)	5 (41.7)	7 (58.3)	—	—	—	—

[a] Tumor nodules on 12 days after inoculation were examined visually after total hepatectomy (see Fig. 22).
[b] Drug was injected immediately before tumor inoculation into the mesenteric vein.
[c] Drug in aqueous formulation was followed by Lipiodol, both injected into the mesenteric vein.

Further, the angiotensin II-induced hypertensive condition may be suitable for i.v. injection for eradication of these arterially fed tumors [29,30]. Systematic clinical evaluation should be pursued.

Clinical Effect of SMANCS in Human Cancer

Therapeutic Effect on Primary and Metastatic Liver Cancer

A lipid lymphographic contrast medium, Lipiodol Ultrafluid (Lipiodol)®, has been found to be selectively retained in the site of hepatic tumors and other malignant solid tumors when it is injected arterially in our clinic as described previously [22,24,25,26,30–32]. Selective retention of Lipiodol in malignant solid tumors pemits anticancer drugs dissolved in Lipiodol to diffuse out into the surrounding tumor tissues and tumor cells by the EPR effect described earlier (see Figs. 20 and 21). Lipiodol adhere frequently on the luminal surface of vascular wall.

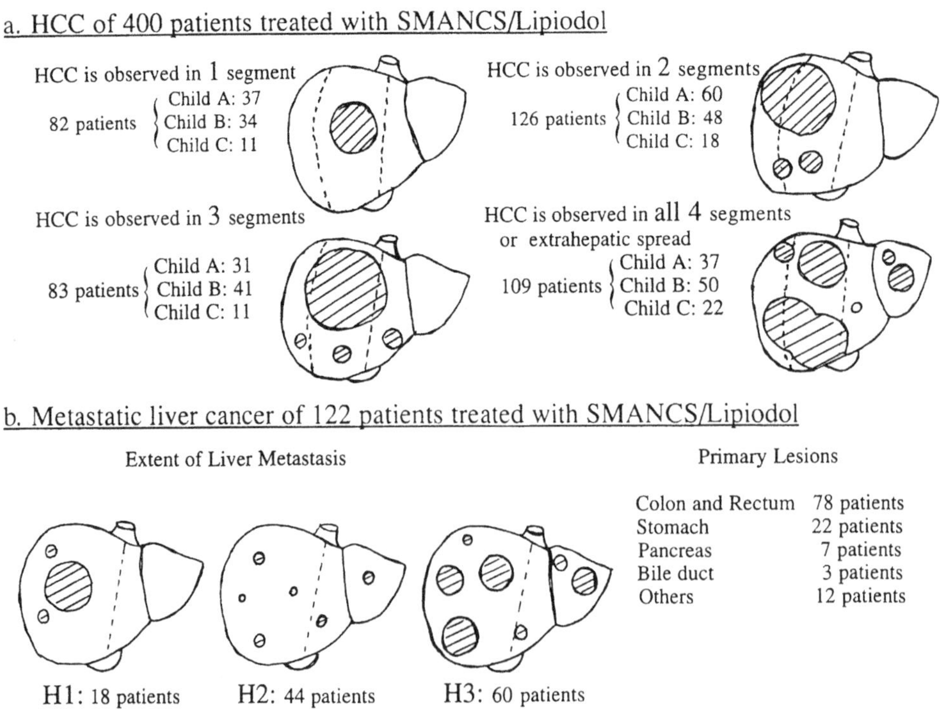

FIG. 23a,b. Patients with tumors in the liver who underwent SMANCS therpay. **a** Primary liver cancer patient with hepatocellular carcinoma (HCC); tumor size and intrahepatic spread. **b** Patients with metastatic liver cancer; size and spread of tumor

The dose regimen required for optimal targeted chemotherapy to liver cancer with Lipiodol as a carrier has already been established in our clinic. Briefly, SMANCS or other anticancer drugs must be dissolved or well dispersed in Lipiodol, and the drugs must be slowly released from Lipiodol while Lipiodol remains selectively in the tumor (see Fig. 19). The oily formulation must be homogeneous without separating into drug and Lipiodol, even if it is left for a long time (Fig. 19). SMANCS dispersed in Lipiodol (SMANCS/Lipiodol) is the first product of this kind and has been used as a first-choice drug for hepatoma.

We describe here the effect of SMANCS/Lipiodol on unresectable, mostly highly advanced, hepatocellular carcinoma (HCC) and metastatic liver cancer using this arterial injection therapy. The adverse effects of this treatment modality were examined clinically.

Patients

Patients had clinically established primary hepatoma (hepatocellular carcinoma, HCC) or metastatic liver cancer on the basis of hepatic angiography, X-ray computer tomography (CT), and ultrasound, with or without increased serum alpha-fetoprotein (AFP) or carcinoembryonic antigen (CEA) levels.

Four hundred patients with HCC (341 men and 59 women between 18 and 84 years of age; mean, 60 years) and 122 patients with metastatic liver cancer (86 men and 36 women between 25 and 85 years of age; mean, 58.4 years) were treated essentially by arterial injection of SMANCS/Lipiodol. For these patients, the extent of HCC and metastatic liver cancer, the state of liver cirrhosis classified by Child's definition in HCC, and the primary sites of metastatic liver cancer are shown in Fig. 23. All patients or their families gave consent to the treatment protocol.

Formulation of Oily Anticancer Agents and Administration of SMANCS/Lipiodol

SMANCS/Lipiodol was prepared by suspending 1 mg of SMANCS in 1 ml of Lipiodol by sonication. SMANCS is partially soluble in Lipiodol and is homogeneous. Selective or superselective catheterization of the celiac artery was performed through the femoral artery under fluoroscopy by using Seldinger's method, confirming the feeding artery of the tumor with the help of a contrast medium, usually a nonionic iodine compound. After identification of the catheter tip located in the feeding artery, SMANCS/Lipiodol was administered via the hepatodistal artery, to be delivered into the feeding artery of the tumor, most often the hepatic, proper hepatic, or common hepatic artery.

To 445 patients, a total of 2080 injections of the drug were given (1543 times for 400 patients with HCC, and 537 times for 120 patients with metastatic liver cancer). Injections were given via the celiac artery (267 times), the common hepatic artery (468 times), the proper hepatic artery (1000 times), or the left or right hepatic artery (345 times). Because the safety of injection of oily anticancer agents into the gastroduodenal artery has not been established, the proper hepatic artery or its distal artery was recommended. The volume of the drug given in each patient depended on the size of the tumor and the artery through which it was administered; the volume ranged from 1 to 12 ml, with a mean of 3.8 ml for HCC and 3.5 ml for metastatic liver cancer. Ideally, about 0.3 mg/cm^2 of cut area of tumor was infused in a small- to medium-size hepatoma. The velocity of this dosing is also very important; i.e., the drug should be infused via syringe very slowly. A slow, intermittent (2- to 3-s interval) pulsating push of the syringe plunger is recommended. Usually 4 ml may be infused in $1 \sim 3$ min, ideally, though it could take longer time. The frequency of administration for HCC varied from 1 to 12 times, with a mean of 4 times, and for metastatic liver cancer it varied from 1 to 25 times with a mean of 4.4 times.

The administration of the oily drug was performed once every 3–5 weeks until evaluation by X-ray CT scan showed grade IV [31,32] was achieved (grade IV in HCC patients means the posttreatment CT at 4–7 days after SMANCS/Lipiodol shows the entire area of HCC to be filled with Lipiodol; this is referred as type A staining). In metastatic liver cancer, posttreatment CT scan of patients after 3–5 days shows the entire rim of the tumor is filled with Lipiodol staining; this type of staining is referred as type B staining [32]. A small metastatic liver cancer may sometime appear similar to HCC staining (type A) [32]. Thereafter, a CT scan was taken once every 1–3 months. If the staining grade showed decrease to grade II or I because of washout, additional doses were necessary to attain grade IV again. The endpoint was judged by the finding that Lipiodol staining was no longer increased in tumor on CT after subsequent injection of oily anticancer agents.

Evaluation of Treatment Response

The tumor response was evaluated by changes in the tumor markers (ATP and CEA) and in the size of tumor. Tumor size was evaluated both by angiogram and X-ray CT scan taken 3–4 days after injection for metastatic liver cancer or 5–7 days after for HCC. For detailed evaluation of tumor size, the root of the product of the longest diametar (a) and the perpendicular diameter (b) ($\sqrt{a \times b}$)of each tumor in the CT scan is used. In addition, duration of survival after the first injection of SMANCS/Lipiodol was calculated by the Kaplan–Meier method. The log-rank test was used for analysis of survival.

Radiological Evaluation

The entire tumor area of HCC was stained by the uptake of Lipiodol when enough volume was injected. In HCC having a central necrosis, Lipiodol uptake was not observed in these portions. In metastatic liver cancer, Lipiodol retained mainly in the periphery of the tumor, described earlier as type B staining, which resulted in a ringlike appearance in about 60% of cases of metastatic liver cancer. In a hypervascular tumor, as judged by angiogram, a thick-walled, ringlike appearance was observed, while a thin ringlike appearance was seen in hypovascular tumors.

Almost all Lipiodol was washed out from the parenchyma of the liver in about 24 h after arterial administration of oily anticancer agents, so the selective staining of Lipiodol in tumor become apparent in about 24 after injection; it is usually retained for 2–3 months in HCC and for 2–4 weeks in metastatic liver cancer (see Table 6). In some cases it is confirmed even after 2 months with high antitumor activity.

Antitumor Activity: Hepatocellular Carcinoma

Serum AFP levels were 100 ng/ml or more (AFP-positive criteria) in 238 of the 400 patients, less than 100 ng/ml in 149 patients, and unknown in 13 patients. Values of AFP, measured for 222 of 238 AFP-positive patients as follow-up marker, were decreased in 209 (94% of 222) after arterial injection therapy and increased in 13 (6%).

The size of tumors that were measurable, in 322 of 400 patients with HCC, showed that the tumor was regressed in 308/322 (96%; size regression to less than 50% of the original occurred in 178 and from 100% to 50% of the original size in 130 patients); no change was observed in 7 (2%) and continued tumor growth in 7 (2%). The degree of reduction in tumor size and the time required for reduction are shown in Fig. 24. Almost all tumors were reduced to less than 50% of the initial size after 1 year from the initial administration of the drug. In 48 of 60 patients, whose tumor shrunk to 10% or less of the initial size, further follow-ups

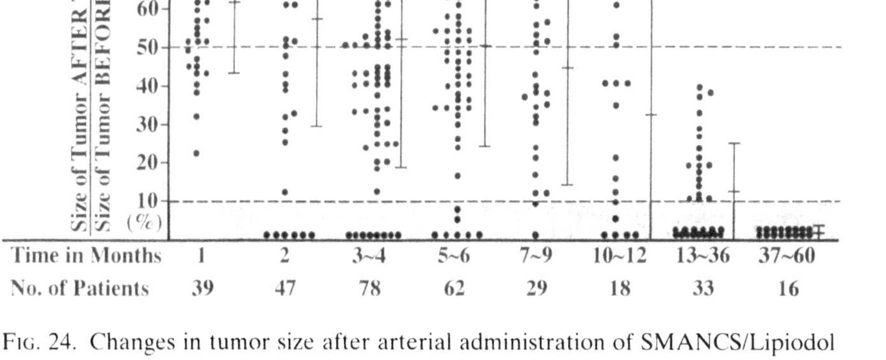

FIG. 24. Changes in tumor size after arterial administration of SMANCS/Lipiodol

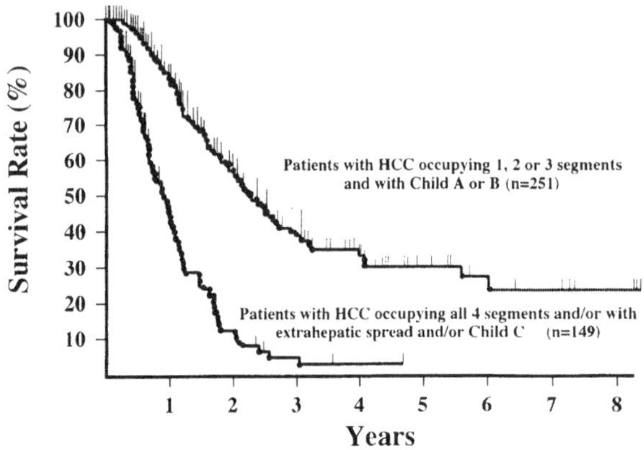

FIG. 25. Survival rates of patients with unresectable HCC treated with SMANCS/Lipiodol: Child A or B class patients with 1–3 segmental metastasis vs. Child C patients with 4 segmental intrahepatic spread

as to tumor size, were carried out for from 1 year to as long as 9 years (average, 3 years), and they revealed that none of these tumors regrew and thus they were considered to be cured.

The 50% survival period of all 330 patients with unresectable HCC was 544 days. In 192 patients who were good candidate for this therapy, excluding patients with cirrhosis of Child C classification and with tumors spreading to all four segments of the liver, or those patients with extrahepatic tumor spread, the 1-, 2-, and 5-year survival rates were 85%, 52%, and 34%, respectively (Fig. 25).

Antitumor Activity: Metastatic Liver Cancer

Elevated serum CEA levels were observed in 95 of the 122 patients with unresectable metastatic liver cancer; CEA levels were not elevated in 22 and were unknown in 5 patients. Changes in CEA were analyzed for 92 of 95 CEA-positive patients. The assay of serum CEA level showed decrease in 75 (82%), no change in 3 (3%), and increase in 14 (15%). Size of metastatic liver cancer size was measured in 99 of the 122 patients: regression was seen in 60 (61%) of these 99 patients, no change in 14 (14%), and increase in 25 (25%). For all 122 patients with unresectable metastatic liver cancer who were given arterial injection therapy of SMANCS/Lipiodol as described before, their 1-, 2-, and 5-year survival rates were 72%, 38%, and 28%, respectively (Fig. 26).

Side Effects

The major adverse side effect of arterial injection therapy of SMANCS/Lipiodol was fever, which occurred in 47% of the patients; this usually lasted 1–2 days and was controlled by appropriate medications. About 12% of the patients reported

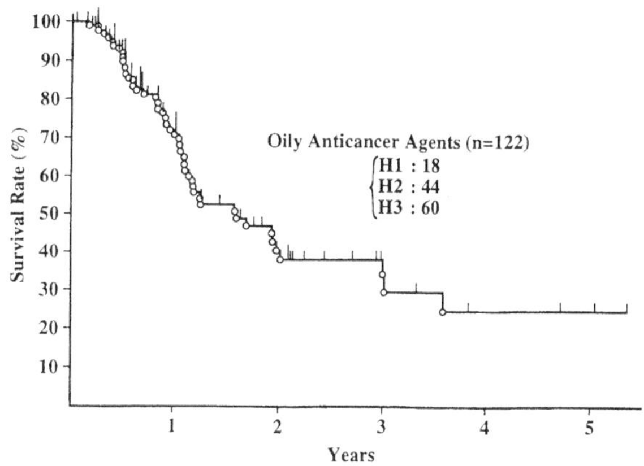

FIG. 26. Survival rates of patients with unresectable metastatic liver cancer patients who were treated with SMANCS/Lipiodol

mild temporary pain in the epigastrium shortly after the injection. The pain occurred immediately after the arterial injection and generally eased or disappeared within 15 min.

Liver function tests showed elevated levels of both aspartate aminotransferase and alanine aminotransferase in 11% and 8% of the patients, respectively. The total bilirubin level rose in 12% of the patients and fell in 9%. These changes were all mild and transient. There was virtually no bone marrow suppression nor life-threatening or lethal complications. Some patients exhibited shock during or shortly after arterial injection because of the procedure, SMANCS, contrast medium/iodine, or an unknown cause; they could be controlled by dexamethasone or another steroid. The frequency of hypotension/shock seems to vary from one hospital to another, and the velocity of arterial infusion seems important (faster infusion tended to cause more shock state). The incidence of shock may be less than 5%, which could be managed with a steroid hormone.

Therapeutic Effect on Renal Cell Carcinoma

As for renal cell carcinoma, SMANCS/Lipiodol was used first for a patient with nonresectable advanced renal cell carcinoma. She had a large mass ($16.5 \times 9.5\,cm^2$) in the left kidney with lymph node metastases. Extensive spread of the tumor and her general condition made tumor resection impossible. SMANCS/Lipiodol was injected 14 times from February 4, 1983 to May 8, 1986, with a total dose of 86.4 mg SMANCS. After treatment with SMANCS/Lipiodol, the tumor shrunk dramatically and lymph node metastases disappeared; the patient's general condition improved so that resection of the tumor became possible. The patient is still alive and well 10 years after the initial administration of the drug [33]. Following this experience, many patients were then treated by the arterial injection of SMANCS/Lipiodol at Gunma University, and a definite antitumor activity against renal cell carcinoma was confirmed clinically and histologically [34].

Because blood flow in the renal artery is very fast the viscosity and drug concentration of SMANCS/Lipiodol should be higher (SMANCS, up to 2mg/ml), and the interval to subsequent injections should be shorter than that with HCC. It was also noted that when the primary tumor in the kidney was treated many metastatic nodules in the lung disappeared [34], indicating that an immunological mode of action may be important in addition to the direct action of SMANCS to tumor cells.

Effect of SMANCS/Lipiodol on Other Solid Tumors and Metastatic Lymph Nodes from Various Tumor Origins

Selective retention of Lipiodol in a tumor occurs by the characteristic nature of the newly formed vessels, neovasculature developed in the tumor, and slow clearance (EPR effect), as described earlier. Although the extent and distribution pattern of the neovasculature in the tumor varied from one tumor to another, neovasculature will develop in all cancers. Therefore, it is thought that the selective retention of Lipiodol might occur in all solid cancers. Clinically, selective retention of Lipiodol in cancer was observed in all cancers tested such as bronchogenic/lung cancer, gastric cancer, cancer of the uterus, cancer of the gallbladder, and cancer of the pancreas [26]. Additionally, this selective retention was also observed in metastasized lymph nodes, meaning that the metastasized lymph nodes were nourished by the arterial blood flow as described, not by or least not entirely by the lymphatic flow [28]. Arterial injection therapy using SMANCS/Lipiodol against these various types of cancer is thought to be more effective than any other chemotherapy described previously. Tumor size reduction and prolonged survival were observed in patients with these unresectable cancers.

Briefly, as for advanced unresectable bronchogenic cancer, tumor size reduction was observed in more than 80% of the patients. And the 1-year survival rate was more than 50%. The dosage of SMANCS/Lipiodol in bronchogenic cancer is 0.5–1.5 ml. However, one needs to be cautious as to whether the spinal artery is far enough from the bronchial artery; overflow of SMANCS/Lipiodol into the spinal artery could possibly cause paralysis of the lower trunk.

In Fig. 27, treatment of a patient having unresectable advanced pancreatic cancer with liver metastases with SMANCS/Lipiodol is presented.

Effect of SMANCS/Lipiodol for Pleural or Peritoneal Carcinomatosis

Very often lung cancers or cancers in the digestive organs result in dissemination of cancer cells into the pleural or peritoneal cavity. This is more frequently

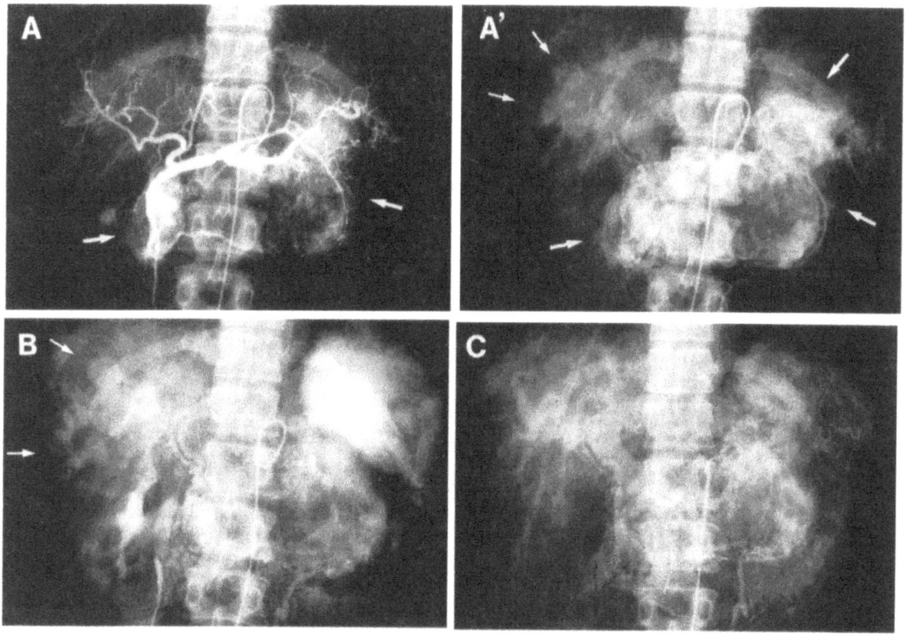

FIG. 27. Treatment with arterial injection therapy of SMANCS/Lipiodol against unresectable advanced pancreatic cancer in 31-year-old woman. Angiography before treatment: **A**, arterial phase; **A'**, venous phase. **B**, Two months after initial injection of the drug, enlargement of liver metastases was observed. **C** One year after the treatment, shrinkage of the pancreatic tumor and liver metastases was observed. These tumors did not regrow until the patient died from multiorgan failure caused by biliary infection and pneumonia 2 years and 2 months after the initial injection of SMANCS/Lipiodol; she had severe diabetes mellitus

observed at the terminal stage or after surgery of the respective organs and is a very formidable condition to treat.

It was found that intracavital injection of SMANCS/Lipiodol resulted in the retention of SMANCS for a long time at much higher concentration in the cavity [35]. This indicates that this macromolecular drug, SMANCS, is not readily absorbed from the capillary blood vessels because of its high molecular weight and is not freely transferred into the blood vessels. Thus, we attempted to inject SMANCS/Lipiodol into the peritoneal and the pleural cavities for the treatment of peritoneal and pleural carcinomatosis. Of the seven patients who received intracavital injections of the drug, all responded. Five patients showed complete response (CR; from Papanicolaou smear class IV to class I), and 2 patients responded partially (PR). No serious side effects were encountered after one or two injections [36].

Concluding Remarks

We have described the various biochemical and pharmacological characteristics of SMANCS, as well as preclinical and clinical data. SMANCS is a new class of anticancer agent of polymer conjugate, and its high tumor-targeting efficiency is further augmented by formulating it in lipid contrast medium (Lipiodol) to the greatest extent. It is truly novel in the targeting method utilizing polymer and lipid contrast medium as a carrier, and among all therapeutic efficacy in various cancers, the present results for hepatoma patients are unprecedented and the side effect is minimal. It seems that treatment for other cancers including renal cancer or cancer of the gallbladder, cholangiocarcinoma, etc., as well as carcinomatosis of either pleural or peritoneal origins can be highly effective, and future investigation is warranted. Oral administration (not described here) was shown to be effective in rodents (B-16 melanoma) (see chapter by F. Suzuki and M. Kobayashi, this volume), but no human data are available.

Although the dose regimen for primary liver cancer is fairly well established, that for renal or other cancer needs to be determined. A higher viscosity and higher drug concentration are suggested for renal cancer, and a smaller volume for lung cancer. The interval of drug administration for hepatoma is usually once a month via hepatic artery during the initial 2–4 months, while metastatic liver cancer needs more frequent treatment and perhaps a higher concentration of the drug may be preferable. Combination with other drugs, especially the use of oily anticancer agents including vinblastin, carboquone derivative, mitomycin C, carmustine, and aclarubicin/doxorubicin, has been pursued (mostly by Konno [30]) and needs further exploitation.

Use of SMANCS/Lipiodol for imaging purposes, either diagnosis of nodular spread, dose determination, or follow-up in general, is invaluable, and it is even used as a standard for tumor detection. All these facts are ideal requirements for tumor drug targeting in future.

Whatever unique mechanisms or pharmacokinetic or targeting superiority a new drug exhibits, it must contribute to improving or sustaining patient' quality of life and life span irrespective of tumor size reduction (to satisfy so-called response criteria). In this connection SMANCS/Lipiodol by use of enforced targeting contributes to both quality of life and prolongation of life span of patients with hepatocellular carcinoma [21,30,31] (see Fig. 26). Therefore, this method should be easily extended to other types of solid tumors such as metastatic liver cancer and cancers of the kidney, lung, pancreas, gallbladder, etc. Thus, we look for future advances in cancer therapy using this strategy.

References

1. Maeda H, Takeshita J, Yamashita A (1980) Lymphotropic accumulation of an antitumor antibiotic protein neocarzinostatin. Eur J Cancer 16:723–731

2. Maeda H, Takeshita J, Kanamaru R (1979) A lipophilic derivative of neocarzinostatin: a polymer conjugation of an antitumor protein antibiotic. Int J Pept Protein Res 14:81–87
3. Maeda H, Ueda M, Morinaga T, Matsumoto T (1985) Conjugation of poly(styrene-co-maleic acid) derivatives to the antitumor protein neocarzinostatin: pronounced improvements in pharmacological properties. J Med Chem 28:455–461
4. Maeda H, Matsumoto T, Konno T, Iwai K, Ueda M (1984) Tailor-making of protein drugs by polymer conjugation for tumor targeting: a brief review on SMANCS. J Protein Chem 3:181–193
5. Takeshita J, Maeda H, Kanamaru R (1982) In vitro mode of action, pharmacokinetics, and organ specificity of poly(maleic acid-styrene)-conjugated neocarzinostatin, SMANCS. Gann 73:278–284
6. Maeda H, Takeshita J, Kanamaru R, Sato H, Katoh J, Sato H (1979) Atimetastatic and antitumor activity of a derivative of neocarzinostatin: an organic solvent- and water-soluble polymer-conjugated protein. Gann 70:601–606
7. Kobayashi A, Oda T, Maeda H (1988) Protein binding of macromolecular anticancer agent SMANCS: characterization of poly(styrene-co-maleic acid) derivatives as an albumin binding ligand. J Bioact Compat Polym 3:319–333
8. Maeda H (1979) Assay of proteolytic enzymes by fluorescence polarization technique. Anal Biochem 92:222–227
9. Maeda H (1978) Assay of an antitumor protein neocarzinostatin and its antibody by fluorescence polarization. Clin Chem 24:2139–2144
10. Oda T, Morinaga T, Maeda H (1986) Stimulation of macrophages by polyanions and its conjugated proteins and effect on cell membrane. Proc Soc Exp Biol Med 181:9–17
11. Takeshita J, Maeda H, Koike K (1980) Subcellular action of neocarzinostatin. Intracellular incorporation, DNA breakdown and cytotoxicity. J Biochem (Tokyo) 88:1071–1080
12. Maeda H, Aikawa S, Yamashita A (1975) Subcellular fate of protein antibiotic neocarzinostatin in culture of a lymphoblastoid cell line from Burkitt's lymphoma. Cancer Res 35:554–559
13. Sakamoto S, Maeda H, Ogata J (1979) An uptake of fluorescein isothiocyanate-labeled neocarzinostatin into the cancer and normal cells. Experientia (Basel) 35:1233–1234
14. Maeda H, Matsumoto M (1979) Cytotoxic effect of neocarzinostatin on human lymphoid cells. Tohoku J Exp Med 128:313–323
15. Oda T, Maeda H (1987) Binding to and internalization by cultured cells of neocarzinostatin and enhancement of its action by conjugation with lipophilic styrene-maleic acid copolymer. Cancer Res 47:3206–3211
16. Oda T, Sato F, Maeda H (1987) Facilitated internalization of neocarzinostatin and its lipophilic polymer conjugate, SMANCS, into cytosol in acidic pH. J Natl Cancer Inst 79:1205–1211
17. Oka K, Miyamoto Y, Matsumura Y, Tanaka S, Suzuki F, Maeda H (1990) Enhanced intestinal absorption of a hydrophobic polymer-conjugated protein drug, smancs, in a oily formulation. Pharm Res (NY) 7:852–855
18. Matsumura Y, Maeda H (1986) A new concept for macromolecular therapeutics in cancer chemotherapy: mechanism of tumoritropic accumulation of proteins and the antitumor agent smancs. Cancer Res 46:6387–6392
18a. Maeda H (1991) SMANCS and polymer-conjugated macromolecular drugs: advantages in cancer chemotherapy. Adv Drug Deliv Rev 6:181–202

19. Maeda H, Matsumura Y (1989) Tumoritropic and lymphotropic principle of macromolecular drug. Crit Rev Ther Drug Carrier Syst 6:193–210
20. Hirayama S, Sato F, Oda T, Maeda H (1986) Stability of high molecular weight anticancer agent SMANCS and its transfer from oil-phase to water-phase. Jpn J Antibiot 39:815–822
21. Maeda H (1994) Polymer conjugated macromolecular drugs for tumor specific targeting. In: Domb AJ (ed) Polymeric site-specific pharmacology. Wiley, New York, pp 95–116
22. Konno T, Maeda H, Iwai K, Tashiro S, Maki S, Morinaga T, Mochinaga M, Hiraoka T, Yokoyama I (1983) Effect of arterial administration of high-molecular-weight anticancer agent SMANCS with lipid lymphographic agent on hepatoma: a preliminary report. Eur J Cancer Clin Oncol 19:1053–1065
23. Kimoto A, Konno T, Kawaguchi T, Miyauchi Y, Maeda H (1992) Antitumor effect of SMANCS on rat mammary tumor induced by 7, 12-dimethylbenz[a]anthracene. Cancer Res 52:1013–1017
24. Iwai K, Maeda H, Konno T (1984) Use of oily contrast medium for selective drug targeting to tumor: enhanced therapeutic effect and X-ray image. Cancer Res 44:2115–2121
25. Iwai K, Maeda H, Konno T, Matsumura Y, Yamashita R, Yamasaki K, Hirayama S, Miyauchi Y (1987) Tumor targeting by arterial administration of lipid: rabbit model with VX2 carcinoma in the liver. Anticancer Res 7:321–328
25a. Maeda H (1986) SMANCS (in Japanese). Zusetsu-Rinsho [Gann] series (Prog Cancer Clin). 1:132–137
26. Konno T, Maeda H, Iwai K, Maki S, Tashiro S, Uchida M, Miyauchi Y (1984) Selective targeting of anti-cancer drug and simultaneous image enhancement in solid tumors by arterially administered lipid contrast medium. Cancer (Phila) 54:2367–2374
27. Yamasaki K, Konno T, Miyauchi Y, Maeda H (1987) Reduction of hepatic metastases in rabbits by administration of an oily anticancer agent into the protal vein. Cancer Res 47:852–855
28. Ohtsuka N, Konno T, Miyauchi Y, Maeda H (1987) Anticancer effects of arterial administration of the anticancer agent SMANCS with Lipiodol on metastatic lymph nodes. Cancer (Phila) 59:1560–1565
29. Li CJ, Miyamoto Y, Kojima Y, Maeda H (1993) Augmentation of tumor delivery of macromolecular drugs with reduced bone marrow delivery by elevating blood pressure. Br J Cancer 67:975–980
30. Konno T (1990) Targeting cancer chemotherapeutic agents by use of Lipiodol contrast medium. Cancer (Phila) 66:1897–1903
31. Konno T, Maeda H (1987) Targeting chemotherapy of hepatocellular carcinoma. In: Okuda K, Ishak KG (eds) Neoplasms of the liver. Springer, Berlin Heidelberg New York, pp 343–352
32. Maki S, Konno T, Maeda H (1985) Image enhancement in computerized tomography for sensitive diagnosis of liver cancer and semiquantitation of tumor selective drug targeting with oily contrast medium. Cancer (Phila) 56:751–757
33. Noda S, Konno S, Tanaka J, Yamada M, Yoshitake N (1990) Treatment of renal cell carcinoma with intra-arterial administration of SMANCS dissolved in Lipiodol. Anticancer Res 10:709–716
34. Kobayashi M, Imai K, Sugihara S, Maeda H, Konno T, Yamanaka H (1991) Tumor-targeted chemotherapy with lipid contrast medium and macromolecular anticancer drug (SMANCS) for renal cancer. Urology 37:288–294

35. Kimura M, Konno T, Oda Maeda H, Miyauchi Y (1993) Intracavitary treatment of malignant ascitic carcinomatosis with oily anticancer agents in rat. Anticancer Res 13:1287–1292
36. Kimura M, Konno T, Yamashita R, Oda T, Nagamitsu A (!994) Intracavital chemotherapy for peritoneal and pleural carcinomatosis with lipid-formulated anticancer agents. Oncol Rep 1:313–315
37. Maeda H, Miyamoto Y (1994) SMANCS approach—oily formulations of protein drugs for arterial injection and oral administration. In: de Boer AG (ed) Drug absorption enhancement, concept, possibilities, limitations and trends. Harwood Academic, Chur, pp 221–247

Supplementary Materials: Publications on Neocarzinostatin or Related Subjects, in English, 1960–1981

Compiled by SHIGEHIRO MATSUMOTO[1], HIROSHI MAEDA[2], and NAKAO ISHIDA[3]

Summary. The following list of publications is incorporated in this monograph because earlier publications are not accessible to many modern computer search systems or have been forgotten because they are too old. Therefore, we want to acknowledge those who dedicated their efforts to clarifying the uniqueness or unknown mystery of NCS at an early stage.

I. Reviews

1. Lega SS, Von Haff DD, Rozencweig M, Abraham D, Slavink M, Muggia M (1976) Neocarzinostatin (NSC 157365): a new cancerostatic compound. Oncology (Basel) 33(5,6):265–270
2. Ishida N, Ebina T (1977) Neocarzinostatin (NCS). Jpn J Antibiot 30(suppl):93–103
3. Maeda H (1981) Neocarzinostatin in cancer chemotherapy (review). Anticancer Res 1:175–186

II. Isolation, Purification, Chemistry, Physicochemistry, and Derivatization

1. Ishida N, Kumagai K, Miyazaki K, Ito M (1960) Carzinostatin, a new antitumor substance. Gann 51:Suppl 56
2. Shoji J (1961) Preliminary studies on the isolation of carzinostatin complex and its characteristics. J Antibiot (Tokyo) Ser A 14:27–33

[1] Kayaku Co. Ltd., 2-8-16 Funado, Itabashi-ku, Tokyo, 174 Japan
[2] Department of Microbiology, Kumamoto University School of Medicine, 2-2-1 Honjo, Kumamoto, 860 Japan
[3] Sendai Institute of Microbiology, 2-5-22 Ichiban, Aoba-ku, Sendai, Miyagi, 980 Japan

3. Ishida N, Miyazaki K, Kumagai K, Rikimaru M (1965) Neocarzinostatin, an antitumor antibiotic of high molecular weight. Isolation, Physicochemical properties and biological activities. J Antibiot (Tokyo) Ser A 18(2):68–76
4. Maeda H, Kumagai K, Ishida N (1966) Characterization of neocarzinostatin. J Antibiot (Tokyo) Ser A 19(6):253–259
5. Maeda H, Ishida N (1967) Conformational study of antitumor protein and a deaminated derivative. Biochim Biophys Acta 147:597–599
6. Maeda H, Koyanagi T, Ishida N (1968) C-Terminal amino acid of neocarzinostatin. Biochim Biophys Acta 160:249–251
7. Kitame S, Maeda H, Kumagai K, Ishida N (1968) Tryptic digestion of an antitumor protein: neocarzinostatin. J Antibiot (Tokyo) 21(7):460–462
8. Maeda H, Meienhofer J (1970) On the tryptophan of content of neocarzinostatin. Int J Pept Protein Res 2(2):135–136
9. Maeda H, Meienhofer J (1970) Amino and carboxy-terminal sequences of neocarzinostatin, an antitumor polypeptide antibiotic. Fed Eur Biochem Soc Lett 9:301–303
10. Maeda H, Glaser CB, Meienhofer J (1970) Detection of disulfide bonds on peptide maps. Fed Proc 29:278
11. Meienhofer J, Czombos J, Maeda H (1971) Reduction of disulfide bonds in peptides and proteins by dithiothreitol in liquid ammonia. J Am Chem Soc 93(12):3080–3081
12. Glaser CB, Maeda H, Meienhofer J, Niall HD (1972) Partial sequence of the tyrosine region of neocarzinostatin. Experientia (Basel) 28(7):772–773
13. Kudo K, Kikuchi M, Ishida N (1972) Biogenesis of an antitumor antibiotic neocarzinostatin. Antimicrob Agents Chemother 1(4):289–295
14. Meienhofer J, Maeda H, Glaser CB, Czombos J, Kuromizu K (1972) Primary structure of neocarzinostatin, an antitumor protein. Science 178:875–876
15. Maeda H, Glaser CB, Czombos J (1972) Structural studies on neocarzinostatin, an antitumor polypeptide antibiotic. Prog Pept Res 2:295–306
16. Maeda H, Shiraishi H, Onodera S, Ishida N (1973) Conformation of antibiotic protein, neocarzinostatin, studied by plane polarized infrared spectroscopy, circular dichroism and optical rotatory dispersion. Int J Pept Protein Res 5:19–26
17. Maeda H (1973) Chemical reactivity and state of tryptophan residues in neocarzinostatin under physiological condition. J Antibiot (Tokyo) 26(12):776–777
18. Samy TSA, Atreyi M, Maeda H, Meienhofer J (1974) Selective tryptophan oxidation in the antitumor protein neocarzinostatin and effects and biologial activity. Biochemistry 13(5):1007–1014
19. Maeda H (1974) Chemical and biological characterization of succinyl neocarzinostatin. J Antibiot (Tokyo) 27(5):303–311
20. Maeda H (1974) Preparation of succinyl neocarzinostatin. Antimicrob Agents Chemother 5(3):354–355
21. Maeda H, Glaser BC, Kuromizu K, Meienhofer J (1974) Structure of the antitumor protein neocarzinostatin amino acid sequence. Arch Biochem Biophys 164(2):379–385
22. Maeda H, Glaser CB, Czombos J, Meienhofer J (1974) Structure of the antitumor protein neocarzinostatin. Purification, amino acid composition, disulfide reduction, and isolation and composition of tryptic peptides. Arch Biochem Biophys 164(2): 369–378
23. Maeda H, Takeshita J (1975) Degradation of neocarzinostatin by blood sera in vitro and its inhibition by DFP and N-ethylmaleimide. Gann 66:523–526
24. Sieker LC, Jensen LH, Samy TS (1976) Neocarzinostatin: an antitumor protein. A preliminary X-ray diffration study. Biochem Biophys Res Commun 68(2):358–362

25. Samy TS, Raso V (1976) Radioimmunoassay of neocarzinostatin, an antitumor protein. Cancer Res 36(12):4378–4381
26. Samy TS (1977) Neocarzinostatin: effect of modification of side chain amino and carboxyl groups on chemical and biological properties. Biochemistry 16(25): 5573–5578
27. Maeda H, Kuromizu K (1977) Spontaneous deamination of a protein antibiotic, neocarzinostatin, at weakly acidic pH. Conversion to a homologus inactive preneocarzinostatin due to change asparagine 83 to asparatic acid 83 accompanied by conformational and biological alterations. J Biochem (Tokyo) 81(1):25–35
28. Samy TS, Hu JM, Meienhofer J, Lazarus H, Johnson RK (1977) A facile method of purification of neocarzinostatin, an antitumor protein. J Natl Cancer Inst 58(6): 1765–1768
29. Maeda H (1978) Assay of an antitumor protein and its antibody by fluorescence polarization. Clin Chem 24(12):2139–2144
30. Burger RM, Peisach J, Horwitz SB (1978) Effect of light and oxygen on neocarzinostatin stability and DNA cleaving activity. J Biol Chem 253(14):4830–4832
31. Maeda H, Takeshita J, Kanamaru R (1979) A lipophilic derivative of neocarzinostatin: a polymer conjugation of antitumor protein antibiotic. Int J Pept Protein Res 14:81–87
32. Napier M, Holmquist B, Strydom D, Gordberg I (1979) Neocarzinostatin: special characterization and separation of a non-protein chromophore. Biochem Biophys Res Commun 89(2):635–642
33. Iseki S, Koide Y, Ebina T, Ishida N (1980) Biological activities and physicochemical properties of pre-neocarzinostatin and UV-irradiated neocarzinostatin. J Antibiot (Tokyo) 33(1):110–113
34. Koide Y, Ishii F, Hasuda K, Koyama Y, Edo K, Takamine S, Kitame F, Ishida N (1980) Isolation of a non-protein component and a protein component from neocarzinostatin (NCS) and their biological activities. J Antibiot (Tokyo) 33(3):347–351
35. Edo K, Katamine S, Kitame F, Ishida N, Koide Y, Kusano G, Nozoe S (1980) Naphthalenecarboxylic acid from neocarzinostatin (NCS). J Antibiot (Tokyo) 33(3):347–351
36. Edo K, Iseki S, Ishida N, Horie T, Kusano G, Nozoe S (1980) An electron spin resonance study of a spin adduct of the non-protein component (NPC) of neocarzinostatin. J Antibiot (Tokyo) 33(12):1586–1589
37. Sheridan RP, Gupta RK (1981) Electron spin resonance detection of free radical in the mercaptan-activation and UV-inactivation of neocarzinostatin. Biochem Biophys Res Commun 99(1):213–220
38. Jung G, Kohnlein W, Luders G (1981) Biological activity of the antitumor protein neocarzinostatin coupled to a monoclonal antibody by N-succinimidyl 3-(2-pyridyldithio)-propionate. Biochem Biophys Res Commun 101(2):599–606
39. Napier MA, Holmquist B, Strydom DJ, Goldberg IH (1981) Neocarzinostatin chromophore: purification of the major active form and characterization of its spectral and biological properties. Biochemistry 20(19):5602–5608
40. Kappen LS, Napier MA, Goldberg IH (1980) Role of chromophore and nonprotein in neocarzinostatin action. Proc Natl Acad Sci USA 77(4):1970–1974
41. Napier MA, Kappen LS, Goldberg IH (1980) Effect of nonprotein chromophore removal on neocarzinostatin action. Biochemistry 19:1767–1773
42. Povirk LF, Goldberg IH (1980) Binding of the nonprotein chromophore of neocarzinostatin to deoxyribonucleic acid. Biochemistry 19:4773–4780

43. Kappen LS, Goldberg IH (1980) Stabilization of neocarzinostatin nonprotein chromophore activity by interaction with apoprotein and with HeLa cells. Biochemistry 19:4786–4790
44. Hatayama T, Goldberg IH (1980) Deoxyribonucleic acid sugar damage in the action of neocarzinostatin. Biochemistry 19:5890–5898
45. Suzuki H, Miura K, Kumada Y, Takeuchi T, Tanaka N (1980) Biological activities of non-protein chromophores of antitumor protein antibiotics: auromomycin and neocarzinostatin. Biochem Biophys Res Commun 94(1):255–261
46. Tatsumi K, Bose KK, Ayres K, Strauss BS (1980) Repair of neocarzinostatin-induced deoxyribonucleic acid damage in human lymphoblastoid cells: possible involvement of apurinic/apyrimidinic sites as intermediates. Biochemistry 19:4767–4772
47. Lewis RS, Jung G, Kohnlein W (1980) A model for the activation and inactivation of neocarzinostatin, an antitumor protein. Biochim Biophys Acta 608:138–146
48. Jung G, Lewis RS, Kohnlein W (1980) The generation of pre-neocarzinostatin and antagonism of neocarzinostatin-induced DNA strand scission in vitro. Biochim Biophys Acta 608:147–153
49. Jubg G, Kohnlein W (1981) Neocarzinostatin: controlled release of chromophore and its interaction with DNA. Biochem Biophys Res Commun 98(1):176–183
50. Sheridan RP, Gupta RK (1981) Electron spin resonance detection of free radicals in the mercaptan-activation and UV-inactivation of neocarzinostatin. Biochem Biophys Res Commun 99(1):213–220
51. Napier MA, Goldberg IH, Hensens OD, Dewey RS, Liesch JM, Albers-Schonberg G (1981) Neocarzinostatin chromophore: presence of a cyclic carbonate subunit and its modification in the structure of other biologically active forms. Biochem Biophys Res Commun 100(4):1703–1712
52. Povirk LF, Dattagupta N, Warf BC, Goldberg IH (1981) Neocarzinostatin chromophore: binds to deoxyribonucleic acid by intercalation. Biochemistry 20:4007–4014
53. Napier MA, Holmquist B, Strydom DJ, Goldberg HI (1981) Neocarzinostatin chromophore: purification of the major active form and characterization of its spectral and biological properties. Biochemistry 20:5602–5608

III. Cytotoxicity, Mode of Actions, Pharmacokinetics, and Pharmacodynamics

1. Kumagai K, Miyazaki K, Rikimaru M, Ishida N (1962) The cytotoxicity effect of carzinostatin on experimental tumor. J Antibiot (Tokyo) Ser A 15:154
2. Kumagai K (1962) Antitumor activity of carzinostatin. J Antibiot (Tokyo) Ser A 15:53–59
3. Kumagai K, Miyazaki K (1963) The effect of carzinostatin on mouse ascitic leukemia SN-36. J Antibiot (Tokyo) Ser A 16:55
4. Nishikawa T, Kumagai K, Kudo A, Ishida N (1965) Cytological study on the effect of neocarzinostatin on sarcoma 180 in vivo. J Antibiot (Tokyo) Ser A 18(5):223–227
5. Kumagai K, Maeda H, Ishida N (1966) Biological activity of neocarzinostatin and its derivatives. Antimicrob Agents Chemother: 546–550
6. Ono Y, Watanabe Y, Ishida N (1966) Mode of action of neocarzinostatin: inhibition of DNA synthesis and degradation of DNA in *Sarcina lutea*. Biochim Biophys Acta 119:46–58

7. Bradner TB, Hutchison DH (1966) Neocarzinostatin (NSC-69856): an antitumor antibiotic effective against ascitic leukemia L-1210 in mice. Cancer Chemother Rep 50(1,2):78–84
8. Kumagai K, Ono Y, Nishikawa T, Ishida N (1966) Cytological studies on the effect of neocarzinostatin on HeLa cells. J Antibiot (Tokyo) Ser A 19(2):69–74
9. Ono Y, Ito Y, Maeda H, Ishida N (1968) Mode of action of neocarzinostatin: requirement of protein synthesis for neocarzinostatin-induced DNA degradation in *Sarcina lutea*. Biochim Biophys Acta 155:616–618
10. Kikuchi M, Sasaki Y, Ishida N (1969) Variation of *Shigella flexneri* sensitive to neocarzinostatin and actinomycin D. J Antibiot (Tokyo) 22(8):390–391
11. Sairenji T, Yamaguchi J, Katagiri S, Hinuma Y (1970) Effect of neocarzinostatin on the cultured Burkitt's lymphoma cells. Gann 61:451–460
12. Tsuruo T, Satoh H, Ukita T (1971) Effect of the antitumor antibiotic neocarzinostatin on DNA synthesis in vitro. J Antibiot (Tokyo) 24(7):423–429
13. Kudo K, Saito N, Ohtsuki K, Kikuchi M, Ishida N (1972) Colistin-mediated action of neocarzinostatin, actinomycin and mikamycin of *Esherichia coli*. J Antibiot (Tokyo) 24:907–910
14. Kawai Y, Katoh A (1972) Differential inhibition of DNA and RNA synthesis by neocarzinostatin in cultured Burkitt's lymphoma. J Natl Cancer Inst 48:1535–1538
15. Sakamoto K, Sakka M (1973) Effect of neocarzinostatin on V-79 Chinese hamster cells in vitro. Eur J Cancer 9:725–728
16. Shamoto M, Kawai Y, Kato A (1973) Electron microscopic observation of neocarzinostatin-treated Burkitt's lymphoma. Experientia (Basel) 29(2):197–198
17. Sakamoto K, Sakka M (1973) Survival of clonogenic cells of a murine epithelioma treated with neocarzinostatin. Eur J Cancer 9:829–831
18. Beerman TA, Goldberg IH (1974) DNA strand scission by the antitumor protein neocarzinostatin. Biochem Biophys Res Commun 59(4):1254–1261
19. Takahashi I, Sakato J, Mikouchi H, Moriwaki H, Kitajima K, Irino S, Hiraki K (1974) Antitumor activity of neocarzinostatin, effect of Rausher leukemia in mice. Acta Med Okayama 28(4):271–276
20. Tatsumi K, Nakamura T, Wakisaka G (1974) Damage of mammalian cell DNA in vivo and in vitro induced by neocarzinostatin. Gann 65:459–461
21. Samy TS, Atreyi M, Maeda H, Meienhofer J (1974) Selective tryptophan oxidation in the antitumor protein neocarzinostatin. Biochemistry 13(5):1007–1014
22. Nakazawa A, Tamada T (1974) Induction of colicin E-1 synthesis by neocarzinostatin and bleomycin. J Antibiot (Tokyo) 27(12):984–986
23. Sawada H, Tatsumi K, Sasada M, Shirakawa S, Nakamura T, Wakisaka G (1974) Effect of neocarzinostatin on DNA synthesis in L-1210 cells. Cancer Res 34:3341–3346
24. Nakazawa A, Tamada T (1974) Induction of colicin E1 synthesis by neocarzinostatin and bleomycin. J Antibiot (Tokyo) 27(12):984–986
25. Schaeppi U, Menninger F, Fleischmen RW, Bogden AE, Schein PS, Coony DA (1974) Toxicity of neocarzinostatin (NSC-69856): an antitumor antibiotic with radiomimetic and antigenic characteristics. Cancer Chemother Rep (Suppl) 54(1):43–52
26. Toriyama K, Fujita H, Ishida N (1975) Absorption, distribution, excretion of neocarzinostatin (NCS) in mice after oral administration. J Antibiot (Tokyo) 28(1):64–72
27. Ohtsuki K, Ishida N (1975) Neocarzinostatin-induced breakdown of deoxyribonucleic acid in HeLa cells. J Antibiot (Tokyo) 28(2):143–148
28. Ohtsuki K, Ishida N (1975) Mechanism of DNA degradation induced by neocarzinostatin in *Bacillus subtilis*. J Antibiot (Tokyo) 28(3):229–236

29. Takahashi I, Mikochi H, Sakato J, Nakanishi T, Toki H (1975) Effect of neocarzinostatin on hemolytic plaque-forming cell production in mice. Acta Med Okayama 29(4):327–328
30. Maeda H, Aikawa S, Yamashita A (1975) Subcellular fate of protein antibiotic neocarzinostatin in culture of a lymphoid cell line from Burkitt's lymphoma. Cancer Res 35:554–559
31. Ebina T, Ohtsuki K, Ishida N (1975) Specific G2 block in HeLa S-3 cells by neocarzinostatin. Eur J Cancer 2:155–158
32. Ebina T, Ishida N (1975) Inhibition of formation of microtubular paracrystal in HeLa-S3 cells by neocarzinostatin. Cancer Res 35:3705–3709
33. Maeda H, Takeshita J (1975) Degradation of neocarzinostatin by blood sera in vitro and its inhibition by diisopropyl fluorophosphate and N-ethylmaleimide. Gann 66(5):523–527
34. Abe I, Sato S (1976) Beta-glucuronidase activity of Yoshida ascites hepatomas of different drug sensitivity and its change after treatment of host rats with various anticancer agents. Gann 67(1):55–65
35. Ishida R, Nishimoto T, Takahashi T (1976) In vitro release of thymine from DNA by neocarzinostatin. Biochem Biophys Res Commun 68:256–261
36. Rao AP, Rao PN (1976) The cause of G2-arrest in Chinese hamster ovary cells treatment with anticancer drugs. J Natl Cancer Inst 57(5):1139–1143
37. Maeda H, Takeshita J (1976) Inhibitors of proteolytic enzymes prevent the inactivation by blood of the protein antibiotic neocarzinostatin and its succinyl derivative. J Antibiot (Tokyo) 29(1):111–112
38. Beerman TA, Goldberg IH, Kappen LS, Poon R, Suzuki H (1976) Molecular basis of action of cytotoxic antibiotics. Adv Enzyme Regul 14:207–225
39. Sarma DS, Rajalakshmi S (1976) Studies on the interaction of neocarzinostatin with rat liver DNA in vivo and in vitro. Biochem Pharmacol 25(7):789–792
40. Maeda H, Yamamoto N, Yamashita A (1976) Fate and distribution of [^{14}C]succinyl neocarzinostatin in rats. Eur J Cancer 12(11):865–870
41. Ono T, Nakabayashi T (1976) Electron microscopic observations of structures appearing in *Trypanosoma gambiense* after treatment with neocarzinostatin. Biken J 19(4):171–177
42. Rao AP, Rao PN (1976) The cause of G2-arrest in Chinese hamster ovary cells treated with anticancer drug. J Natl Cancer Inst 57(5):1139–1143
43. Sasada M, Sawada H, Nakamura T, Uchino H (1976) Caffeine potentiation on lethality of L-1210 cells treated with neocarzinostatin. Gann 67(3):447–449
44. Goldberg IH, Kappen LS, Beerman TA (1977) The nature and mechanism of the damage in DNA by a protein antibiotic. Adv Enzyme Regul 16:239–254
45. Maeda H, Sakamoto S, Ogata J (1977) Mechanism of accumulation of the antitumor protein antibiotic neocarzinostatin in bladder tissue. Antimicrob Agents Chemother 11(6):941–945
46. Beerman TA, Poon R, Goldberg IH. (1977) Single-stranded nicking of DNA in vitro by neocarzinostatin and its possible relationship to the mechanism of durg action. Biochim Biophys Acta 475(2):294–306
47. Beerman TA, Goldberg IH (1977) The relationship and strand-scission and DNA synthesis inhibition in HeLa cells treated with neocarzinostatin. Biochim Biophys Acta 475(2):281–293
48. Poon R, Beerman TA, Goldberg IH (1977) Characterization of DNA strand breakage in vitro by the antitumor protein neocarzinostatin. Biochemistry 16(3):486–493

49. Kappen LS, Goldberg IH (1977) Effect of neocarzinostatin-induced strand scisson on the template activity of DNA polymerase I. Biochemistry 16(3): 479–485
50. Ebina T, Satake M, Ishida N (1977) Inhibition of surface immunoglobulin central capping of Daudi cells and cell spreading of HeLa cells by neocarzinostatin. Cancer Res 37(12):4423–4429
51. Tatsumi K, Nishioka H (1977) Effect of DNA repair synthesis on antibacterial and mutagenic activity of an antitumor protein, neocarzinostatin. Mutat Res 48(2):195–203
52. Lazarus H, Raso V, Samy TS (1977) In vitro inhibition of human leukemia cells (CCFR-CEM) by agarose-immobilized neocarzinostatin. Cancer Res 37(10): 3731–3736
53. Cozzarelli NR (1977) The mechanism of action of DNA Synthesis. Annu Rev Biochem 46:641–668
54. Raso V (1977) Radioimmunoassay for methotrexate, citrovorum factor, neocarzinostatin, and actinomycin. Cancer Treat Rep 61(4):585–590
55. Maeda H, Ichimura H, Satoh H, Ohtsuki K (1978) Evaluation of succinyl neocarzinostatin in vivo. J Antibiot (Tokyo) 31(5):468–472
56. Kappen LS, Goldberg IH (1978) Neocarzinostatin induction of DNA repair synthesis in HeLa cells and isolated nuclei. Biochim Biophys Acta 52(3):481–489
57. Hatayama T, Goldberg IH, Takeshita M, Grollman AP (1978) Nucleotide specificity in DNA scission by neocarzinostatin. Proc Natl Acad Sci USA 75(8): 3603–3607
58. Sasada M, Sawada H, Nakamura T, Uchino H (1978) DNA repair in human leukemic leukocytes treated with neocarzinostatin. Gann 69(3):407–412
59. Sausville EA, Horwitz SB (1978) Inhibition of SV40 DNA synthesis by campthothecin and neocarzinostatin. Mol Pharmacol 14(6):1156–1166
60. D'Andrea AD, Haseltine WA (1978) Sequence-specific cleavage of DNA by the antitumor antibiotic neocarzinostatin and bleomycin. Proc Natl Acad Sci USA 75(8):3608–3612
61. Nakazawa I, Ouchi E, Isida N, Wagai K (1978) Effect of amphotericin B on the incorporation of neocarzinostatin into human gastric cancerous tissues. Tohoku J Exp Med 124(1):97–98
62. Kappen LS, Goldberg IH (1978) Gaps in DNA induced by neocarzinostatin near 3′- and 5′-phosphoryl termini. Biochemistry 17(4):729–734
63. Kuo MT, Samy TS (1978) Effect of neocarzinostatin on mammalian nuclei: release of nucleosome. Biochim Biophys Acta 518(1):186–190
64. Ishida R, Takahashi T (1978) Role of 2-mercaptoethanol in in vitro DNA degradation by neocarzinostatin. Cancer Res 38:2617–2620
65. Shimoyama M, Kimura K (1979) Effect of blood, ascites, and tumor cell density on cytocidal action of neocarzinostatin. Gann 70(2):165–171
66. Lowenthal IS, Parker IM, Greenblatt DJ, Brown BL, Samy TS (1979) Pharmacokinetic analysis of neocarzinostatin in normal and tumor-bearing rodents. Cancer Res 39(5):1547–1551
67. Satake M, Ebina T, Ishida N (1979) Effect of neocarzinostatin on microtubules. Gann 70(4):545–548
68. Comis RL, Griffin TW, Raso V, Ginsberg SJ (1979) Pharmacokinetics of the protein antitumor antibiotic neocarzinostatin after bolus injection in humans. Cancer Res 39(3):757–761

69. Tatsumi K, Tashima M, Shirakawa S, Nakamura T, Uchino H (1979) Enhancement by caffeine of neocarzinostatin cytotoxicity in murine leukemia L1210 cells. Cancer Res 39(5):1623–1627
70. Maeda H, Matsumoto M (1979) Cytotoxic effect of neocarzinostatin on human lymphoid cells. Tohoku J Exp Med 128:313–323
71. Yamashita A, Kitawaki Y, Miyamoto M, Maeda H (1979) Suppression of graft-versus-host reaction by preincubation of the graft with an antitumor protein neocarzinostatin. Immunopharmacology 1:255–265
72. Sakamoto S, Maeda H, Ogata J (1979) An uptake of fluorescein isothiocyanate-labeled neocarzinostatin into the cancer and normal cells. Experientia (Basel) 35:1223–1234
73. Maeda H, Takeshita J, Kanamaru R, Sato H, Katoh J, Sato H (1979) Antimetastatic and antitumor activity of a derivative of neocarzinostatin: an organic solvent- and water-soluble polymer-conjugated protein. Gann 70:601–606
74. Ishida R, Nishimoto T, Takahashi T (1979) DNA strand scission by neocarzinostatin and its relation to the inhibition of cell cycle traverse and DNA synthesis. Cell Struct Funct 4:235–250
75. Maeda H, Takeshita J, Yamashita A (1980) Lymphotropic accumulation of an antitumor antibiotic protein, neocarzinostatin. Eur J Cancer 16:723–731
76. Takeshita J, Maeda H, Koike K (1980) Subcellular action of neocarzinostatin: Intracellular incorporation, DNA breakdown and and cytotoxicity. J Biochem (Tokyo) 88:1071–1080
77. Ohtsuki K, Ishida N (1980) The biological effect of a nonprotein component removed from neocarzinostatin (NCS). J Antibiot (Tokyo) 33:744–750
78. Ohtsuki K, Koike T, Sato T, Koyama K, Ishida N, Satake M (1980) Neocarzinostatin and microtubule-associated kinases. J Antibiot (Tokyo) 33:1590–1593
79. Maeda H, Sano Y, Takeshita J, Iwai Z, Kosaka H, Marubayashi T, Matsukado Y (1981) A pharmacokinetic simulation model for chemotherapy of brain tumor with antitumor protein antibiotic, neocarzinostatin: theoretical considerations behind a two-compartment model for continuous infusion via an internal carotid artery. Cancer Chemother Pharmacol 5:243–249
80. Ohtsuki K, Sato T, Koike T, Yokoyama K, Ishida N (1981) The inhibitory mechanism of in vitro protein phosphorylation by a nonprotein chromophore removed from neocarzinostatin. Biochim Biophys Acta 673(2):147–156

IV. Clinical Effects and Studies

1. Takahashi M, Toriyama K, Maeda H, Kikuchi M, Kumagai K, Ishida N (1969) Clinical trials of a new antitumor polypeptide: neocarzinostatin. Tohoku J Exp Med 98(2): 273–280
2. Hiraki K, Kamimura O, Takahashi I, Nagao T, Kitajima K, Irino S (1973) Neocarzinostatin, a new chemotherapeutic approach to acute leukemia. Nouv Rev Fr Hematol 131(4):445–451
3. Toki H (1976) Absence of anti-neocarzinostatin (NCS) antibody production in leukemia patients treated with NCS. Acta Med Okayama 30(2):135–142
4. Wiepnik PH (1976) The therapy of acute leukemia in the adult: a progress report. Haematol Bluttransfus 19:271–284

5. Kitajima K, Toki H, Sanada HA (1976) Case of smoldering acute leukemia: long survival duration of 3 years and 9 months after the diagnosis. Acta Med Okayama 30(5):311–319
6. Rivera G, Howarth C, Aur RJ, Pratt CB (1978) Phase I study of neocarzinostatin in children's cancer. Cancer Treat Rep 62(12):2105–2107
7. Sakamoto S, Maeda H, Matsumoto T, Ogata J (1978) Experimental and clinical studies on the formation of antibody to neocarzinostatin, a new protein antibiotic. Cancer Treat Rep 62(12):2063–2070
8. Griffin TW, Comis RL, Lokich JJ, Blum RH, Canellos GP (1978) Phase I and preliminary phase II study of neocarzinostatin. Cancer Treat Rep 62(12):2019–2025
9. Ohnuma T, Nogeire C, Cutter J, Holland JF (1978) Phase I study with neocarzinostatin: tolerance to two-hour infusion and continuous infusion. Cancer (Phila) 42(4):1670–1679
10. Kimura I (1978) Clinical investigations of neocarzinostatin in Japan. Recent Results Cancer Res 63:252–260
11. Van Echo AD, Wiernik PH (1978) Phase I study of neocarzinostatin in patients with acute leukemia. Cancer Treat Rep 62(9):1363–1365
12. Sakamoto S, Ogata J, Ikegami K, Maeda H (1978) Effects of systemic administration of neocarzinostatin, a new protein antibiotic, on human bladder cancer. Cancer Treat Rep 62(3):453–454
13. Comis RL, Griffin T, Raso V, Ginsberg SJ (1978) Neocarzinostatin: initial clinical and pharmacological studies in the United States. Recent Results Cancer Res 63:261–271
14. Seltzer SE, Griffin T, D'Orsi C, Tryka F, Herman PJ (1978) Pulmonary reaction associated with neocarzinostatin therapy. Cancer Treat Rep 62(8):1271–1272
15. Sakamoto S, Ogata J, Ikegami K, Maeda H (1980) Chemotherapy for bladder cancer with neocarzinostatin: evaluation of systemic administration. Eur J Cancer 16:103–113
16. Okada K, Kitade F, Kawashima Y, Yamada S, Sakuramoto K, Okajima K (1981) Clinical evaluation of preoperative combined chemotherapy with neocarzinostatin (NCS) and 5-fluorouracil (combined NF therapy) against gastric cancer. Bull Osaka Med Sch 27(2):62–78

Subject Index

Kolibri 360 GmbH
Ferdinand-Jühlke-Straße 7
99095 Erfurt, Deutschland
produktsicherheit@kolibri360.de